MARINE WILDLIFE AND TOURISM MANAGEMENT

Insights from the Natural and Social Sciences

To George, Gus and Will
and
To Jette

MARINE WILDLIFE AND TOURISM MANAGEMENT

Insights from the Natural and Social Sciences

Edited by

James Higham and Michael Lück

www.cabi.org

CABI is a trading name of CAB International

CABI Head Office
Nosworthy Way
Wallingford
Oxfordshire OX10 8DE
UK

CABI North American Office
875 Massachusetts Avenue
7th Floor
Cambridge, MA 02139
USA

Tel: +44 (0)1491 832111
Fax: +44 (0)1491 833508
E-mail: cabi@cabi.org
Website: www.cabi.org

Tel: +1 617 395 4056
Fax: +1 617 354 6875
E-mail: cabi-nao@cabi.org

A catalogue record for this book is available from the British Library, London, UK.

Library of Congress Cataloging-in-Publication Data

Marine wildlife and tourism management : Insights from the natural and social sciences /
James Higham and Michael Lück (eds).
    p. cm.
  Includes bibliographical references and index.
  ISBN: 978-1-84593-345-6 (alk. paper)
  1. Marine ecotourism -- Management. I. Higham, James E.S. II. Lück, Michael,
1966--III. Title.
  G156.5.M36M385 2007
  910.68--dc22

2007024193

ISBN: 978 1 84593 345 6

Typeset by SPi, Pondicherry, India.
Printed and bound in the UK by Biddles Ltd, King's Lynn.

Contents

Contributors

Lars Bejder, Murdoch University (Australia)
Throughout the last decade Dr Bejder has carried out research on various aspects of cetacean biology, ecology and conservation. His particular field of expertise falls into two main categories, the social behaviour of cetaceans and impact assessments of human activities on cetaceans. He has a strong interest in developing quantitative techniques to analyse and unravel the social structure of complex animal systems. His research on the impacts of whale and dolphin watching on targeted animals has resulted in changes in management and regulations within the industry in both New Zealand and Australia. He has written peer-reviewed scientific publications and book chapters on dolphin behaviour, social structure and conservation.

Carl Cater, Griffith University (Australia)
Dr Carl Cater is a Lecturer in Tourism at Griffith University, Queensland, Australia. His research centres on the experiential turn in tourism and the subsequent growth of special interest sectors. He is a fellow of the Royal Geographical Society, a qualified pilot, diver, mountain and tropical forest leader, and maintains an interest in both the practice and pursuit of sustainable outdoor tourism activity. He is co-author (with Dr Erlet Cater) of *Marine Ecotourism: Between the Devil and the Deep Blue Sea* (CAB International, 2007). E-mail: C.Cater@griffith.edu.au

Erlet Cater, The University of Reading (UK)
Dr Erlet Cater is Senior Lecturer in Tourism and Development in the Department of Geography, The University of Reading. Joint editor of *Ecotourism: A Sustainable Option* (1994) and Advisory Editor for *The Encyclopaedia of Ecotourism*, she is an advisor for the Society and Environment Forum of the RGS-IBG and Coral Cay Conservation and judged the British Airways *Tourism for Tomorrow Award* for several years. She is on the International Editorial Boards of *Tourism Geographies*, the *Journal of Ecotourism* and *Tourism in Marine Environments*.

Rochelle Constantine, University of Auckland (New Zealand)

Dr Rochelle Constantine's research on the effects of tourism on dolphins began in 1994 in the Bay of Islands, New Zealand, where she continues to run a research project primarily on the bottlenose dolphins of Northland. She has been involved in research on a variety of species including dusky, common and Hector's dolphins and is also part of the South Pacific Whale Research Consortium focusing mainly on the ecology of the humpback whales of Oceania. Rochelle has a strong interest in applied behavioural ecology and believes that high-quality research is vital to help develop a sustainable whale-watching industry. E-mail: r.constantine@auckland.ac.nz

Philip Dearden, University of Victoria (Canada)

Philip Dearden is a Professor of Geography in the University of Victoria. He leads the Marine Protected Areas Research Group at UVic and has research interests ranging from seagrass ecology through to coral reef monitoring and diver surveys in South-east Asia. He is particularly interested in understanding MPA use patterns, zoning and developing incentive-based approaches to conservation and advises the World Bank, Asian Development Bank, UN, IUCN and national governments in Asia on protected area management. He is the Leader of Canada's national MPA Working Group for the Ocean Management Research Network and Co-Chair of Parks Canada's NMCA Marine Science Network. E-mail: pdearden@office.geog.uvic.ca

John Dobson, University of Wales Institute (UK)

John Dobson is a Senior Lecturer in the Cardiff School of Management at the University of Wales Institute, Cardiff. His research interests included examining the potential for tourism to contribute towards the conservation of vulnerable and endangered species, the growth and management of shark-based tourism and the ethics of wildlife tourism. E-mail: JDobson@uwic.ac.uk

David Duffus, University of Victoria (Canada)

Dave Duffus is a wildlife scientist who began his career in the Canadian Prairies working on wetland conservation. Subsequent to that, he began work in the mid 1980s on marine mammals and the then sprouting eco-tourism management issues surrounding recreational whale watching. Since 1992, he has been on the faculty at the University of Victoria where he directs the Whale Research Lab in the Geography Department. Over the last 15 years, the graduate student scientists have worked in a variety of locations and a broad number of issues in marine mammal science and management. The current focus of Duffus' research is on ecological function and process in a grey whale foraging area on the west coast of Vancouver Island.

Markus Dyck, Department of Environment, Government of Nunavut (Canada)

Markus Dyck currently works for the Department of Environment, Government of Nunavut, from the decentralized community of Iglulik. During the last 10 years he has examined the possible effects of tourism on the polar bears of western Hudson Bay and he is involved in various aspects of polar bear research (mark–recapture studies, behavioural ecology, etc.) in the Canadian Arctic. His current research interests focus on paternity and mate selection in polar bears, as well as the possible latitudinal effects of climate change across various trophic levels throughout Nunavut.

Ursula Ellenberg, University of Otago (New Zealand)

Ursula Ellenberg's interest in human–wildlife interactions was first sparked while studying behavioural ecology and wildlife management at the University of Victoria, British Columbia, Canada. But it was during work as a backcountry guide on Svalbard, Norwegian Arctic, that she gained first-hand experience of the impacts human visitors may have on wildlife. This motivated her to study Arctic Terns as a side project during her second Arctic summer. After completing a Master's degree at the University of Kiel, Germany, she joined a seabird ecology project in Northern Chile, where she focused on the impact of human activity on Humboldt penguins. Ursula is currently a PhD student at the Department of Zoology, University of Otago, New Zealand, where she is investigating the effects of human disturbance on Yellow-eyed and Snares penguins.

Paul H. Forestell, Long Island University (USA)

Paul H. Forestell is Professor of Comparative Psychology and Animal Behavior at Long Island University in New York, and Vice President of the Pacific Whale Foundation in Hawaii. Born in Canada and educated at the University of New Brunswick, he received his PhD from the University of Hawaii in 1988. For 30 years, he has studied whales and dolphins throughout the Pacific. He has gained international recognition for his understanding of the impacts of a rapidly expanding whale- and dolphin-watching industry, and his development of formal training programmes for naturalists and educators. He has conducted whale-watching workshops in Hawaii, Australia, Japan and Ecuador, and continues to study bottlenose and tucuxi dolphins in Costa Rica, and humpback whales in Ecuador and Australia. E-mail: Paul.Forestell@liu.edu

Brian Garrod, University of Wales Aberystwyth (UK)

Brian Garrod is a Senior Lecturer and Head of Tourism in the Institute of Rural Sciences at the University of Wales Aberystwyth, UK. He trained originally as an economist, specializing in environmental and natural-resource economics, and was awarded a PhD from the University of Portsmouth in 1993 on the subject of fisheries bioeconomics. His research interests encompass ecotourism, nature-based tourism and heritage tourism, and he has published four books and more than 25 research articles in these areas. He is Book Reviews Editor for the *Journal of Heritage Tourism*, an Associate Editor of the *Journal of Ecotourism*, and on the editorial boards of *Tourism in Marine Environments* and the *International Journal of Sustainable Development*. He has worked as a consultant to various public bodies, including the UN World Tourism Organization (UNWTO), the Organization for Economic Cooperation and Development (OECD), the Countryside Agency and the Welsh Development Agency. E-mail: bgg@aber.ac.uk

Wiebke F. Hendry, Green Concrete Films

Wiebke F. Hendry completed her Master's degree in Marine Science at the University of Otago, New Zealand, focusing on whale watching and tourism in the

American Northwest Pacific. After working as a wildlife guide she studied Natural History Filmmaking and Communication. She now runs her independent filmmaking business Green Concrete Films.

James Higham, University of Otago (New Zealand)

James Higham is Professor of Tourism at the School of Business, University of Otago, New Zealand. His research centres on understanding and managing tourist interactions with marine wildlife, particularly as it relates to colonial nesting seabirds and marine mammals. He serves on the editorial boards of the *Journal of Sustainable Tourism*, *Tourism in Marine Environments* and the *Journal of Ecotourism*. He is the editor of *Critical Issues in Ecotourism: Understanding a Complex Tourism Phenomenon* (Elsevier, 2007) and co-editor (with Professor C. Michael Hall) of *Tourism, Recreation and Climate Change* (Channel View Publications, 2005). E-mail: jhigham@business.otago.ac.nz

Bruce McKinlay, Department of Conservation (New Zealand)

Bruce McKinlay is Technical Support Officer, Terrestrial Ecosystems, for the Otago Conservancy, New Zealand Department of Conservation *Te Papa Atawhai*. Bruce has been involved in all aspects of Yellow-eyed penguin conservation for more than 20 years, including research, policy development, species recovery plan formulation, community liaison and hands-on fieldwork.

Harvey Lemelin, Lakehead University (Canada)

Dr Lemelin is an Assistant Professor with the School of Outdoor Recreation, Parks and Tourism at Lakehead University in Thunder Bay, Canada. He has published extensively on the socio-economic and socio-environmental dimensions of wildlife tourism (i.e. polar bear tourism) in Churchill, Manitoba and elsewhere. Extensive field research conducted with First Nations in Canada complements his current research projects examining tourism, forestry, wildlife management and natural resources management strategies in Northern Ontario. E-mail: rhlemeli@lakeheadu.ca

Michael Lück, Auckland University of Technology (New Zealand)

Michael currently holds the position of Associate Professor in the School of Hospitality and Tourism, AUT University in Auckland, New Zealand. He is an active member of the New Zealand Tourism Research Institute (NZTRI), where he is Associate Director, and responsible for the development of a marine tourism research programme. Michael's research interests are in the wider area of marine tourism, with a focus on marine wildlife tourism and interpretation and education. He is also interested in ecotourism, sustainable tourism, the impacts of tourism, aviation and gay tourism. He has developed a keen interest in innovative and alternative teaching and assessment methods. He has published in international academic journals, and contributed to various books. He is co-editor of two books on ecotourism (in print), two books on marine tourism (in press), and he is the overall editor of the *Encyclopedia of Tourism and Recreation in Marine Environments* (CAB International, in press). Michael is the founding editor-in-chief of the academic journal *Tourism in Marine Environments* and Associate Editor of the *Journal of Ecotourism*. E-mail: mlueck@aut.ac.nz

David Lusseau, Dalhousie University (Canada)
David Lusseau is currently Izaak Walton Killam Postdoctoral Fellow in the Department of Biology at Dalhousie University, Canada. He obtained his BSc in marine biology from Florida Tech in 1996. He then obtained a PhD in Zoology from the University of Otago, New Zealand, in 2003 studying the impacts of tourism on bottlenose dolphins in Fiordland. David is a member of the IUCN Species Survival Commission Cetacean Specialist Group and regularly advice the International Whaling Commission on whale-watching issues. He has been nominated for a 2008 Pew Fellowship for achievements in marine conservation. E-mail: d.lusseau@dal.ca

Patrick T. Maher, University of Northern British Columbia (Canada)
Patrick T. Maher is an Assistant Professor in the Outdoor Recreation and Tourism Management program at the University of Northern British Columbia. Prior to this he has taught at universities in both Canada and New Zealand. His research interests include tourism in polar regions, marine and nature-based tourism, social and behavioural dimensions of outdoor recreation and tourism, and outdoor and experiential education. He is a Fellow of both the Royal Geographical Society and the Explorer's Club, and has acted in editorial and reviewing capacities for a number of international journals and conferences. E-mail: maherp@unbc.ca

Chris Malcolm, Brandon University (Canada)
Chris Malcolm is a wildlife researcher who has worked on the development of whale watching management using both scientific and social science approaches. He spent 10 years working in and with the whale-watching industry in British Columbia, during its period of rapid growth in the 1990s, and is at present collaborating with Fisheries and Oceans Canada to recommend management for beluga watching in Churchill, Manitoba. Since 2002, he has been on the faculty in the Department of Geography at Brandon University. He currently has students working on a variety of wildlife research, including the development of a GIS-based seabird habitat use model in Clayoquot Sound, British Columbia, and VHF telemetry to study spring dispersal and summer habitat use by northern pike in south-western Manitoba. E-mail: Malcolmc@Brandonu.ca

Marc L. Miller, University of Washington (USA)
Marc L. Miller is Professor in the School of Marine Studies and Adjunct Professor in the School of Aquatic and Fishery Sciences and the Department of Anthropology at the University of Washington (Seattle, Washington, USA). He has published widely on the topics of coastal tourism and recreation management, marine protected area management, fisheries management, integrated coastal zone management and marine environmental ethics and aesthetics. Professor Miller has co-edited three proceedings (1991, 1998 and 2002) of Congresses on Coastal and Marine Tourism. He served as an Associate Editor of *Coastal Management* from 1983 until 1993, and of the *Sage Qualitative Research Methods Series* since it began in 1986. He also serves as the Research Commentary and Notes Editor of *Tourism in Marine Environments* and a Coordinating Editor for *Annals of Tourism Research*. E-mail: mlmiller@u.washington.edu

Sue Muloin, James Cook University (Australia)

Sue Muloin is the Student Equity Officer at James Cook University in Cairns, North Queensland, Australia. Sue has previously been a lecturer at Newcastle University, New South Wales, in the Department of Leisure and Tourism where she developed and coordinated subjects in ecotourism and wildlife tourism. She has also been a senior research officer at the Hunter Valley Research Foundation in Newcastle, New South Wales. Her research interests include wildlife tourism (especially marine wildlife), ecotourism, indigenous tourism and access and equity in tourism.

David Newsome, Murdoch University (Australia)

David Newsome is a senior lecturer in the School of Environmental Science at Murdoch University, Perth, Western Australia. David holds degrees in botany, soil science and geomorphology. His principal research interests are geotourism, human–wildlife interactions and the biophysical impacts of recreation and tourism. His research and teaching, and the activities of his research group, focus on the sustainable use of landscapes and the assessment and management of recreational activity in protected areas. He is the lead author of the recently published books *Natural Area Tourism: Ecology, Impacts and Management* and *Wildlife Tourism* and co-editor of *Geotourism*, a book which lays the foundation for the emergence of geotourism as a distinct discipline within the area of natural area tourism. E-mail: D.Newsome@murdoch. edu.au

Kate Rodger, Murdoch University (Australia)

Kate Rodger is a Research Fellow in the School of Environmental Science at Murdoch University, Perth, Western Australia. Her main research interests include the environmental and social impacts of human–wildlife interactions, identifying and minimizing visitor impacts and linking the sociology of science with recreation in natural areas. Currently, she is furthering her experience as a researcher working in the fields of visitor monitoring and management as well as wildlife tourism.

Philip Seddon, University of Otago (New Zealand)

Philip Seddon completed a PhD on Yellow-eyed penguins in 1988, and subsequently worked on both African and King penguins while based at the University of Cape Town, South Africa. Nine years in the Middle East took him away from penguin research, until 2001, when he returned to New Zealand and New Zealand penguins. He is currently Director of the Wildlife Management programme at the Department of Zoology, University of Otago, where his research interests include the restoration and conservation management of native species in New Zealand. He is a member of the IUCN/SSC Reintroduction Specialist Group, and the Recreation Ecology Research Network (RERN). E-mail: philip.seddon@stonebow. otago.ac.nz

Eric J. Shelton, University of Otago (New Zealand)

Eric J. Shelton teaches interpretation and visitor management at the University of Otago and has had a lifelong interest in Yellow-eyed penguins and their conservation. He currently serves as a Trustee on an environmental NGO, the Yellow-eyed Penguin Trust, and represents the Tourism Department of the University of Otago on the Yellow-eyed Penguin Consultative Group. He is committed to the notion that well-managed tourism can enhance species and habitat conservation. E-mail: eshelton@business.otago.ac.nz

Karen N. Topelko, British Columbia Ministry of Environment (Canada)

Karen N. Topelko completed her Bachelor of Business Administration (Honours) at Wilfrid Laurier University in 1994, and her BA (Honours) at the University of Victoria in 2003. In 2007, she completed her MA at the University of Victoria. Her graduate research focused on the social and environmental impacts of recreational use of coral reefs in a marine park in Thailand. She has co-authored several technical reports on marine park management, and co-authored a paper on the shark-watching industry in the *Journal of Ecotourism*. Currently, she is a Policy Analyst with the British Columbia Ministry of Environment, providing advice in support of intergovernmental coastal and marine initiatives.

Heather Zeppel, James Cook University (Australia)

Heather Zeppel is a Senior Lecturer in Tourism, School of Business at James Cook University in Cairns, North Queensland, Australia. She lectures on Tourism and the Environment, Australian Ecotourism and Wildlife Tourism Management, Tourism Issues in Developing Countries and Tourism Analysis. Her research interests include indigenous tourism, ecotourism, wildlife tourism and environmental best practice in tourism. She is the author of *Indigenous Ecotourism: Sustainable Development and Management* (CAB International, 2006). E-mail: heather.zeppel@jcu.edu.au

Jackie Ziegler, University of Victoria (Canada)

Jackie Ziegler has had a lifelong passion for the ocean and the marine life it contains, which led to an undergraduate degree in Marine and Freshwater Biology from the University of Guelph. It was during this time that Jackie was introduced to the field of conservation biology and the beginning of her current research interests. She is now pursuing a graduate degree at the University of Victoria focusing on Marine Protected Areas and the impacts of marine wildlife ecotourism on endangered species.

List of Tables

List of Figures

List of Boxes

List of Appendices

Acknowledgements

The preparation and publication of this book has benefited greatly from the direct support and influence of colleagues, friends and family. Rebecca Stubbs, Claire Parfitt and Sarah Hulbert (CABI) provided all the assistance required of a supportive publisher. Needless to say we are indebted to the scholars who contributed their original work to this volume: Brian Garrod, Bruce McKinley, Carl Cater, Chris Malcolm, Dave Duffus, David Lusseau, David Newsome, Eric Shelton, Erlet Cater, Harvey Lemelin, Heather Zeppel, Jackie Ziegler, John Dobson, Karen Topelko, Kate Rodger, Lars Bejder, Marc Miller, Markus Dyck, Paul Forestell, Pat Maher, Philip Dearden, Phil Seddon, Rochelle Constantine, Sue Muloin, Ursula Ellenberg and Wiebke Hendry. Their professionalism and efficiency have greatly eased the steps in the process of bringing this volume together. Our research interests in the field of marine wildlife and tourism management have been stimulated by a number of colleagues in New Zealand and overseas. Discussions and collaborations with David Lusseau (Dalhousie University), Lars Bejder (Murdoch University), Mark Orams (Sir Peter Blake Trust), Rochelle Constantine (Auckland University), Phil Seddon, Wiebke Hendry, Eric Shelton (University of Otago), David Newsome (Murdoch University), Ross Dowling (Edith Cowan University), Erlet Cater (University of Reading), Dave Duffus and Phil Dearden (University of Victoria) have always been extremely valuable.

The latter part of this project coincided with a period when I (James) was on sabbatical leave from the University of Otago. The support of Professor George Benwell (Pro Vice-Chancellor, Commerce, and Dean, School of Business) and Professor Alan MacGregor (former Dean, School of Business) was crucial to this period of Research and Study Leave (RSL). My colleagues at the Department of Tourism (University of Otago) have provided me with a supportive, genuinely collegial and enjoyable place of work. They include Andrea Valentine, Anna Carr, Brent Lovelock, Caroline Orchiston, David Duval, David Scott, Diana Evans, Donna Keen, Eric Shelton, Hazel Tucker, Jan Mosedale, Michael Hall, Monica Graham, Neil Carr, Richard Mitchell, Tara Duncan and Teresa Leopold. My sabbatical would not have been possible without their strong support. My participation in the Australian National Wildlife Tourism Conference (August

2006) in Perth, Western Australia, was extremely opportune in the context of this book and for that I thank Ross Dowling (FACET). I also benefit greatly from the support of my family and friends. Most particularly the support of my immediate family, Linda, Alexandra, Katie and George, has as always been instrumental.

Equally, I (Michael) am fortunate to enjoy the continued support of colleagues at the Auckland University of Technology. In particular, Linda O'Neill (Head, School of Hospitality & Tourism), Simon Milne (Director, New Zealand Tourism Research Institute) and Rob Allen (Pro Vice-Chancellor, Learning and Teaching, and Dean, Faculty of Applied Humanities) not only supported my endeavours enthusiastically, but also gave me the freedom to engage in my research interests. It is a great pleasure to work in a supportive team, and I would like to sincerely thank Alice Gräupl, Arno Sturny, Carolyn Nodder, Charles Johnston, Hamish Bremner, Jill Poulston, Kathy Slater, Nathaniel Dobbin and Roberto Altobelli. Any achievement is much easier to accomplish when you have a loving and caring home environment. My parents, Heidi and Siegfried, never cease to lend me moral support and encouragement, even though they are literally at the other end of the world. Andreas and Tanja are always welcoming and great to discuss ideas with. My 'family away from home', Carol and Colin, are incredible and always welcome me to their home. My heartfelt thanks go to Neil Gussey, who is my rock, and who not once complained about my many evenings and weekends working, or my frequent times away from home when travelling. I dedicate this book to Jette, with the sincere hope that all the wildlife discussed in these pages will be there for you to enjoy.

James Higham   **Michael Lück**
Dunedin, New Zealand   Auckland, New Zealand

1 Marine Wildlife and Tourism Management: In Search of Scientific Approaches to Sustainability

J.E.S. HIGHAM AND M. LÜCK

Introduction

The marine environment encompasses two-thirds of the surface of the 'blue planet' (Lück, 2007a). From inshore environments, such as estuaries, lagoons, atolls and reef systems, mud flats and mangroves, to the pelagic environments of the open oceans, the marine environment has become, albeit relatively belatedly (Orams, 1999), a major venue for tourism and recreation. Many marine environments, such as the North Atlantic Gulf Stream and the Antarctic convergence, boast high biomass and fantastic arrays of wildlife. Marine wildlife ranges from the complex ecologies of the Great Barrier Reef (Coral Sea) – coral reefs support over 25% of all known marine species (International Coral Reef Information Network, 2002) – to the Southern Ocean, where one link in the food chain is all that separates the smallest one-cell organisms from the largest animal on earth (see Maher, Chapter 16, this volume).

It is remarkable, then, that nature-based marine tourism has so recently become the subject of tourist attention. While marine environments have long been, and continue to be, venues for exploration, subsistence, transport and communication, merchant trade and conflict, recreation and tourist attention have relatively recently turned to the pursuit of marine experiences. Excursions to coastal resorts in Great Britain date to the 1850s, and beach holidays to the 1930s, following the unveiling of the bikini on the cover of *Vogue* magazine in 1929. The phenomenon of holidays at Mediterranean and Caribbean coastal and island resorts and destinations dates from the 1950s (Bramwell, 2004), and cruise shipping, exclusively the domain of the rich and famous in the early 20th century, has experienced a renaissance since the 1990s (Lück, 2007b).

In recent decades the spatial expression of marine tourism has expanded far beyond coastal resorts and the beach as a setting for leisure and recreation.

©CAB International 2008. *Marine Wildlife and Tourism Management: Insights from the Natural and Social Sciences* (eds J.E.S. Higham and M. Lück)

1

Forming an important part of this process, appreciation of and demand for marine wildlife experiences are recent developments. Viewing whales in the wild, for example, originally dates to the early 1950s, but the growth of commercial whale watching, along with other forms of non-consumptive wildlife-based marine tourism, has burgeoned since the 1980s (Hoyt, 2000). The scuba phenomenon and, as a consequence, a growing appreciation of the need to protect fragile marine ecologies also date to the 1980s (Bennett *et al.*, 2003).

Perhaps as a consequence of this belated development context a full appreciation and adequate conservation of marine environments remains largely unfulfilled. While the establishment of terrestrial national parks, initially Yellowstone National Park (USA), dates to the 1870s, the designation of marine protected areas (MPAs) remains a work in progress. For example, over one-third of New Zealand's land area has been designated for conservation (much of it in a system of national parks), yet less than 1% of New Zealand's extensive marine environments has been incorporated into a system of marine reserves (Department of Conservation, 2007). New Zealand's first marine reserve (Cape Rodney – Okakari Point Marine Reserve) was established in 1975 and was one of the world's first no-take marine reserves. There are now 28 marine reserves established in New Zealand waters with the majority initiated by applications lodged by groups such as the indigenous *tangata whenua*, conservation groups, fishers, divers and marine science interest groups (Department of Conservation, 2007).

However, 99% of the total area designated as New Zealand marine reserves lies in two extremely remote offshore island groups: Kermadec Island to the far north of New Zealand and the Auckland Islands to the far south. The Department of Conservation (2007) notes that 'of New Zealand's total marine environment, just 0.3% is protected in marine reserves'. This situation still exists despite the Department of Conservation's intention to incorporate 15% of New Zealand's marine environments into MPAs. Canada, which boasts the longest coastline of any nation (244,000 km), adopted an 'Oceans Action Plan for Present and Future Generations' as recently as 2005. This document notes in its foreword: 'Our oceans are important and represent an opportunity to make a greater contribution to our well-being and to benefit from the protection of critical marine environments' (Government of Canada, 2005, p. 3).

Prior to the Romantic movement of the 19th century, wilderness areas in Europe and North America were seen as cursed and chaotic wastelands (Oelschlager, 1991). In many respects marine environments are still seen in discriminatory terms. Many see marine environments as threatening, unpredictable and dangerous, not to mention home to some of the world's last great and least understood predators (see Dobson, Chapter 3, this volume). Although efforts to protect the megafauna and the great predators of terrestrial environments are well established, again, protection of their marine counterparts is belated and not so well advanced. The large-scale hunting of whale populations continued unopposed until the 'Save the Whales' campaigns of the 1970s (Barstow, 1986; Dalton and Isaacs, 1992), and in some countries the slaughter and exploitation of whales and dolphins continue even today.

Diversity of Marine Tourism

Yet, despite infuriatingly slow progress towards a new marine environmental paradigm, tourist interests in marine experiences and the growing diversity of marine tourism is perhaps evidence of the emergence of such a paradigm. Tourism activities that are set in coastal and marine environments have evolved far beyond the traditional passive leisure experiences of the classic resort holiday. While the traditional beach holiday remains a contemporary mass tourism phenomenon (Bramwell, 2004), marine tourism now extends beyond beach activities to a wide spectrum of activities, such as scuba-diving and snorkelling, windsurfing, jet skiing, fishing, sea kayaking, visits to fishing villages, marine parks and aquaria, sailing and motor yachting, maritime events and races, and the cruise ship industry, among others (Lück, 2007a). This list makes no specific mention of the tourists and their activities that are the focus of this book – those who specifically access marine environments to observe and appreciate marine wildlife.

Such has been the pace of growth in demand for marine tourism that visitor numbers, development of private sector tourism businesses and issuing of permits and consents, as well as outfitting of private recreational interests, have forged ahead of legislative and management responses aimed at sustainability. As such, marine recreation and tourism has, at least for the time being, been added to a lengthy list of interests that essentially treat the marine environment as a common pool resource to be exploited or otherwise used in the interests of personal gain or other reward.

Marine Environments: A Common Pool Resource

Today, as in the past, the vast majority of the global human population lives in close proximity to coastal areas. According to Burke *et al.* (2002), more than 350 million people live within 50 km of the coast in South-east Asia. Historically this has been due to the high biomass of riverine, estuarine and other littoral environments, making coastal areas strategically important in terms of the diversity and relative abundance of annual and seasonal subsistence resources. While this remains critically important in many parts of the world, coastal areas additionally offer strategic advantages in terms of communications, transport, commercial development of marine resources, indigenous claims for exclusive access to traditional marine resources, lifestyle, recreation and tourism. Thus, in terms of tourism as well as all other forms of human use and exploitation of the marine environment, the neritic (inshore) and pelagic (ocean) environments of the world remain, and have become, an increasingly contested, common pool resource.

Consequently, there exist manifold examples of resource use conflict in the marine context. The plunder of marine resources continues unabated in many parts of the world. Stocks of large fish species such as bluefin tuna (*Thunnus thynnus*) (one of the most prized fish at risk of overfishing), and long-lived species such as orange roughy (*Hoplostethus atlanticus*) and Patagonian toothfish (*Dissostichus eleginoides*) have been exploited to the very brink of collapse (Ellis, 2003). Efforts to establish Southern Ocean fishing rights and catch sizes led to the Convention for the Conservation of Antarctic Marine Living Resources (CCAMLR) in 1980. However, a sustainable toothfish industry remains dubiously improbable. The seabird by-catch of longline fisheries and the indiscriminate destruction of non-target species such as sea lions and diving birds by drift and set nets remain unresolved. However, it should be acknowledged that the 'tragedy of the commons' (Hardin, 1968) also applies in many instances of marine wildlife-based tourism development.

Whale Watch: The Vanguard of Marine Wildlife Tourism

Commercial whale watching dates to the 1950s and originated in Baja California/ Mexico and Hawaii (Tilt, 1987). Since then whale watching has proliferated into boat, land and airborne interactions with all 83 species of whales, dolphins and porpoises (Hoyt, 2000). The phenomenal growth in popularity of whale watching post-dates the mid-1980s. Whale and dolphin activities in Australia and New Zealand became major tourist activities from the late 1980s (Orams, 1999) and since then similar activities have become commercially available in destinations such as Indonesia, Hong Kong, Fiji, Tonga and the Solomon Islands (Lück, 2007a). In Asia, similar patterns of growth have occurred in the 1990s. Although in 1994, no whale watching whatsoever took place in Taiwan, in 1998, 30,000 people engaged in whale watching in that country. Despite a whale- and dolphin-hunting industry that is both highly visible and highly contentious internationally, whale and dolphin experiences have also become big business in Japan. An average annual growth rate of 37.6% between 1991 and 1998 demonstrates the rapid ascension of the whale-watching phenomenon in Japan. By 1998, more than 100,000 people sought whale- and dolphin-watching experiences in Japan, and spent nearly US$33 million in doing so (Hoyt, 2000).

Such rates and patterns of growth may be viewed as encouraging in terms of conservation, but they also raise intriguing questions relating to resource use conflicts. Thus, while whale hunting continues to be practised in Japan and Norway, these countries have also seen the development of significant whale-watching industries in recent years. Norway formally objected to the International Whaling Commission's moratorium which was set in 1986, and therefore never stopped hunting Minke whales. By contrast, whale watching began in Iceland in 1991, and within 3 years the number of whale watchers rapidly approached 10,000 per annum. By 1999, this number exceeded 30,000 and in 2001 over 60,000 people engaged in whale-watching activities (E. Hoyt, 2001; World Wide Fund for Nature, 2003). In 2002/03, the year in which the Icelandic government announced its intention of resuming scientific whaling, this period of rapid growth in whale watching came to an abrupt end (Higham and Lusseau, 2008). Thus, it is timely for researchers in the social science disciplines to address complex resource use issues such as the impacts of whale hunting (be it commercial, scientific or traditional/indigenous) on the whale-watching industry.

Of course, in the intervening years a diverse range of other marine wildlife viewing experiences have also grown in prominence, each associated with different global, national and regional environmental, resource conflict and conservation issues. These include viewing wading and migratory birds, marine mammals (from cetaceans and pinnipeds to polar bears), coral reef ecologies, species of great albatross, penguins and sharks. With each new manifestation of tourist engagements with marine wildlife come new and unique conservation and tourism management challenges.

Seeking the Insights of Natural Science

Tourist interactions with marine mammals on a regular basis can have detrimental effects on both focal animals and the health of local animal populations. While some science has paid attention to the impacts of tourism on wild animals, a comprehensive understanding of impacts is incomplete. Constantine (1999, p. 14) states that since 'the development of commercial dolphin watching and seal watching is a relatively new occurrence in most places, information on the effects of tourism on these animals is limited'.

The management of tourist interactions with cetaceans is a case in point. A range of rigorous publications concerning possible impacts on cetaceans in different regions and contexts has emerged (Finley *et al.*, 1990; J. MacGibbon, New Zealand, 1991; Gordon *et al.*, 1992; Corkeron, 1995; Williams *et al.*, 2002) but a comprehensive *understanding* of those impacts does not yet exist.

Hearing is the primary sense of cetaceans (Higham and Lusseau, 2004). They use vocalizations not only to communicate and maintain group cohesion (Janik and Slater, 1998), but also to locate prey and navigate using echolocation (Popper, 1980). Vocalization patterns are altered by the presence of tour boats. In the case of Humpback whales in Hawaii, the presence of boats has been found to affect song phase and unit duration (Norris, 1994). The production of an 'alarm signal', as well as an increase in silence time, in belugas and narwhals has been related to the presence of boats (Finley *et al.*, 1990). An increase in whistling rate in different species of dolphins has also been linked to the maintenance of group cohesion during interactions with boats (Scarpaci *et al.*, 2000; Van Parijs and Corkeron, 2001).

So clearly, it is difficult to assess the impact of human activities on marine mammals because they live in a different environment and use their senses differently from humans (Higham and Lusseau, 2004). Strict methodologies are necessary to interpret responses to anthropogenic impacts objectively. Several short-term studies have shown a variety of responses. Most studies have focused on behavioural changes depending on the presence and density of boats. In most cases, schools of animals tend to tighten when boats are present (e.g. Blane and Jaakson, 1995; Barr, 1996; Novacek *et al.*, 2001). Some species show signs of active avoidance. Responses range from changes in movement patterns (Edds and MacFarlane, 1987; Salvado *et al.*, 1992; Campagna *et al.*, 1995; Bejder *et al.*, 1999; Novacek *et al.*, 2001), to increases in dive intervals (Baker *et al.*, 1988; Baker and Herman, 1989; Blane, 1990; J. MacGibbon, 1991; Janik and Thompson, 1996), and increases in swimming speed (Blane and Jaakson, 1995; Williams *et al.*, 2002). These signs of avoidance can be a result of not only the presence of boats, but also the manoeuvring of boats including sudden changes in vessel speed or rapid approaches (J. MacGibbon, 1991; Gordon *et al.*, 1992; Constantine, 1999).

The presence and density of boats (Briggs, 1985; Kruse, 1991; Barr, 1996) and the distance between boats and individuals (Corkeron, 1995) can also affect the frequency or occurrence of behaviours. Humpback whales in Alaska have been seen reacting to vessels up to 4 km away from their pod (Baker *et al.*, 1988). In addition, the behavioural state of cetacean groups interacting with tourist vessels can be affected and changed (Ritter, 1996; Constantine and Baker, 1997; Lusseau, 2003). For example, interactions with boats led to a decrease in resting behaviour in spinner dolphins in Hawaii (Würsig, 1996); resting behaviour seems to be the most sensitive state to boat interactions (Lusseau, 2003).

More and more studies show that the navigation of vessels interacting with animals is a key parameter in the intrusiveness of interactions (Novacek *et al.*, 2001; Lusseau, 2002; Williams *et al.*, 2002). The more boats are manoeuvred unpredictably and erratically, the more animals tend to try to elude them. The observed avoidance strategies are similar to typical anti-predator responses (Howland, 1974). For decades many species of marine mammals have associated the presence of a boat following them at close range with the harpooning, distress and death of members of the pod. It is therefore not surprising that whales and dolphins employ anti-predator techniques when a vessel targets them directly, especially when the vessel attempts to out-manoeuvre or impair their movement. Of course, much of this research relates to the impacts of boat-based whale watching rather than land-based or airborne tourist activities, which have been the subject of significantly less research attention to date.

Unfortunately, most studies have examined only one aspect of complex impact problems and few studies have gathered data that can address the long-term impacts associated with tourist disturbance of wild animals. Increasingly, studies are being based on long-term observations, or are designed to capture a temporal element of analysis (Würsig, 1996; Constantine, 1999). Some studies have, for example, been able to relate changes in habitat use as well as avoidance of previously preferred areas to an increase in boat traffic (Baker *et al.*, 1988; Salden, 1988; Corkeron, 1995; Lusseau, 2002). Studies with long-term elements of analysis have now become a priority. To date, perhaps only the work of Bejder *et al.* (2006a) adequately meets this need.

Measuring and understanding biological significance

It is generally recognized that one critical but largely unresolved issue centres on the consequences of observed marine mammal avoidance responses. The biological consequences of increased dive times, decreased blow intervals, changes in travel directions, disruption of important behaviours and increases in aggressive behaviours are not adequately understood. It is necessary to relate the effects of the responses observed to standardized parameters such as the energetic budget of the species to assess their biological significance (Higham and Lusseau, 2004).

Moreover, observing the impacts of tourism on the behavioural budget of different populations offers the opportunity to scientifically link observational data to energetic budget (Lusseau, 2003). The behavioural budget of a population is directly linked to its energetic budget (Lusseau, 2002). It is therefore possible to assess the energetic cost of avoiding interactions with boats by observing the changes in the proportion of time engaged in different behavioural states (e.g. resting, socializing and feeding). New analytical techniques are opening this avenue of research and will afford more rigorous insights into the likely biological significance of observed responses (Lusseau, 2003). For some this means that a precautionary approach should be applied to the management of cetacean-watching activities until the real extent of the problem is understood scientifically (see Shelton and McKinley, Chapter 12, this volume). Clearly, there exists an urgent need to encourage and act upon good research in the natural sciences to inform the sustainable management of tourist interactions with marine wildlife species.

Developing Insights into the Social Science of Wildlife Tourism

The complex relationship between different human activities and resource utilities also raises a number of intriguing questions which need to be addressed by social scientists. Apart from the work of Herrera and Hoagland (2006), little is known about the social and economic opportunities and opportunity costs of whaling, whale watching, tourism and tourist boycotts of destinations where whale hunting continues. In instances where potential tourists are discouraged from visiting particular destinations because of whaling activities, an important question arises as to the net economic impact of such decisions.

Higham and Lusseau (2008), in their call for empiricism to address these issues, raise a range of timely research questions. They ask: Do tourists, both actual and latent, respond to the national stance on whaling of a country where they may otherwise choose to engage in whale watching? Do they respond differently to commercial, scientific and traditional/aboriginal whaling? Does whaling undertaken by indigenous communities actually add to the cultural mosaic that makes destinations unique and attractive? Do tourists engage in whale watching in a country that hunts whales to promote the prospects of whale watching becoming an exclusive alternative to killing whales of any species? Rodger *et al.* (2007) highlight the need to better understand the interface between visitors and wildlife. They note that an understanding of the social and environmental contexts of wildlife tourism generally must make a critical contribution to the sustainability of wildlife viewing.

Responding Effectively to Good Science

The engagement of scientists from both the natural and social science discip-lines is, however, only a first step. The effectiveness of good science ultimately rests with the ability for policy makers and resource managers to respond to research, and apply the insights achieved by the scientific community in mean-ingful and effective ways. Here again lies a barrier to sustainable tourist–wildlife interactions in marine contexts. In their study of wildlife tours in Australia, Rodger *et al.* (2007) specifically address the place of science and monitoring in wildlife tourism businesses. Their recent results demonstrate low levels of engagement of scientists in protecting the wildlife of interest to tours. They conclude that 'given the centrality of science to sustainability, mechanisms for increasing this involvement particularly in impact research, through partner-ships and other means, are critical for the long-term sustainability of this indus-try' (Rodger *et al.*, 2007, p. 160).

The management of tourist interactions with wild dolphin populations in various parts of the world demonstrates a lack of acknowledgement of the find-ings of scientists, and a high degree of policy and effective management paraly-sis. Data collection dating back over 20 years at Shark Bay (Western Australia), where low-level commercial tourism brings groups of tourists into interactions with bottlenose dolphins (*Tursiops* sp.), was recently published in the December 2006 (Vol. 20, No. 6) issue of *Conservation Biology* (Bejder *et al.*, 2006a). This article provides rich historical insights into the development of dolphin-based tourism over time. Specifically, the unique Shark Bay data set allows detailed interrogation of the long-term impacts of vessel activity in the vicinity of bottlenose dolphins. Bejder *et al.* (2006a) present data generated over three phases: a pre-tourism phase, through the establishment of one commercial dolphin-watching operation, and two commercial operations. Through all of these phases research activity was constant. Their data collection also affords the comparative analysis of dolphin behaviour in zones where interactions with tourists take place and control (non-tourism) sites. They report that

> A nonlinear logistic model demonstrated that there was no difference in dolphin abundance between periods with no tourism and periods in which one operator offered tours. As the number of tour operators increased to two, there was a significant average decline in dolphin abundance . . . approximating to a decline of one per seven individuals.
>
> (Bejder *et al.*, 2006a, p. 1793)

Their research also identified a divergence in the tourism and control site data sets based on an analysis of patterns of dolphin avoidance and reduced female reproductive success.

The authors conclude that where 'small, closed, resident, or endangered cet-acean populations' are exposed to such impacts, the consequences are likely to be serious. These findings are derived from a site of low levels of recreational and commercial tourism activity. In contrast, Higham and Hendry (see Chapter 19, this volume) report on whale watching in the San Juan Islands (USA) where it is not uncommon to witness more than 100 commercial and private boats following a group of 25–30 cetaceans (Kind-Keppel *et al.*, 1999). Bejder *et al.* (2006a) highlight both the critical need for good science and the considerable challenge of the sustainable management of tourist–wildlife interactions. Where a large fleet of vessels seeks interactions with small, closed or endangered cetacean populations – not an uncommon scenario when tourist demand runs ahead of appropriate and comprehensive management response – the situation is particularly pressing.

So what, precisely, has been the response to the findings from Shark Bay published by Bejder *et al.* (2006a) in *Conservation Biology*? Both the Department of Conservation and Land Management (CALM) and the Marine Parks and Reserves Authority (MPRA) considered the research findings and the options to reduce the exposure of dolphins to tour vessels and provided advice to the Western Australian Minister of Environment. After careful consideration and consultation with CALM, MPRA, the existing licence holders, other dolphin researchers and stakeholders, Mark McGowan, the Minister for the Environment decided, among other things, to reduce the number of commercial dolphin-watching licences from two to one and to introduce a moratorium on any increase in research vessel activity in the affected area (Western Australian Environment Ministry Media Statement, 2006, see Box 1.1).

The Minister for the Environment clearly stated that the Shark Bay tourism industry (including dolphin provisioning at Monkey Mia) was almost entirely based on dolphin experiences and the withdrawal of one licence was a necessary sacrifice for the long-term sustainability of tourism in the area. An expression of interest

---

Box 1.1. Western Australian Environment Ministry Media Statement (26 June 2006) relating to sustainable tourist–dolphin interactions at Shark Bay. (From Western Australian Environment Ministry Media Statement, 2006.)

Long-term sustainability central to Monkey Mia decision (26 June 2006)

The Monkey Mia dolphin population will be given a lifeline, following a decision by Environment Minister Mark McGowan to reduce the number of commercial boat tour licences in the area. Mr McGowan announced today that he would reduce the number of licences issued to marine-based wildlife interaction tour operators in the Monkey Mia Bay from two to one, in the interests of the dolphin populations in the area. 'I will extend the two existing licences – which expire on June 30 – for another three months while an expression of interest process is undertaken to determine a new sole licensee,' he said. The Minister made the decision after carefully considering wide-ranging advice on the best manner in which to license and manage tour boat activities into the future. 'Unfortunately, the research shows that both dolphin populations – the Red Cliff Bay dolphins and the Monkey Mia beach dolphins – are being affected by the tour boat activities,' he said. 'A study by Murdoch University researcher Dr Lars Bejder has found that the Red Cliff Bay dolphins have been using the area frequented by the tour vessels less and less. The same study also found that females exposed to the vessels had lower reproductive success than the females with less exposure. The new licence will strictly limit the number and time of dolphin interactions, as well as minimising engine and propeller impacts of tour vessels. I will also introduce a moratorium on any increased research vessel activity within the Red Cliff Bay area and seek a review of the operations of private and commercial fishing vessels.'

process was subsequently undertaken to determine a sole commercial operator (Naturebase, 2006).

The Second Australian National Wildlife Tourism Conference which was hosted by Wildlife Tourism Australia (WTA) and the Forum Advocating Cultural and Eco-Tourism (FACET) took place in Fremantle, Western Australia, during 13–15 August 2006, soon after the aforementioned ministerial decision. The conference explored issues surrounding the development and long-term sustainable management of wildlife tourism and succeeded in highlighting and exploring a range of key issues that are central to the sustainability of wildlife-based tourism. Minister McGowan's statement on dolphin-based tourism at Shark Bay was both timely and topical, and provided much basis for discussion at the conference.

Two clear conclusions to emerge from the conference were that: (i) it is only with rigorous scientific research that we can begin to understand the complex relationship that prevails when tourists engage with wild animals (individual animals or populations of animals); and (ii) managers must be responsive to the outcomes of rigorous science. These conclusions were clearly articulated in a series of resolutions which were discussed at the closing session of the conference, and drafted in full (with post-conference delegate input via e-mail) following the conference. The conference resolutions included the following statement:

> The conference delegates endorse and support the decision by Western Australian Minister for the Environment, Mark McGowan, to reduce the number of commercial boat tour licences in Shark Bay in response to research into the impacts of tour boat activities on dolphins.
>
> (FACET, 2006, n.p.)

In Shark Bay, the dolphin-watching tourism industry is licensed and controlled, yet measurable impact over a relatively brief period has been documented (Bejder *et al.*, 2006a). If the findings at this site of low-level tourism are extrapolated to the many high-level tourism sites around the world (e.g. killer whales in British Columbia, Canada (Williams *et al.*, 2002), bottlenose dolphins in the Bay of Islands (Constantine, 1999) and Port Stephens, Australia (Allen, 2005)), one might conclude that cetacean-based tourism may not be as low-impact as previously presumed. Given the scarcity of studies with adequate controls or longevity to fully evaluate tourism impacts, a cumulative impact, like that detected in Shark Bay, could go unnoticed for many years, perhaps decades. This case clearly reinforces the need for responsive and proactive management.

However, despite the concerted efforts of various stakeholders with interests in Shark Bay, challenges remain. While dolphin-viewing permits in Shark Bay have been reduced to one, nothing can legally prevent other commercial operators or private boat owners operating vessels in the area, including the control site that has previously been used by mutual agreement exclusively for research purposes. Thus, it is possible that despite the best of intentions, the recent development at Shark Bay may inadvertently result in an expanded spatial range of tourism operations and, therefore, an expanded range of tourism impacts as well as the loss of comparative data from tourism and non-tourism (control) sites.

Meanwhile, on the east coast of Australia, the New South Wales state government adopted the new National Parks and Wildlife Amendment (Marine

Mammals) Regulations 2006 to apply in that state from 2 June 2006. The main features of the amendment include:

- Minimum approach distances in line with the new Australian guidelines;
- New penalty infringement notices (AUS$300) for any recreational and commercial vessel breaching the regulations;
- New operating rules for vessels and aircraft;
- Provision for the minister to declare approach distances for special interest marine mammals.

In a subsequent letter to all commercial operators and other tourism stakeholders, emerged the following:

> In recognition of the importance of the commercial marine mammal observation tour industry to regional economies and the role of industry in educating the public about marine mammals, the Minister for the Environment has asked the Department of Environment and Conservation to investigate a *closer approach distance to whale and dolphin calves for commercial marine mammal observation tour operators* than that prescribed in the Regulation. [emphasis added]
>
> (S.J. Allen, 2006, Sydney, personal communication)

Such a move has no doubt mystified the research community given that any such decision would fly in the face of a significant weight of scientific research that confirms the importance of approach direction, speed and distance in terms of the impacts of tourism upon focal animals (Baker *et al.*, 1988; Corkeron, 1995; Ritter, 1996; Würsig, 1996; Constantine and Baker, 1997; Lusseau, 2003), with animals engaging in resting behaviour most likely to be disturbed by approach distance (Lusseau, 2003).

In recent years, there has been a call for site- and species-specific research into the impacts of tourist interactions with various species of marine wildlife. This call is echoed in various chapters in this volume (see Seddon and Ellenberg, Chapter 9, this volume). However, the urgency of the dolphin-viewing situation perhaps argues in support of the case for careful management in respect to some clear and consistent impact issues that are now well documented in the scientific literature (Higham and Lusseau, 2004). Well-researched sites provide clear indications that dolphin-based tourism should be subject to close management (Lusseau, 2003) to limit interactions, as well as allow a degree of both spatial and temporal relief from anthropogenic interference. Meanwhile, at Port Stephens (New South Wales), where no fewer than 17 dolphin-watching boats operate, the likelihood of medium- or long-term sustainability must be brought into question.

Similarly at Kealakekua Bay (Hawaii) spinner dolphins (*Stenella longirostris*) come inshore in the middle of the day to rest, making them a likely target for observation by visitors on boats or kayaks or in the water (Driscoll-Lind and Östman-Lind, 1999). Barber's (1993) land-based observational research demonstrated shorter resting periods for animals exposed to swimmers and to boat traffic. Fortunately in this case, the State of Hawaii Department of Land and Natural Resources (DLNR) has in recent months moved to establish a temporary human exclusion area (HEA) to protect the critical resting areas of spinner dolphins in Kealakekua Bay. It is intended that after a 1-year trial period, DLNR will implement a more permanent management protocol, which may include the continued use of an HEA.

The urgency is apparent

'Manage it or lose it' is the conclusion drawn by Bejder in the delivery of his paper at the Second Australian National Wildlife Tourism conference in August 2006 (Bejder *et al.*, 2006b). This is a conclusion that could apply to many forms of tourist engagements with marine wildlife populations, hence the title of this book. In numerous instances of tourist–wildlife interactions, it has proved that voluntary codes of practice and self-regulation do not work in the absence of limits applied to the issuing of commercial operator permits, frequency and duration of interactions and numbers of vessels and/or visitors interacting with animals. Numerous sites worldwide, many mentioned in this chapter, where boat-based interactions with cetaceans take place, such as Port Stephens (New South Wales), Shark Bay (Western Australia), Bay of Islands and Doubtful Sound (New Zealand), San Juan Islands (USA), Kealakekua Bay (Hawaii), Puget Sound (Canada), Moray Firth (Scotland) and Baja (USA/Mexico), all point towards the need for careful visitor management. Despite an expanding body of research that demonstrates the urgency of careful management, little or nothing is happening. The challenge clearly remains to turn scientific knowledge (where it exists) into management actions.

However, it is also important to recognize that tourism is often seen as a pariah and is treated as an easy target for those with concerns for sustainable resource management (see Shelton and McKinlay, Chapter 12, this volume). Concerns for the impacts of human activities upon marine wildlife do not relate exclusively to tourism, indeed some would argue that the impacts of tourism pale alongside the more immediate and in many cases terminal consequences of, for example, fisheries by-catches. In recent times it has also been interesting to note members of the scientific community responding to the impacts of their own research (and that of other researchers) on focal animals. In 2006, the collection of biopsy samples from bottlenose dolphins in Doubtful Sound (New Zealand) by a team of Auckland University marine biologists resulted in protests from fellow scientists. Increasingly, marine scientists are seeking new approaches to the mitigation of research impacts (Lusseau, 2003). Simultaneously, calls in New Zealand to ban recreational set netting in selected inner harbour and in-shore habitats to protect the engendered Hector's dolphin (*Cephalorhynchus hectori*) have largely fallen upon deaf ears.

Thus, it seems that decisions – and instances of apparent indecision – tend to be based on economics and politics which often work against, rather than for, interests in sustainability. In tourism, as in these other areas, there remain considerable barriers to effective planning, the establishment of clear management objectives, positive incentives for good research and management responsiveness to good science. The application of science to marine tourism, as well as other big system issues, remains deeply challenging to social and political systems. Furthermore, the challenge of integrating science into complex systems to accommodate medium- to long-term future timeframes is a challenge that continues to remain outstanding.

Conclusion

It has been noted previously that the Second Australian National Wildlife Tourism Conference (Fremantle, Western Australia, 13–15 August 2006) concluded with a declaration that included a range of research and management priorities. Among them were some that bear considerable relevance to the central point of emphasis in this chapter. They included to:

- Conduct research to support identification, evaluation and monitoring of environmental impacts associated with wildlife tourism
- Review legislation relating to wildlife tourism, with a view to achieving 'uniform' national regulations and focusing more on positive outcomes
- Develop specific sustainability indicators for wildlife tourism to ensure identification and management of priority environmental impacts
- Build better coordination and cooperation in data collection mechanisms and systems
- Develop and promote broad uptake of national guidelines for managing impacts, especially of sensitive interaction types/species
- Undertake long-term research and monitoring involving sensitive species/ interactions and integrate this with management
- Prioritise research on species and sites of most concern in relation to impact management

(FACET, 2006)

This book seeks to underscore the urgent need for scientific approaches to first understanding and then managing tourist interactions with marine wildlife. It draws upon the work of leading natural and social scientists whose work serves the interests of sustainable wildlife-based marine tourism.

Thus, from within the natural science disciplines of marine biology, environmental science, behavioural ecology, conservation biology and wildlife management come chapters that provide insights into the effects of human disturbance on marine wildlife, understanding impacts that tourists may have upon wild animals, and management approaches to mitigating impacts that may in the long term be biologically significant. Equally from the social science disciplines of geography, sociology, management and social anthropology are drawn chapters that explore demand for marine wildlife experiences, the benefits that visitors derive from their experiences, ethical and legislative contexts and management issues that arise when tourists interact with populations of wild animals in coastal and marine environments.

This book inevitably, perhaps preferably, poses more questions than it answers. Selected chapters provide rigorous scientific insights that should inform the management of wildlife tourism; others raise challenges and articulate important research questions that may be taken up by researchers in the natural and social science disciplines. In both cases, the fundamental aim is to advance an understanding of the complexities of marine wildlife and tourism management, while seeking to gather further momentum behind the advancement and uptake of scholarly research serving this important field.

References

Allen, S.J. (2005) Management of bottlenose dolphins (*Tursiops aduncus*) exposed to tourism in Port Stephens, New South Wales, Australia. MS thesis, Graduate School of the Environment, Macquarie University, Sydney, Australia.

Baker, C.S. and Herman, L.M. (1989) *Behavioural Responses of Summering Humpback Whales to Vessel Traffic: Experimental and Opportunistic Observations*. Final Report to the National Park Service, Alaska Regional Office, Anchorage, Alaska.

Baker, C.S., Perry, A. and Vequist, G. (1988) Humpback whales of Glacier Bay, Alaska. *Whalewatcher* Fall, 13–17.

Barr, K. (1996) Impacts of tourist vessels on the behaviour of dusky dolphins (*Lagenorhynchus obscurus*) at Kaikoura. MSc thesis, Department of Marine Sciences, University of Otago, Dunedin, New Zealand.

Barstow, R. (1986) Non-consumptive utilization of whales. *Ambio* 15(3), 155–163.

Bejder, L., Dawson, S.M. and Harraway, J.A. (1999) Responses by Hector's dolphins to boats and swimmers in Porpoise Bay, New Zealand. *Marine Mammal Science* 15(3), 738–750.

Bejder, L., Samuels, A., Whitehead, H., Gales, N., Mann, J., Connor, R., Heithaus, M., Watson-Capps, J., Flaherty, C. and Krützen, M. (2006a) Decline in relative abundance of bottlenose dolphins (*Tursiops* sp.) exposed to long-term anthropogenic disturbance. *Conservation Biology* 20(6), 1791–1798.

Bejder, L., Whitehead, H., Samuels, A., Mann, J., Connor, R., Gales, N., Heithaus, M., Watson-Capps and Flaherty, C. (2006b) *Decline in Relative Abundance of Bottlenose Dolphins Exposed to Long-Term Disturbance*. The 2nd Australian Wildlife Tourism Conference, Fremantle, Australia, 13–15 August 2006.

Bennett, M., Dearden, P. and Rollins, R. (2003) The sustainability of dive tourism in Phuket, Thailand. In: Landsdown, H., Dearden, P. and Neilson, W. (eds) *Communities in SE Asia: Challenges and Responses*. University of Victoria, Centre for Asia Pacific Initiatives, Victoria, British Columbia, pp. 97–106.

Blane, J.M. (1990) Avoidance and interactive behaviour of the Saint Lawrence beluga whale (*Delphinapterus leucas*) in response to recreational boating. MA thesis, University of Toronto, Canada.

Blane, J.M. and Jaakson, R. (1995) The impact of ecotourism boats on the Saint Lawrence beluga whales. *Environmental Conservation* 21(3), 267–269.

Bramwell, B. (2004) Mass tourism, diversification and sustainability in southern Europe's coastal regions. In: Bramwell, B. (ed.) *Coastal Mass Tourism: Diversification and Sustainable Development in Southern Europe*. Channel View Publications, Clevedon, UK, pp. 1–31.

Briggs, D.A. (1985) *Report on the Effects of Boats on the Orcas in the Johnstone Strait from July 11, 1984 – September 1, 1984*. UCSC, Santa Cruz, California.

Burke, L., Selig, L. and Spalding, M. (2002) *Reefs at Risk in Southeast Asia*. Available at: http://www.wri.org/reefsatrisk/reefsatriskseaisa.html

Campagna, C., Rivarola, M.M., Greene, D. and Tagliorette, A. (1995) *Watching Southern Right Whales in Patagonia*. UNEP, Nairobi.

Constantine, R. (1999) Effects of Tourism on Marine Mammals in New Zealand. Science for Conservation, 106, Department of Conservation, Wellington, New Zealand.

Constantine, R. and Baker, C.S. (1997) Monitoring the commercial swim-with-dolphin operations in the Bay of Islands. *Science and Research Series* 104. Department of Conservation, Wellington, New Zealand.

Corkeron, P.J. (1995) Humpback whales (*Megaptera novaeangliae*) in Hervey Bay, Queensland: behaviour and responses to whale-watching vessels. *Canadian Journal of Zoology* 73, 1290–1299.

Dalton, T. and Isaacs, R. (1992) *The Australian Guide to Whale-watching*. Weldon Publishing, Sydney, Australia.

Department of Conservation (2007) *Marine Reserves in New Zealand*. Available at: http://www.doc.govt.nz/templates/summary.aspx?id = 33776

Driscoll-Lind, A. and Östman-Lind, J. (1999) Harassment of Hawaiian spinner dolphins by the general public. *MMPA Bulletin* 17, 8–9.

Edds, P.L. and MacFarlane, J.A.F. (1987) Occurrence and general behaviour of balaenopterid cetaceans summering in the Saint Lawrence Estuary, Canada. *Canadian Journal of Zoology* 65, 1363–1376.

Ellis, R. (2003) *The Empty Ocean*. Island Press, Washington, DC.

Finley, K.J., Miller, G.W. and Davis, R.A. (1990) Reactions of belugas, *Delphinapterus leucas*, and narwhals, *Monodon monoceros*, to ice-breaking ships in the Canadian high Arctic. *Canadian Bulletin of Fisheries and Aquatic Science* 224, 97–117.

Forum for the Advancement of Cultural and Ecotourism (FACET) (2006) *Conference Resolutions*. The 2nd Australian National Wildlife Tourism Conference, 13–15 August 2006, Fremantle, Western Australia. Available at: www.facet.asn.au

Gordon, J., Leaper, R., Hartley, F.G. and Chappell, O. (1992) Effects of whale-watching vessels on the surface and underwater acoustic behaviour of sperm whales off Kaikoura, New Zealand. *Science and Research Series*, 52. Department of Conservation, Wellington, New Zealand.

Government of Canada (2005) *Canada's Oceans Action Plan: For Present and Future Generations*. Communications branch, Fisheries and Oceans Canada. Ottawa, Ontario. Available at: www.dfo-mpo.gc.ca

Hardin, G. (1968) The Tragedy of the Commons. *Science* 162, 1243–1248.

Herrera, G.E. and Hoagland, P. (2006) Commercial whaling, tourism and boycotts: an economic perspective. *Marine Policy* 30, 261–269.

Higham, J.E.S. and Lusseau, D. (2004) Ecological impacts and management of tourist engagements with Cetaceans. In: Buckley, R. (ed.) *Environmental Impacts of Ecotourism*. CAB International, Wallingford, UK, pp. 173–188.

Higham, J.E.S. and Lusseau, D. (2008) Slaughtering the goose that lays the golden egg: are whale-watching and whaling mutually exclusive? *Current Issues in Tourism* 11(1), in press.

Howland, H.C. (1974) Optimal strategies for predator avoidance: the relative importance of speed and manoeuvrability. *Journal of Theoretical Biology* 47, 333–350.

Hoyt, E. (2000) *Whalewatching 2000: Worlwide Numbers, Expenditures, and Expanding Socioeconomic Benefits*. Available at: http://www.ifaw.org/press/whalewatching2000.html

Hoyt, E. (2001) Whale watching 2001. *Report to IFAW and UNEP*. London.

International Coral Reef Information Network (2002) *About Coral Reefs*. Available at: http://www.coralreef.org/coralreefinfo/about.html

Janik, V.M. and Slater, P.J.B. (1998) Context-specific use suggests that bottlenose dolphin signature whistles are cohesion calls. *Animal Behaviour* 56, 829–838.

Janik, V.M. and Thompson, P.M. (1996) Changes in surfacing patterns of bottlenose dolphins in response to boat traffic. *Marine Mammal Science* 12, 597–602.

Kind-Keppel, J., Nikolay, A., Muloin, S. and Otis, R. (1999) Whale watchers' attitudes towards boats accompanying killer whales (*Orcinus orca*). Paper presented at the 1999 Northeast Regional Animal Behaviour Society Annual Meeting, C.W. Post Campus, Long Island University, New York.

Kruse, S. (1991) The interactions between killer whales and boats in Johnstone Strait, B.C. In: Norris, K.S. and Pryor, K. (eds) *Dolphin Societies: Discoveries and Puzzles*. University of California Press, Berkeley, California, pp. 149–159.

Lück, M. (2007a) Nautical tourism development: opportunities and threats. In: Lück, M. (ed.) *Nautical Tourism: Concepts and Issues*. Cognizant Communication Corp., Elmsford, New York (in press).

Lück, M. (2007b) The cruise ship industry: curse or blessing? In: Lück, M. (ed.) *Nautical Tourism: Concepts and Issues*. Cognizant Communication Corp., Elmsford, New York (in press).

Lusseau, D. (2002) The effects of tourism activities on bottlenose dolphins (*Tursiops* spp.) in Fiordland. PhD thesis, University of Otago, Dunedin, New Zealand.

Lusseau, D. (2003) The effects of tour boats on the behavior of bottlenose dolphins: using Markov chains to model anthropogenic impacts. *Conservation Biology* 17, 1785–1793.

MacGibbon, J. (1991) *Responses of Sperm Whales (Physeter macrocephalus) to Commercial Whale Watching Boats off the Coast of Kaikoura*. Department of Conservation, Wellington, New Zealand.

Naturebase (2006) Available at: http://www.naturebase.net/tourism/pdf_files/mm_eoi_guidelines_ 04072006.pdf

Norris, T. (1994) Effects of boat noise on the acoustic behavior of humpback whales. *Journal of the Acoustic Society of America* 96(5–2), 3251–3257.

Novacek, S.M., Wells, R.S. and Solow, A.R. (2001) Short-term effects of boat traffic on bottlenose dolphins, *Tursiops truncatus*, in Sarasota Bay, Florida. *Marine Mammal Science* 17(4), 673–688.

Oelschlager, M. (1991) *The Idea of Wilderness: From Prehistory to the Age of Ecology*. Yale University Press, New Haven, Connecticut.

Orams, M.B. (1999) *Marine Tourism: Development, Impacts and Management*. Routledge, London.

Popper, A.N. (1980) Sound emission and detection by Delphinids. In: Herman, L.M. (ed.) *Cetacean Behaviour: Mechanisms and Functions*. Wiley, New York, pp. 1–52.

Ritter, F. (1996) Abundance, distribution and behaviour of cetaceans off La Gomera (Canary Islands) and their interaction with whale-watching boats and swimmers. Diploma thesis, University of Bremen, Germany.

Rodger, K., Moore, S.A. and Newsome, D. (2007) Wildlife tours in Australia: characteristics, the place of science and sustainable futures. *Journal of Sustainable Tourism* 15(2), 160–179.

Salden, D.R. (1988) Humpback whales encounter rates offshore of Maui, Hawaii. *Journal of Wildlife Management* 52(2), 301–304.

Salvado, C.A.M., Kleiber, P. and Dizon, A.E. (1992) Optimal course by dolphins for detection avoidance. *Fishery Bulletin* 90, 417–420.

Scarpaci, C., Bigger, S.W., Corkeron, P.J. and Nugegoda, D. (2000) Bottlenose dolphins (*Tursiops truncatus*) increase whistling in the presence of swim-with-dolphin tour operations. *Journal of Cetacean Research and Management* 2(3), 183–185.

Tilt, W.C. (1987) From whaling to whalewatching. Paper presented at the 52nd North American Wildlife and Natural Resources Conference, Washington, DC.

Van Parijs, S.M. and Corkeron, P.J. (2001) Boat traffic affects the acoustic behaviour of Pacific humpback dolphins, *Sousa chinensis*. *Journal of the Marine Association UK* 81, 533–538.

Western Australian Environment Ministry Media Statement (2006) Available at: http://www.media statements.wa.gov.au/media/media.nsf/news/ 958A19167C70F7934825719900206D69

Williams, R., Trites, A.W. and Bain, D. (2002) Behavioural responses of killer whales (*Orcinus orca*) to whale-watching boats: opportunistic observations and experimental approaches. *Journal of Zoology* 256, 255–270.

World Wide Fund for Nature (2003) *Whale Watching: A Future for Whales?* WWF Report. Available at: http://www.panda.org

Würsig, B. (1996) Swim-with-dolphin activities in nature: weighing the pros and cons. *Whalewatcher* 30(1), 11–15.

I

Demand for Marine Wildlife Tourism

2 Marine Wildlife Tours: Benefits for Participants

H. ZEPPEL AND S. MULOIN

Introduction

Wildlife-based tours in marine and coastal areas provide a range of psychological, educational and conservation benefits for visitors encountering marine animals (Higham, 1998; Orams, 2000; Schänzel and McIntosh, 2000; Lück, 2003; Finkler and Higham, 2004; Mayes *et al.*, 2004; Hughes and Saunders, 2005; Tisdell and Wilson, 2005; Andersen and Miller, 2006). This chapter reviews and evaluates benefits for participants on marine wildlife tours. The focus is on non-consumptive, free-ranging marine wildlife tourism where visitors can view, photograph, feed, and swim with, or assist in research on, marine animals in their natural habitats. Other broader participants include the marine tour operators, coastal and island communities in marine areas and researchers studying marine wildlife and/or tourists. Most research on marine wildlife tourism addresses environmental impacts on sea animals, industry compliance with codes of conduct and managing visitor interactions with marine species. However, this chapter reviews studies that primarily focus on tourist experiences of marine and coastal wildlife in Australia, New Zealand, Scotland and western Canada/USA. Much of this research on marine wildlife tourism is site or species specific and limited to one type of encounter. There is a need for more systematic, in-depth evaluation of marine wildlife tourism experiences and educational programmes to identify techniques that increase tourist benefits and knowledge, promoting attitude shifts and lifestyle changes (Samuels *et al.*, 2003). In addition, both on-site and longer-term conservation behaviours that benefit marine wildlife and marine environments need to be explored. This chapter introduces marine wildlife tours and visitor benefits from marine wildlife encounters, then critically reviews the psychological, educational and conservation benefits of tourist participation in a range of marine wildlife experiences.

Marine Wildlife Tours

Marine wildlife tourism is defined as 'any tourist activity with the primary purpose of watching, studying or enjoying marine wildlife' (Masters, 1998). It includes marine wildlife-watching holidays, wildlife boat trips in marine or estuarine areas, guided island or coastal walks, observing marine life from land viewpoints, visiting marine or coastal nature reserves, participating in a marine life study tour or conservation holiday, and visiting marine wildlife visitor centres and marine aquaria. Marine wildlife includes 'flora and fauna that live in the coastal and maritime zone and are dependent on resources from the marine environment' (Masters, 1998). There are around 30,000 marine species, compared to 10,000 bird species (Wildlife Extra, 2006). This chapter focuses on mobile free-ranging marine animals such as marine mammals, sharks, fish, rays, turtles and seabirds. Marine mammals, in particular, are a key tourism attraction (Birtles et al., 2001; Stokes et al., 2002; Orams, 2003, 2005; Higham and Lusseau, 2004). Popular marine mammals include dolphins (Orams, 1997a; Hughes, 2001), whales and porpoise (i.e. cetaceans); dugong and manatee (Sorice et al., 2006); and seals and sea lions (i.e. pinnipeds) (Barton et al., 1998; Booth, 1998; Kirkwood et al., 2003; Scarpaci et al., 2005). Other marine wildlife of tourist interest includes whale sharks and other shark species (Dobson, 2006); fish and rays; sea turtles (Wilson and Tisdell, 2001); and penguins, albatross, gannet and other seabirds. Worldwide, 500,000 divers a year now feed, photograph and swim with sharks (Topelko and Dearden, 2005). Nesting or rookery areas for seabirds and marine turtles (Higham, 1998, 2001; Schänzel and McIntosh, 2000; Tisdell and Wilson, 2002) and haul-out areas for seals and sea lions (Orsini and Newsome, 2005) also attract visitors.

In Australia in 1999, there were over 70 marine species targeted for marine tourism, from whales (e.g. Humpback, southern right and dwarf minke), dolphins, turtles, sea lions and seals, to penguins, fish, sharks (e.g. reef, grey nurse, great white and whale sharks), rays, sea dragons and cuttlefish. Fish species promoted as an attraction to divers on the Great Barrier Reef include potato cod, Queensland grouper, giant Maori wrasse, moray eels, anemone fish and scorpion fish. Penguins, seals and whales are a visitor attraction in Antarctica (Birtles et al., 2001). Dolphin and fish-feeding programmes; swimming with sea lions, whales, dolphins and manta rays; and stingray feeding are other marine wildlife attractions in Australia (Newsome et al., 2005). A survey of 376 marine tourism operators in New Zealand also found viewing marine wildlife was a key attraction, focusing on marine mammals (44%, with 22% on dolphins), sea birds (42%), fish (30%), penguins (18%) and other marine wildlife (16%) (McKegg et al., 1998 cited in Orams, 2003, p. 237). By 2000, there were 75 permits allocated for marine mammal tourism in New Zealand, mainly for dolphin watching and swimming (Orams, 2005). Marine species endemic to New Zealand include Hector's dolphins, Hioho or yellow-eyed penguins and Hooker's sea lions. Sperm whales at Kaikoura and dolphins (dusky, common and bottlenose) along with royal albatross, penguin and gannet rookeries also offer unique marine wildlife experiences (Pearce and Wilson, 1995; Amante-Helweg, 1996).

Marine animals in their natural habitats range from completely wild, habituated to humans in coastal areas, to provisioned or fed (e.g. dolphins, fish, rays, moray eels and sharks). The types of viewing platforms used by tourists to watch free-ranging marine wildlife species include shore-based, boat-based and in-water encounters (Birtles *et al.*, 2001). Onshore viewing of marine wildlife includes rookeries and resting areas (e.g. turtles, seabirds, seals and sea lions) and feeding marine animals in shallow water at the shoreline (e.g. stingrays, fish and dolphins). Vantage points on land also provide for near-shore viewing of marine wildlife such as migrating whales, dolphins, turtles, sharks, seals, sea lions, fish shoals and seabirds. Boat-based tours focus on cetaceans, pinnipeds, sharks, fish and pelagic seabirds, while in-water encounters include tourists swimming, snorkelling or scuba diving with marine animals such as whales, dolphins (Würsig, 1996; Samuels *et al.*, 2003), turtles, dugong, manatee, seals, sea lions, sharks, rays and fish. Some 100,000 visitors a year feed and stroke stingrays off Grand Cayman Island in the western Caribbean (Shackley, 1998) and snorkel with manatees at Crystal River in Florida, USA (Sorice *et al.*, 2006). 'Interactive diving', based on operators feeding sharks, fish and eels, is a growing activity although shark feeding was banned in 2002 in Florida, Hawaii and the Cayman Islands. Diving in cages underwater with great white sharks is a tourist attraction in South Africa and South Australia. Hence, there are now many opportunities for visitors to interact with marine wildlife at sea and near the shore, from passive viewing on land and from boats, to more active swim with activities.

Visitor Benefits from Marine Wildlife Encounters

Visitor benefits from participation in marine wildlife tourism experiences include enhanced psychological, educational and conservation or environmental outcomes (Higham, 1998; Orams, 2000; Schänzel and McIntosh, 2000; Tisdell and Wilson, 2002, 2005; Lück, 2003; Finkler and Higham, 2004; Mayes *et al.*, 2004; Hughes and Saunders, 2005; Andersen and Miller, 2006). Psychological benefits of tourist encounters with marine wildlife include excitement, novelty, intensity and uniqueness promoting personal well-being and quality of life (Muloin, 1998; Schänzel and McIntosh, 2000; Birtles *et al.*, 2002; Curtin, 2005). Marine wildlife tourism results in psychological and sociopsychological outcomes such as learning about wildlife, relaxation in marine areas and personal growth (Muloin, 1998). There are five main types of outcomes or benefits for visitors interacting with marine wildlife: physiological, economic, environmental, social and psychological. Social values influence the type of benefits sought from marine wildlife interactions, such as the trend towards non-consumptive viewing of cetaceans rather than killing whales (Frohoff and Packard, 1995; Bulbeck, 1999; Hoyt, 2003). Kellert (1999) evaluated how Americans perceive marine mammals and key management issues. Psychological benefits are also related to individual needs or motivations, while visitor satisfaction relates to personal benefits received from the marine wildlife encounter. Visitors are intrinsically motivated to seek these personal rewards or benefits through interactions with marine wildlife, while also escaping from their normal routines and daily life.

Visitor motivation and satisfaction with marine tourism experiences provides a basis for market segmentation based on benefits sought (Moscardo, 2000) and also the management of visitors in marine areas. Murphy and Norris (2005) segmented recreational visitors to the Great Barrier Reef according to reef trip benefits based on learning experiences or the desire for socializing and relaxing. Hoyt (2005) also describes how managers can enhance a wide range of potential tourist benefits, natural values or services provided by ecotourism based on whale watching. Muloin (1997, 1998, 2000) identified the need to conduct research on the psychological benefits that tourists derive from wildlife encounters. Curtin (2005) highlighted experiential views of wildlife tourism, including the emotional and physical benefits of wildlife interaction for visitors. These experiences derive from the innate human desire to interact with, and interpret, wildlife; the cultural and anthropomorphic appeal of animals; and urban living motivating visitors to seek physical, emotional and psychological benefits from connecting with nature and wildlife. Schänzel (2004, p. 354) found that visitors at a marine life centre mainly gained psychological benefits such as 'positive moods and emotions, environmental sensitivity, sense of place and species, and affective learning' from hands-on involvement with sea animals.

This research on benefits provides insight into the relationship between tourists and wildlife and 'an understanding of *exactly* what it is about the wildlife interaction that facilitates psychological well-being' (Muloin, 1998, p. 203). The personal benefits of viewing wildlife are the basis for conservation actions (Manfredo and Driver, 2002). On-site benefits of increased understanding or emotional responses to marine wildlife encounters may also lead to off-site benefits such as greater environmental awareness, supporting nature conservation work and protecting endangered species.

Framework for Managing Marine Wildlife Tourism Experiences

This chapter follows the framework devised by Orams (1995a,b, 1999) that measures positive changes in both tourists and the marine environment for effective management of marine tourism (see Table 2.1). Indicators of tourist benefits from marine animal encounters include enjoyment and learning contributing to pro-environmental attitude and behavioural changes, along with conservation benefits for marine environments and marine wildlife. Indicators of conservation benefits include tourists reducing wildlife disturbance, protecting habitats and aiding the viability of marine ecosystems. The framework by Orams (1999) was based on a previous model of experiential education in whale-watching ecotourism programmes in Hawaii (Forestell and Kaufman, 1990; Forestell, 1993). This model focused on the cognitive states or learning of visitors using interpretation in marine settings to reduce impacts and promote pro-environmental behaviours on whale-watching tours. Lück (2003) evaluated the key role of interpretation on 'swim with dolphin' tours in New Zealand, based on models by Forestell and Kaufman (1990) and Orams (1997a,b).

Table 2.1. Indicators for managing marine wildlife tourism experiences. (From Orams 1995a, 1999.)

Tourist
 Satisfaction/enjoyment (closeness/type of interaction)
 Education/learning (knowledge)
 Attitude/belief change
 Behaviour/lifestyle change

Marine environment
 Minimize disturbance
 Improve habitat protection
 Contribute to long-term health and viability of ecosystem

Orams (1999) extended the three-step experiential education sequence of Forestell (1993) into a four-stage sequence of desirable tourist outcomes from marine education programmes. Mayes *et al.* (2004) also adopt a model based on changing attitudes, beliefs, behaviours and actions through wildlife interaction and interpretation with benefits for animals, the environment and visitors. The following sections apply Orams' (1999) key indicators to review the psychological, educational and conservation benefits of visitor participation in marine wildlife tourism experiences.

Visitor satisfaction with wildlife tourism experiences

Recent models identify key elements or factors for satisfying tourist encounters with wildlife. These focus on the type of animal, the setting and the wildlife experience. Reynolds and Braithwaite (2001) highlight the intensity, naturalness and uniqueness of wildlife encounters along with visitor control and wildlife attributes. For Moscardo and Saltzer (2004), key aspects of the visitor–wildlife experience were the perceived interaction with the wildlife, touching/handling the wildlife, perceived naturalness/ authenticity of the encounter, and surprise or novelty of the wildlife experience (see Table 2.2). Positive visitor interactions with wildlife lead to mindful, satisfied and conservation-oriented visitors. Animal attributes, surprise encounters, natural environments and new animals were key features of the best or most satisfying wildlife tourism experiences (Moscardo *et al.*, 2001; Woods and Moscardo, 2003). A survey of 5000 visitors at 15 wildlife sites in Australia and New Zealand found that key features were close viewing of unique wildlife species behaving naturally in natural areas. A knowledgeable guide (19%), wildlife information (18%) and touching or handling wildlife (7%) were less important factors (Moscardo and Saltzer, 2004). On the Great Barrier Reef, visitor satisfaction related to the diversity of reef wildlife (e.g. bright colours and patterns of the reef fish and corals), reef interpretation and naturalness of the encounters with reef wildlife (Moscardo, 2001).

Bentrupperbaumer (2005) reviewed psychological aspects underlying human–wildlife interactions. This included affiliation with animals (i.e. companionship, bonding and emotional attachment); key attributes and human preferences for wildlife (e.g. similarity, aesthetic appeal and rare/endangered); human values for wildlife,

Table 2.2. Key factors in visitor satisfaction with wildlife tourism experiences.

Intensity or excitement of the experience	Variety/large numbers of animals[a]
Authenticity or naturalness of the experience	Natural/pleasant physical setting
Uniqueness of the experience	Quality interpretation
Amount of visitor control over the experience	Knowledgeable staff
Popularity of the species	Clear orientation/structure
Species status, i.e. rare and/or endangered	Quality facilities, lack of crowds
(Reynolds and Braithwaite, 2001)	(Moscardo and Saltzer, 2004)

[a]Wildlife characteristics: large, colourful, rare/unique, dangerous, human-like, presence of infants.

with psychological benefits of wildlife as part of the utilitarian view of animals; types of settings and human activities in wildlife encounters (e.g. photography, feeding and hunting) and marketing images of wildlife. Marine wildlife was not specifically addressed in this study but these findings could apply to marine species. Key factors in the wildlife viewing experience include a natural setting, closeness, information, range or numbers of wildlife and facilities (Pearce and Wilson, 1995). Tremblay (2002) examined wildlife icons as tourist attractions and marketing symbols. Wildlife attributes such as cute, appealing and human-like in appearance or behaviour elicit positive responses from visitors (Moscardo and Saltzer, 2004). The ecological, cultural and economic importance of animals is also part of their appeal. Wildlife viewing is thus an educational activity and an emotional experience; both affective motives and cognitive learning shape tourist experiences of wildlife. The emotional and personal aspects of viewing marine wildlife along with education play a major role in visitor satisfaction and enjoyment of wildlife (Schänzel, 2004).

Table 2.3 lists key factors contributing to tourist enjoyment and satisfaction with marine wildlife tours. Seeing whales, their proximity and whale behaviours were the most important factors on whale-watching tours, followed by other marine wildlife, the coastal scenery or boat trip, and learning about whales (Duffus and Dearden, 1993; Muloin, 1998; Foxlee, 2001; Andersen and Miller, 2006). There were similar responses on whale shark tours (Davis et al., 1997, 2000) and viewing penguins at their nesting areas (Schänzel and McIntosh, 2000; Saltzer, 2003a). Being close to nature, the natural habitat and animal behaviour were important for all marine species.

Experiential aspects such as hearing the Orca 'blow', seeing penguins coming out of the water and stingrays close to shore also heightened visitor satisfaction. Seeing and feeding dolphins, natural scenery, socializing and a learning experience all contributed to visitor enjoyment of shore-based interactions with wild dolphins (Mayes et al., 2004; Smith et al., 2006).

In-water encounters with whale sharks prompted feelings of excitement, adventure, freedom and relaxation (Davis et al., 1997, 2000). Visitor satisfaction

Table 2.3. Key factors for visitor satisfaction with marine wildlife experiences. (Adapted from Moscardo and Saltzer, 2004, p. 179; Moscardo *et al.*, 2004, p. 235.)

Marine wildlife tourism study	Factors contributing to visitor enjoyment/ satisfaction in ranked order

Duffus and Dearden (1993)
   Killer whale-watching tours, British
     Columbia, Canada
	Seeing whales
	Getting close to whales
	Seeing displays of whale behaviour
	Seeing coastal scenery
	Having a naturalist/crewmember
	   answering questions
	Seeing other marine mammals

Andersen and Miller (2006)
   Killer whale-watching tours, Washington
	Seeing Orcas, large numbers of Orcas
	Orca behaviours, proximity to boat
	Learning about whales, wildlife and the
	   area
	Mention of other wildlife
	Hearing the Orca 'blow'
	Boat ride/scenery/weather
	Small boat with personable staff
	Time with family and friends

Muloin (1998) Whale watching, Hervey Bay, Queensland	Sightings of whales Proximity to whales Boat trip Dolphins/other wildlife Whale activity Facilities/amenities Interpretation/education Weather and sea conditions

Foxlee (2001)
   Whale watching, Hervey Bay, Queensland
	Number of whales seen
	Distance from whales
	Whale activity
	Information available about whales
	Information available about other marine
	   life
	Style in which information was presented

Mayes *et al.* (2004)
   Dolphin feeding, Tangalooma
     and Tin Can Bay, Queensland
	Seeing dolphins
	Getting close to nature
	Spending time with family and friends
	Seeing the region
	Feeding dolphins
	Have a learning/education experience

Davis *et al.* (1997, 2000)
   Whale shark tours, Ningaloo Reef,
     Western Australia
	Being close to nature
	Seeing large animals
	Many different types of marine life
	Feeling of excitement
	Learning about the marine environment
	Feeling of adventure
	Underwater scenery
	Diving somewhere new
	Freedom, relaxation, being with friends

Continued

Table 2.3. *Continued*

Marine wildlife tourism study	Factors contributing to visitor enjoyment/satisfaction in ranked order
Lewis and Newsome (2003) Stingray feeding, Western Australia	New experience/unique/different Wild stingrays in natural environment Interaction with stingrays Stingrays close to shore Stingrays gentle and trusting Stingrays beautiful creatures/large size

Schänzel and McIntosh (2000)
  Penguin viewing, Otago Peninsula,
    New Zealand
	Natural habitat and penguin behaviour
	Proximity to the penguins
	Educational opportunities
	Innovative/novel approach (hides,
	  covered trenches)
	Fewer other people present
	Presence of infant penguins

Saltzer (2003a)
  Penguin Place, New Zealand
	Natural environment and penguin
	  behaviour
	Proximity to penguins
	Seeing more penguins
	Tour guide
	Video and other information
	See penguins coming out of the water

from swimming with dwarf minke whales included closeness, total interaction time and the number of whales seen during in-water encounters. The best elements were the minke whales (18%) and minke whale interactions (22%) followed by diving (15%) and seeing other marine life (10%) (Valentine *et al.*, 2004). In contrast, Orams (2000) found that proximity to whales onboard boats (4%) was not a major part of visitor satisfaction compared to whale behaviours and boat features. Muloin (1998) found that visitor satisfaction (78% novices) mainly related to seeing whales in the wild and minimizing impacts through approach regulations. Furthermore, Muloin (1998, 2000) noted a gender difference where women reported a higher level of satisfaction and were more emotional in their responses to whales than men. The novelty of viewing penguins from covered trenches and hides with few other people present (Schänzel and McIntosh, 2000), and the uniqueness of feeding (38%), touching (72%) or viewing stingrays also enhanced visitor enjoyment (Lewis and Newsome, 2003). Social aspects such as spending time with family or friends added to visitor satisfaction on boat tours viewing whales and whale sharks and at shore-based wild dolphin-feeding sessions.

Psychological Benefits of Marine Wildlife Experiences

Psychological benefits or outcomes are what visitors personally gain from wildlife experiences. Tourists derive individual psychological benefits from interacting with both wildlife and people. These benefits include relaxation or stimulation, mood benefits or feelings, and learning benefits (Schänzel and McIntosh, 2000). Table 2.4 presents a range of psychological benefits for visitors viewing whales or whale sharks. Watching Humpback whales at Hervey Bay provided personal benefits for visitors such as thrills and excitement (whales), tranquillity and peacefulness (nature), inspiration, learning and relaxation. Escaping from normal life and enjoying a special experience with family and other people interested in whales were other key benefits (Muloin, 1998, 2000). On killer whale tours, the main visitor benefits were seeing Orcas, the boat ride and coastal scenery, learning about marine wildlife, and being with family/friends (Andersen and Miller, 2006). For visitors on whale shark tours, the key psychological benefits focused on finding and interacting with whale sharks, diving, boat operations and seeing other reef animals. The diving opportunities, boat facilities, staff and service aspects of marine wildlife tours have all increased in importance since the mid-1990s (Birtles *et al.*, 1996; Catlin and Jones, 2006). Davis *et al.* (1997, 2000) also identified adventure, learning, excitement, freedom and relaxation as other psychological benefits for visitors swimming with whale sharks. However, the learning benefits of information presented about whales were secondary to the 'mood' benefits of tourists directly experiencing marine wildlife and nature.

Personal and emotional experiences of marine wildlife

The personal and emotional responses of visitors to marine wildlife, along with fun and enjoyment, are part of the affective domain. Table 2.5 presents the personal and emotional experiences of visitors to whales, dolphins, sea lions, fur seals, penguins and albatross, mainly through close encounters and mediated interactions. There is a recent trend in marine wildlife tourism from viewing marine mammals onboard boats or from land to more active in-water encounters, such as swimming alongside, feeding or touching marine animals (Orams, 2002; Curtin, 2005). Swimming with whales (Valentine *et al.*, 2004), dolphins (Amante-Helweg, 1996; Würsig, 1996; Samuels *et al.*, 2003) and seals (Kirkwood *et al.*, 2003; Scarpaci *et al.*, 2005) in the wild has become an increasingly popular visitor experience since the early 1990s. This desire for close personal encounters has prompted strong emotional reactions to marine wildlife experiences, such as 'peace', 'calm', 'grace' and 'beauty' from swimming with whale sharks (Davis *et al.* 1997, p. 266, cited in Curtin, 2005, p. 4). Tourists on 'swim with dwarf minke whale' tours also elicit emotional responses (10%) such as 'awed', 'humbled/enlightened', 'dream come true' and 'feel closer to whales/nature' (Birtles *et al.*, 2002).

At Penguin Place in New Zealand, tourists reported feelings of pleasure, curiosity, amazement, admiration, fascination, amusement/fun, happiness and sympathy from close encounters with yellow-eyed penguins (Schänzel and McIntosh, 2000). Feelings of wonder, exploration, privilege, affection and simplicity were also elicited. Tourists viewing Royal albatross at Taiaroa Head and on board Monarch Wildlife Cruises in New Zealand used positive words to describe

Table 2.4. Psychological benefits experienced by participants on marine wildlife tours.

Rank	Psychological benefits	Mean	Response (%)
Humpback whale watching, Queensland (Muloin, 1998)[a]			
1	Seeing marine mammals I do not normally see	4.7	
2	Experiencing thrills and excitement from seeing Humpback whales	4.6	
3	Experiencing the tranquillity and peacefulness of nature	4.6	
4	Being inspired by seeing Humpback whales	4.4	
5	Increasing my knowledge about Humpback whales	4.3	
6	Relaxing in a pleasant setting	4.3	
7	Escaping the confinement of my usual routine	4.2	
8	Doing something special with important people in my life	4.2	
9	Telling others about the experience	4.1	
10	Enjoying the facilities and services offered on the boat	4.0	
11	Enjoying the company of people with similar interests	3.5	
Killer whale-watching tours, USA (Andersen and Miller, 2006)[b]			
1	Seeing Orcas/whales		75.4
2	Enjoying the boat ride (outdoors, scenery, islands, weather)		66.7
3	Mention of other (marine) wildlife		42.1
4	Learning about Orcas/whales, wildlife and the area		38.6
5	Time with family and friends		14.9
6	Seeing whales and wildlife in their natural form		8.8
Whale shark tours, Western Australia (Catlin and Jones, 2006)[c]			
1	Other scuba-diving or snorkelling		16.4
2	Seeing, watching, observing or finding the whale shark		15.7
3	Staff, food and operations		15.4
4	Being or swimming with, next to or alongside the whale shark		15.2
5	Other animals, reefs or nature		12.5
6	Being close to the whale shark		5.5
7	Size or number of whale shark(s)		4.3
8	Self experiences or interactions with whale shark		3.3
9	Weather and sea conditions		2.0
10	People/family/friends		1.9

[a]All variables were measured on a 5-point Likert scale with a value of 1 indicating not at all important and a value of 5 indicating very important (*n* = 667 questionnaires).
[b]Top three responses on killer whale tours (*n* = 57).
[c]Best three experiences on whale shark tours (*n* = 539).

Table 2.5. Personal and emotive experiences of visitors participating in marine wildlife tours.

Personal experiences	Total themes[a]	Response (%)
Dwarf minke whale tours, Queensland (Birtles *et al.*, 2002)[a]		
'Awed' by/always remember experience	8	1.8
General personal experience (non-specific)	7	1.5
Pleased/happy/satisfied by experience	6	1.3
Feel closer to/relate to whales	6	1.3
'Humbled/enlightened' by experience	6	1.3
Feel privileged/lucky/fortunate	4	0.9
'Eye opening'/'Horizon broadening'/'See bigger picture'	4	0.9
'Dream come true'/feel closer to nature	4	0.9
Religious/spiritual/life changed by experience	2	0.4
Total – YES – Personal Experience	47	10.1
Sea lion encounters, Western Australia (Orsini and Newsome, 2005)[b]		
Enjoyed viewing sea lions		76
View wild sea lions 'in their natural habitat'		22
View sea lions 'being themselves' without 'human interference'		20
Interactions with sea lions		19
Being close to/swimming with sea lions		12
Opportunity to view sea lions 'freely' without crowds		4
Royal albatross, New Zealand (Saltzer, 2003a,b)		
Albatross (Monarch Cruises): large, magnificent, graceful, beautiful, majestic		63
Albatross (Taiaroa Head): huge, elegant, amazing, impressive, unique, great flyers		73
New Zealand fur seals (Booth, 1998)		
Positive: lovely, beautiful, friendly, graceful, cute		75
Neutral: big, sleepy, natural		14
Negative: aggressive, fat, ugly, useless		11
Penguin place (Schänzel and McIntosh, 2000)[c]		
Appreciation of the conservation effort to save penguins		39
Pleasure from being in natural surroundings		36
Admiration of the animal kingdom		34
Felt amused or had fun, penguins funny, happy and laughed		34
Interest in the life cycle of the penguins		32
Sympathy for the penguins		21
Dolphins, whales, sea lions, penguins (Bulbeck, 2005)[d]	(Antarctica/GAB/Seal Bay/ Monkey Mia)	
Felt good animals being preserved	37/45/75/33	
Felt privileged to have encounter	86/55/66/69	
Felt close to nature	45/82/42/31	
Felt affection for animals	5/27/22/50	
Felt protective towards animals	9/0/17/40	

Continued

Table 2.5. *Continued*

Personal experiences	Total themes[a]	Response (%)
Felt animals trusted/liked me		5/9/9/25
Felt spiritually uplifted		23/36/3/13
Had deeper understanding of meaning of life		14/9/9/4
Feelings towards dolphins (Bulbeck, 2005)[e]		(In wild/Monkey Mia/aquarium)
Fondness or affection		68/84/67
Fun or pleasure		69/58/55
Peace or tranquillity		43/47/27
Comradeship or oneness		15/32/7
Feeling of communication		7/24/2

[a]Total coded themes or elements = 466.
[b]207 questionnaires on sea lion tours.
[c]40 respondents at Penguin Place.
[d]Antarctica/Macquarie Island (*n* = 22), GAB = Great Australian Bight (*n* = 11), Seal Bay (*n* = 65), Monkey Mia (*n* = 48).
[e]Survey question on 'feelings towards dolphins': in wild (*n* = 175), Monkey Mia (*n* = 38), aquarium (*n* = 55).

these birds as 'magnificent, graceful, beautiful, majestic, elegant, and unique' with this emotional response more pronounced for first-time visitors at Taiaroa Head (Saltzer, 2003b,c). New Zealand fur seals were mainly described in positive words, 'friendly, graceful, cute', while some visitors (25%) had either neutral or negative emotional responses to the seals. Stingrays fed by visitors at the shoreline in Hamelin Bay, Western Australia, were described as large, gentle, trusting and beautiful creatures (Lewis and Newsome, 2003). In Australia, visitors at Seal Bay felt good about preserving sea lions (75%), those viewing penguins in Antarctica felt privileged (84%) and gained a deeper meaning of life (14%), others viewing southern right whales from Nullarbor cliffs felt closer to nature (82%) and spiritually uplifted (32%), while tourists interacting with dolphins at Monkey Mia felt affection (50%), protective (40%) and that the dolphins liked them (25%) (Bulbeck, 1999, 2005). Other general feelings elicited about dolphins at Monkey Mia were fondness/affection (84%) peace/tranquillity (32%), comradeship/oneness (32%) and communication (24%), while other dolphins in the wild elicited stronger feelings of fun/pleasure (69%) (Bulbeck, 2005). Hence, for many visitors on whale-watching tours, 'the basis of the participants' enjoyment of the experience is aesthetic and emotional, rather than intellectual' (Neil *et al.*, 1996, p. 186). Schänzel (2004) described the fun and enjoyment derived by visitors from learning experiences provided at the Marine Education Centre in Wellington. The centre had a tidal pool filled with local fish, a rock pool tray to handle marine life and other marine creatures in live habitat displays. Visitor learning at this centre involved both a personal connection and emotional involvement with marine wildlife. A tactile approach and enthusiastic interpreters helped to foster visitor appreciation of local marine life at this centre.

Educational Benefits of Marine Wildlife Experiences

The educational benefits of marine wildlife tours include visitor learning, knowledge and information presented about marine species and marine or coastal environments. These interpretation or education programmes in marine areas involve talks by tour guides, interpreters and rangers onboard boats or at shorelines, along with visitor centres, displays, signs and brochures. This information covers the biology, ecology and behaviours of marine species, best practice guidelines and threats to marine life. This section reviews the education/learning (knowledge), attitude/belief changes and behaviour/lifestyle changes deriving from visitor exposure to marine life education.

Education and learning (knowledge)

Educational experiences were important for visitors on dwarf minke whale tours, at the Mon Repos turtle rookery, on swim with dolphin tours and whale watching (see Table 2.6). Some 14% of visitors highlighted learning about dwarf minke whales, marine life on the Great Barrier Reef, and the educational experience and research conducted about these whales (Birtles *et al.*, 2002). At Mon Repos, visitors learnt about sea turtles at a visitor centre display and during interpretive talks on egg laying or turtle hatchlings. This included knowledge about the life cycle of turtles, their need for protection and current threats they face (Tisdell and Wilson, 2002, 2005). On whale-watching tours in Scotland, one-third of visitors learnt that cetaceans were mammals and about threats to whales from fishing, marine pollution and whaling (Warburton *et al.*, 2000; Parsons *et al.*, 2003). Lück (2003) found that, while most visitors on 'swim with dolphin' tours increased their knowledge of dolphins and wildlife (66–69%), in general and from tour staff, only 29% strongly agreed the dolphin tour was an educational experience. One dolphin operator did not have a guide onboard and most visitors wanted more interpretation about dolphins, the marine environment and threats (Lück, 2003). Visitors on whale-watching tours also wanted more information about the marine environment (Foxlee, 2001). Whale organizations strongly promote the educational and conservation benefits of whale watching by raising public awareness of whales and marine conservation issues (WDCS, 2005; Whales Alive, 2005; McIntyre, 2006). Interpreters or scientists educate visitors about cetacean biology and marine conservation issues onboard most whale- and dolphin-watching boats (Birtles *et al.*, 2002; Russell and Hodson, 2002; Corkeron, 2004; O'Neill *et al.*, 2004; Andersen and Miller, 2006).

On wild dolphin-feeding talks in Queensland, Australia, visitor knowledge about dolphins increased by 81% at Tangalooma Resort on Moreton Island and by 47% at the small seaside town of Tin Can Bay (Mayes *et al.*, 2004). The educational benefits for visitors of the dolphin interpretation and feeding interaction programme at Tangalooma Resort have been well documented by Orams (1994, 1995b,c,d, 1996, 1997a,b, 1999; Orams and Hill, 1998). The site includes a Dolphin Education Centre and a ranger giving nightly talks about dolphin biology and behaviour to both dolphin feeders and observers. Learning

Table 2.6. Educational experiences and visitor learning on marine wildlife tours.

Educational experiences	Total themes[a]	Response (%)
Dwarf minke whale tours, Queensland (Birtles *et al.*, 2002)[a]		
More informed about minke whales	21	4.5
More informed about reef (GBR) and marine life	15	3.2
Increased interest in info on (minke) whales – want to learn more	9	1.9
A learning/educational experience (non-specific)	9	1.9
More informed about/by research	8	1.7
Increased interest in info on marine life – want to learn more	4	0.9
Interested in recent discovery of species	1	0.2
Learned/understood whale-watching guidelines	1	0.2
Total – YES – Educational	68	14.6
Mon Repos Conservation Park, Queensland (Tisdell and Wilson, 2002)		
Sea turtle viewing informative and educational	514	99
Sea turtle visitor centre display		93
Interpretive talks (turtle hatchling behaviour)		90
Interpretive talks (egg-laying process)		87
Life cycles of sea turtles		85
Need to protect sea turtles		82
Information on current threats to sea turtles[b]		78
Amphitheatre		76
Visitor awareness of threats to sea turtles – additional information	282	54
Visitor awareness of threats to sea turtles – first time	163	31
Whale watching, Scotland (Warburton *et al.*, 2000)		
Cetaceans are mammals		36
Fisheries by-catch, threat of over-fishing		32
Threat of marine pollution		31
Commercial whaling		20
Threat of oil spills		5
Excessive boat traffic		4
Swim with dolphin tours, New Zealand (Lück, 2003)[c]		
Teach school courses on conservation of natural resources		72
Dolphin tour staff had good knowledge of dolphins		69
Learn new things/increase my knowledge (general)		66
Learn as much as we can about wildlife		66
Enjoy learning about wildlife on holidays		46
Dolphin tour was an educational experience		29
Learned a lot about dolphins on this tour		17
Learned a lot about other marine life		5

[a]Total number of coded themes/elements = 466.
[b]Prawn trawlers (64%), boat strikes (60%), fox/wild pig predation (59%), turtles harvested (56%), tangled in crab pots (55%), pollution of waterways (53%), eggs collected (52%), goanna predation (45%), natural diseases (37%). Reason to visit Mon Repos: watch sea turtles (78%), study sea turtles (11%), entertain visitors (9%), other (2%).
[c]Responses to 'Strongly agree' only; 733 questionnaires from three 'swim with dolphin' tours (New Zealand).

about the dolphins at Tangalooma motivated Australian tourists, while Japanese tourists wanted to touch and physically interact with dolphins. Language barriers also impeded the Japanese visitors from understanding the dolphin-feeding programme or from adopting more environmentally responsible behaviours (Takei, 1998). Beasley (1997) found that visitors on dolphin tours in Akaroa (New Zealand) and Hong Kong had short-term increases in their knowledge of marine mammals and ocean ecosystems. On Penguin Island (Western Australia), visitors learn about penguins, sea lions and the marine ecology of the area from information on signs, displays, pamphlets and talks by rangers at a Penguin Experience visitor centre that houses orphaned or injured little penguins. Rangers feed the penguins and give scheduled talks about their biology. All tourists increased their knowledge of penguins after visiting this centre from 55% (pre-visit) to between 69% and 74% (post-visit) (Hughes and Saunders, 2005).

Attitude and belief change

Several studies suggest that marine wildlife tours with a strong educational focus can change the pro-environmental attitudes and beliefs of visitors. Table 2.7 summarizes the changes in the environmental attitudes of visitors on whale and dolphin tours. On 'swim with dwarf minke whale' tours, 27% of tourists changed their attitudes to conservation, displaying a greater awareness of whales, marine life, whaling and other human impacts (Birtles *et al.*, 2002). In Scotland, 46% of visitors thought that whale watching had a positive impact mainly by increasing visitor awareness of whales. This included concern for boats or noise affecting whale behaviours (Warburton *et al.*, 2000). In the USA, land-based whale watchers were more concerned than boat tourists about the impacts of noise, boats and kayaks on killer whales (Muloin, 2000; Finkler and Higham, 2004). Tourist desire for close encounters, then, was matched by awareness of impacts on whales.

Visitors on wild dolphin-feeding tours at Tin Can Bay and Tangalooma Resort felt more strongly about conservation (81%), the state of marine areas (66%) and helping out with conservation programmes (52%) after their dolphin experience (Mayes *et al.*, 2004). They also disagreed with keeping dolphins in captivity (59%) and indigenous people hunting dolphins (68% of Australians) while 9% also disagreed with the practice of feeding wild dolphins (Mayes *et al.*, 2004). At the Penguin Experience visitor centre, rangers feed orphaned or injured little penguins and give scheduled interpretive talks about their biology. All visitors had more pro-environmental attitudes after this experience. Exploration-focused visitors held an attitude of responsible conservation based on intrinsic natural values of the area, while recreation-focused visitors moved towards attitudes that valued nature based on its usefulness to humans (Hughes and Saunders, 2005). Education, then, is a key element of managing tourist–wildlife interactions, with the outcome of changing environmental attitudes and the potential to reduce visitor impacts on wildlife and marine areas (Orams, 1995b; Schäenzel, 1998; Townsend, 2003).

Table 2.7. Changes in environmental/conservation attitudes on whale and dolphin tours.

Environmental/conservation attitudes	Total themes[a]	Response (%)
Dwarf minke whale tours, Queensland (Birtles *et al.*, 2002)[b]		
Greater awareness/concern/appreciation of (marine) life/nature	48	10.3
Increased/reinforced conservation awareness	28	6.0
Greater awareness/concern/appreciation of whales	27	5.8
Greater awareness of whaling issues	9	1.9
Increased awareness of human impacts on marine life/nature	6	1.3
Greater awareness of need for whale-watching guidelines	3	0.6
Greater awareness of need for sustainable ecotourism	2	0.4
Greater awareness/appreciation of impacts of humans on whales	2	0.4
Aware that wildlife need not be touched/fed to be enjoyed	2	0.4
Increased awareness of effects of human coastal development	1	0.2
Greater awareness of natural resource exploitation	1	0.2
Total – YES – Conservation attitudes	129	27.7
Whale watching, Scotland (Warburton *et al.*, 2000)		
Whale watching had a positive effect on cetaceans		46
Education and raising awareness of whales		40
Potential interference with individual whales		33
Potential impact on whales from noise pollution		10
Killer whale watching, USA (Finkler and Higham, 2004)		(Land-based/boat-based)
Effects of noise on whales		74/54
Power boats placed in the path of whales		73/56
Disturbance of whales by (other) power boats		69/52
Impacts of kayaks approaching whales		27/18
Dolphin feeding, Queensland (Mayes *et al.*, 2004)[c]		
Felt more strongly about conservation of the environment generally		81
Felt they could make more of a difference to the state of the (marine) environment		66–67
Felt more confident in assisting with conservation programmes		52
Disagreed with indigenous people hunting dolphins (International visitors/Australians)		30/68
Disagreed with keeping dolphins in aquariums		59
Disagreed with feeding wild dolphins		9

[a]Total number of coded themes/elements = 466.
[b]Total: Yes – conservation attitudes (27.7%), Yes – educational (14.6%), Yes – personal experience (10.1%), Yes – other (14.4%), No (33.3%); 527 questionnaires from 52 trips on 5 live-aboard dive boats in 1999/2000.
[c]105 questionnaires (54 Tangalooma, 51 Tin Can Bay) for visitors feeding wild dolphins.

Behaviour and lifestyle changes

Some studies also suggest that marine wildlife tours with a strong educational focus can create longer-term behavioural or lifestyle changes in visitors. These behavioural changes include minimising impacts, donating money and direct actions supporting environmental issues (Moscardo *et al.*, 2004). Table 2.8 presents selected changes in the personal behaviour or lifestyle of visitors during or after a marine wildlife tour. For example, after viewing sea turtles at Mon Repos, Australia, visitors indicated strong support for protecting sea turtles, taking care while using beaches and not buying turtle products overseas. They would also take more care with fishing gear, plastics and lights near beaches (Tisdell and Wilson, 2002, 2005). Howard (2000) found that 74% of visitors (37 out of 50) surveyed 6 months after visiting Mon Repos reported behaviours such as talking to friends/family and teaching people about turtles, removing beach litter, reporting turtle sightings, releasing turtles trapped in nets and volunteering. At the Jurabi Turtle Experience (JTE) tour at Exmouth (Western Australia) nearly all visitors (98%) who joined knew about the code of conduct for viewing nesting sea turtles. Some visitors on this JTE tour still breached aspects of this code, but far less so than independent visitors. Most JTE tour participants avoided sudden movements and stayed low to the ground when viewing turtles (Smith, 2006). Since 2004, volunteer turtle guides and tour operators at Jurabi Turtle Centre increased the level of information and public education about turtle viewing behaviour and conserving marine turtles (Macgregor, 2006). However, 77% of tourist groups still breached the code of conduct by shining lights directly at the turtles, not staying behind turtles and going closer than 3 m, with 51% of breaches disturbing nesting turtles (Waayers *et al.*, 2006).

Visitors on wild dolphin-feeding tours at Tin Can Bay and Tangalooma (56–75%) stated they would remove beach litter, assist in protecting dolphins, decrease water pollution, and tell others about caring for oceans and marine life (Mayes *et al.*, 2004). For on-site behaviours, 25% of Tin Can Bay visitors thought it was acceptable to touch dolphins compared to 3% of visitors at Tangalooma who learnt about human impacts on dolphins at feeding talks. More than half of these dolphin visitors (63%) though were not involved in environmental organizations or activities (Mayes *et al.*, 2004). At Tangalooma Resort, follow-up phone interviews conducted with visitors 2–3 months after the dolphin experience also found longer-term changes in environmental behaviours. Visitors who participated in the dolphin education programme actively sought dolphin information, picked up beach rubbish, were more involved in environmental issues and donated money to environmental organizations (Orams, 1996, 1997b). The Tangalooma education programme also reduced inappropriate on-site behaviours such as visitors touching dolphins (Orams and Hill, 1998). Barney *et al.* (2005) also reported that American college students with higher education or knowledge had a more scientific attitude about dolphins,

Table 2.8. Changes in personal behaviour or lifestyle on or after a marine wildlife tour.

Behaviour or lifestyle changes	Response (%)
Mon Repos Conservation Park, Queensland (Tisdell and Wilson, 2002)	
Take personal action to conserve sea turtles	87
Other respondents (partner, family, children) protecting sea turtles	81
Take care while using beaches used by sea turtles for nesting	75
Do not buy or consume tortoiseshell products, eggs, meat, soups (overseas)	73
Switch off lights near beaches	68
Be more careful disposing of plastics	62
Take care with fishing gear	47
Jurabi turtle experience, Western Australia (Smith, 2006)[a]	
Aware of code of conduct for viewing nesting turtles (independent visitors/JTE tour)	72/98
Walked below high tide (independent visitors – beach only/ JTE display/JTE tour)	84/89/29
Closer than 15 m to turtle digging pit (independent visitors – beach/display; JTE tour)	75/86/23
Did not stay behind the turtle (independent visitors – beach only JTE display/JTE tour)	69/100/12
Avoid sudden movements (JTE tour)	94
Staying low (JTE tour)	68
Dolphin feeding, Queensland (Mayes *et al.*, 2004)[b]	
Remove beach litter that could harm dolphins	75
Assist in the protection of whales and dolphins where possible	64
Decrease their contribution to water pollution	60
Tell others about the need to care more for our oceans and wildlife	56
Become more involved in marine conservation issues	23
Touching dolphins is okay (Tin Can Bay/Tangalooma)	25/3
Dolphin education and feeding Tangalooma, Queensland (Orams, 1996)[c]	(DEP visitors/control)
Get more information on dolphins	41/13
Picked up rubbish from beaches	65/44
Become more involved in environmental issues	32/6
Made a donation to an environmental organization	23/11

[a]Independent visitors who visited the beach only and independent visitors who also saw the JTE interpretive display. JTE tour (*n* = 42), independent visitors and JTE display (*n* = 29), independent visitors – beach only (*n* = 25).
[b]Definite responses only, 105 questionnaires (54 Tangalooma, 51 Tin Can Bay) for visitors feeding wild dolphins.
[c]DEP = Dolphin Education Programme, visitors = 104, control group (pre-DEP) = 110 (phone interviews).

were more environment-friendly and less likely to engage in harmful activities such as touching, feeding or boating near wild dolphins. However, most of these studies measure intention to act rather than actual behaviours (e.g. JTE turtle tour) and rely on self-reporting by visitors, with no longer-term studies of changes over 1–5 years.

Conservation Benefits of Marine Wildlife Experiences

The conservation benefits gained from wildlife tourism include: (i) wildlife management and research; (ii) finances for conservation of species; (iii) socio-economic benefits; and (iv) education of visitors potentially leading to more conservation-focused behaviour and support (Higginbottom and Tribe, 2004). The conservation outcomes for marine wildlife and marine environments aim to: (i) minimize disturbance; (ii) improve habitat protection; and (iii) contribute to long-term health and viability of ecosystems (Orams, 1995a, 1999). Table 2.9 presents conservation appreciation and actions by visitors on marine wildlife tours. At Mon Repos, these included minimizing threats to turtles; talking about sea turtles, reporting mistreated, sick or injured sea turtles; protecting sea turtles as an ancient/unique species; and donating money to conserve sea turtles (Tisdell and Wilson, 2001a,b, 2002, 2005).

Visitors on whale-watching tours indicated that they would do more to protect whales within Australia (80%), supported a global ban on commercial whaling (78%) and would report stranded or injured whales (73%) (Wilson and

Table 2.9. Conservation appreciation or actions by participants on marine wildlife tours.

Conservation appreciation or actions	Response (%)
Mon Repos Conservation Park, Queensland (Tisdell and Wilson, 2002)[a]	
Take more action to minimize threats to sea turtles	98
Talk about sea turtles at Mon Repos to friends and relatives	98
Increased desire to protect sea turtles as unique species	90
Report poaching or mistreatment of sea turtles	88
Take more personal action to conserve sea turtles	87
Report the sighting of sick or injured sea turtles	66
Protect sea turtles because they are an ancient species	66
Contribute more money for sea turtle conservation	40
Protect sea turtles because they have recreational value	32
Protect sea turtles because they can generate income	23
Humpback whale watching, Queensland (Wilson and Tisdell, 2003)[b]	
Take more action to protect whales in Australia	80
Complete worldwide ban on whaling	78
Report stranding of whales and injured/mistreated whales	73
Whale watching, Scotland (Rawles and Parsons, 2005)	
Regularly recycled items	83
Bought cosmetic/hygiene items not tested on animals	73[c]

[a]Subscribe to a newsletter with updates on sea turtle conservation work; form a 'friends of sea turtles' group; more access to (translated) material on sea turtles, current threats and conservation measure; ban photography.
[b]702 questionnaires from whale watchers in Hervey Bay, Queensland.
[c]Definite responses only, 105 questionnaires (54 Tangalooma, 51 Tin Can Bay) for visitors feeding wild dolphins.

Tisdell, 2003). The conservation actions of visitors on whale-watching tours in Scotland included regular recycling, purchasing items not tested on animals, using energy-saving bulbs and other devices, became members of (47%) or did voluntary work for (27%) environmental or animal welfare organizations, and bought organic/environment-friendly products (Rawles and Parsons, 2005). Visitors on Australian whale-watching cruises also exhibited similar conservation actions up to 4 months later (Ballantyne *et al.*, 2006). The International Fund for Animal Welfare supports whale watching as it fosters visitor appreciation of marine conservation and profiles threats to whales (McIntyre, 2006). Dolphin-feeding visitors at Tangalooma also recorded a strong commitment towards informing others about conservation; using energy-saving devices; donating time/money to wildlife conservation or environmental organizations; and joining a dolphin or mammal stranding group (Mayes *et al.*, 2004). Moreover, live-aboard divers in Thailand were more likely to participate in reef conservation or monitoring project, particularly if negative reef impacts were observed (Dearden, 2006). Conservation messages delivered on tours, brochures and displays at seven marine wildlife attractions in New Zealand (e.g. dolphin, whale, penguin, albatross and shorebirds) also highlighted hunting of whales; protection of marine mammals and migratory birds; predator management and eradication; the Southern Ocean Whale Sanctuary; recreational set netting/fisheries by-catch; marine pollution/urban waste management; and Maori environmental values. Some 54% of visitors on Dolphin Watch Marlborough, where tourists could participate in collecting and recording data about dolphin sightings and behaviours, stated their wildlife experience had affected their environmental values and actions (Higham and Carr, 2003).

Visitors to Seaworld Australia also stated they would support wildlife conservation by actions such as recycling (20%), giving money (14%), supporting wildlife networks (13%), looking after animal habitats (8%), reducing pollution (8%), conserving energy/water (6%) and cleaning up waterways (5%). Education (26%), respecting wildlife (12%) and awareness of wildlife (6%) were other conservation-minded actions reported by visitors to Seaworld (Saltzer, 2001). In sum, personal encounters with marine wildlife linked with education programmes were more likely to generate conservation appreciation and action by visitors. However, again most of these studies measure intention to act rather than actual conservation behaviours and rely on self-reporting by visitors, with no longer-term studies of conservation actions by visitors.

Visitor benefits from marine wildlife tours

This chapter has identified a range of psychological, educational and conservation benefits for visitors on marine wildlife tours. The on-site benefits of increased understanding or emotional responses to marine wildlife encounters can lead to off-site benefits such as greater environmental awareness, supporting nature conservation work and protecting endangered species. Empirical studies of marine wildlife tourism experiences were assessed against the framework devised by Orams (1995a, 1999) measuring positive changes in both tourists and the marine environment for effective management of marine wildlife tourism operations and sites. Tourist benefits from marine animal encounters include enjoyment and learning contributing to pro-environmental attitude and behavioural changes, and the longer-term intention to engage in conservation actions that benefit marine wildlife and environments. Therefore, marine wildlife tours with a strong educational focus and interpretation programme can create attitude, behaviour or lifestyle changes in visitors (Ballantyne et al., 2007). This review of visitor benefits from guided encounters with marine wildlife thus supports the framework developed by Orams (1999) for managing marine tourism experiences and also the experiential education sequence model in marine ecotourism programmes (Forestell, 1993). Key factors for tourist enjoyment and satisfaction were seeing marine wildlife, their proximity and behaviours, the natural setting, other marine wildlife and learning about marine life. The psychological benefits of visitors interacting with marine wildlife include relaxation or stimulation, mood benefits or feelings and learning benefits (Muloin, 1998, 2000; Schänzel and McIntosh, 2000). The learning benefits obtained from information about marine wildlife reinforced the emotional benefits of directly experiencing marine animals in their natural habitats. There is some evidence that personal benefits and visitor satisfaction differ according to gender, level of previous experience (Muloin, 1998) and type of wildlife encounter such as boat, land-based or in-water activities with marine life (Birtles et al., 2002; Finkler and Higham, 2004).

Quality educational experiences are also important for visitors to increase their short-term knowledge of marine species. Marine wildlife tours with a strong educational focus changed the pro-environmental attitudes, beliefs and behaviour of visitors. On whale and dolphin tours, tourists changed their attitudes to conservation, displaying a greater knowledge of cetaceans and awareness of threats to marine life. Other changes in the personal behaviour of visitors on a guided tour of turtle nesting beaches included better overall adherence to minimal impact guidelines at the Jurabi turtle experience. Visitors interacting with sea turtles at Mon Repos and dolphins at Tangalooma also adopted short-term pro-environmental behaviours (up to 4 months later) such as cleaning up beaches, recycling and donating money to wildlife groups. Other conservation benefits were enhanced appreciation of marine wildlife and engaging in actions to reduce human threats or impacts on wildlife (Howard, 2000). Close proximity to marine wildlife during in-water encounters or shore-based feeding interactions with dolphins magnified these environmental and personal benefits. The level or intensity of the encounter with marine wildlife needed to change tourist attitudes was linked to direct, close contact with animals more than passive viewing from a boat or on land. The quality of marine wildlife interpretation also influenced conservation outcomes and other environmentally responsible behaviours as reported by visitors. Therefore, visitor interactions with marine wildlife increased environmental awareness, changed attitudes, modified on-site and some longer-term behaviours and benefited marine conservation.

These psychological, educational and conservation benefits for visitors, however, depend on sound management of marine animal encounters and interpretation programmes that integrate knowledge with the emotional aspects

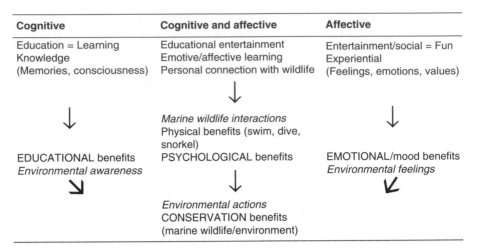

Fig. 2.1. Benefits for participants on marine wildlife tours. (From Schänzel, 2004.)

of observing marine wildlife (see Fig. 2.1). The benefits for participants on marine wildlife tours are realized when the affective (emotional) benefits and excitement of seeing unique marine life are integrated with the cognitive (education) benefits of learning new facts about marine wildlife. Thus, educational entertainment in marine life interpretation needs to include both cognitive and affective aspects of experiential learning (Howard, 2000; Schänzel, 2004). Visitor learning for fun and enjoyment during leisure activities is an important part of tourism experiences (Packer, 2006). Hence, marine wildlife interactions that involve making personal connections with marine animals in a learning context provide a range of psychological/mood and educational benefits. In-water encounters also provide physical benefits from swimming, snorkelling or diving with marine animals. Marine wildlife tourism experiences that increase both environmental awareness and positive feelings are more likely to generate environmental actions resulting in conservation benefits for marine wildlife and the natural environment.

Much of the research on marine wildlife tourism is site or species specific, focused on biological impacts and is limited to one type of encounter (Parsons *et al.*, 2006). This chapter focuses on tourist experiences of marine wildlife in Australia, New Zealand, Scotland and western Canada/USA. There is a need for marine wildlife tourism studies that investigate cultural and environmental values and other cultural perspectives of marine wildlife, such as by Asian tourists. The personal benefits of marine wildlife encounters for Asian visitors may differ from benefits sought by western tourists. The environmental attitudes of visitors in regard to whale watching and commercial or subsistence whaling need further investigation in Iceland, Norway, Japan, Tonga and the Caribbean (Orams, 2001; Hoyt and Hvenegaard, 2002; Higham and Lusseau, 2007). Visitors at aquariums and seaworld parks need to be surveyed about the conservation and educational benefits of marine wildlife encounters at these captive sites (Adelman *et al.*, 2000; Ballantyne, 2007). The personal benefits for staff and operators of marine wildlife tours along with their conservation attitudes and behaviour also need further investigation (Groff *et al.*, 2005). Longer-term studies also need to measure ongoing actual conservation actions of visitors 1–5 years after marine wildlife interactions, beyond self-reported intentions to act environmentally. The wildlife experience itself in a scenic natural area may heighten visitor concern and appreciation for wildlife but behavioural changes may not always ensue. This more in-depth evaluation of visitor benefits from marine wildlife tourism experiences and educational programmes will validate techniques that increase tourist knowledge and promote attitude shifts, lifestyle changes and both on-site and longer-term conservation behaviours that ultimately benefit marine wildlife and marine habitats.

Conclusion

Close personal encounters with selected marine wildlife, especially marine mammals, provide a range of psychological, educational and conservation benefits for visitors. These mediated encounters on wildlife tours motivate visitors to respect marine life, foster environmentally responsible attitudes and behaviours, and benefit marine conservation. Marine wildlife interpretation programmes that highlight biology and human impacts also influence visitor attitudes, beliefs and conservation outcomes. Linking affective and cognitive responses to marine wildlife increases environmental awareness, changes visitor attitudes, modifies intentions to act pro-environmentally, and fosters conservation appreciation and actions by wildlife tourists. Personal benefits for visitors also depend on the intensity and frequency of tourist encounters with marine wildlife and the type of learning experience provided. Visitors differ in their desired mix of psychological, educational and conservation benefits. Therefore, visitor benefits, and potential dissatisfaction with wildlife encounters, need to be considered by the managers and operators of marine wildlife tourism experiences. The challenge is to manage the visitor desire for close interaction with marine wildlife and the need to minimize human impacts on marine animals and marine ecosystems.

References

Adelman, L., Falk, J.H. and James, S. (2000) Assessing the national aquarium in Baltimore's impact on visitors' conservation, knowledge, attitudes and behavior. *Curator* 43(1), 33–61.
Amante-Helweg, V. (1996) Ecotourists' beliefs and knowledge about dolphins and the development of cetacean ecotourism. *Aquatic Mammals* 22, 131–140.
Andersen, M.S. and Miller, M.L. (2006) Onboard marine environmental education: whale watching in the San Juan islands, Washington. *Tourism in Marine Environments* 2(2), 111–118.
Ballantyne, R. (2007) Post-visit 'action resourcing': promoting and supporting visitor adoption of environmentally sustainable behaviours. In: McDonnell, I., Grabowski, S. and March, R. (eds)

Proceedings of the CAUTHE 2007 Conference: Tourism-Past Achievements, Future Challenges. University of Technology, Sydney, Australia [CD-ROM].

Ballantyne, R., Packer, J. and Hughes, K. (2006) Designing 'Jonah' experiences: developing visitor conservation learning through whale watching. In: *Conference Program: The 2nd Australian Wildlife Tourism Conference, 13–15 August 2006.* Promaco Conventions, Western Australia, p. 28.

Ballantyne, R., Packer, J. and Bond, N. (2007) The impact of a wildlife tourism experience on visitors' conservation knowledge, attitudes and behaviour: preliminary results from Mon Repos turtle rookery, Queensland. In: McDonnell, I., Grabowski, S. and March, R. (eds) *Proceedings of the CAUTHE 2007 Conference: Tourism-Past Achievements, Future Challenges.* University of Technology, Sydney, Australia [CD-ROM].

Barney, E.C., Mintzes, J.J. and Yen, C.F. (2005) Assessing knowledge, attitudes, and behavior toward charismatic megafauna: the case of dolphins. *The Journal of Environmental Education* 36(2), 41–55.

Barton, K., Booth, K., Ward, J., Simmons, D.G. and Fairweather, J.R. (1998) *Tourist and New Zealand Fur Seal Interactions along the Kaikoura Coast.* Tourism Research and Education Centre Report No. 9. Lincoln University, New Zealand.

Beasley, I. (1997) Marine mammal tourism: educational implications and legislation. Diploma of Wildlife Management thesis, University of Otago, Dunedin, New Zealand.

Bentrupperbaumer, J. (2005) Human dimensions of wildlife interactions. In: Newsome, D., Dowling, R.K. and Moore, S.A. (eds) *Wildlife Tourism.* Channel View, Clevedon, UK, pp. 82–112.

Birtles, A., Cuthill, M., Valentine, P. and Davis, D. (1996) Incorporating research on visitor experiences into ecologically sustainable management of whale shark tourism. In: Richins, H., Richardson, J. and Crabtree, A. (eds) *Taking the Next Steps: Ecotourism Association of Australia Conference Proceedings.* Ecotourism Association of Australia, Brisbane, Australia, pp. 195–202.

Birtles, A., Valentine, P. and Curnock, M. (2001) *Tourism Based on Free-Ranging Marine Life: Opportunities and Responsibilities.* Wildlife Tourism Research Report No. 11. CRC for Sustainable Tourism, Gold Coast, Queensland, Australia.

Birtles, A., Valentine, P., Curnock, M., Arnold, P. and Dunstan, A. (2002) *Incorporating Visitor Experiences into Ecologically Sustainable Dwarf Minke Whale Tourism in the Northern Great Barrier Reef.* Technical Report No. 42. CRC Reef Research Centre, Townsville, Queensland, Australia.

Booth, K. (1998) The tourist experience of New Zealand fur seals along the Kaikoura coast, New Zealand. In: Kandampully, J. (ed.) *New Zealand Tourism and Hospitality Research Conference 1998: Advances in Research Part 2.* Lincoln University, Canterbury, New Zealand.

Bulbeck, C. (1999) The 'nature dispositions' of visitors to animal encounter sites in Australia and New Zealand. *Journal of Sociology* 35, 129–148.

Bulbeck, C. (2005) *Facing the Wild: Ecotourism, Conservation and Animal Encounters.* Earthscan, London.

Catlin, J. and Jones, R. (2006) Evolution of the whale shark tourism market. In: *Conference Program: The 2nd Australian Wildlife Tourism Conference.* Promaco Conventions, Western Australia, pp. 85–95.

Corkeron, P.J. (2004) Whale watching, iconography, and marine conservation. *Conservation Biology* 18(3), 847–849.

Curtin, S. (2005) Nature, wild animals and tourism: an experiential view. *Journal of Sustainable Tourism* 4(1), 1–15.

Davis, D., Banks, S., Birtles, A., Valentine, P. and Cuthill, M. (1997) Whale sharks in Ningaloo marine park: managing tourism in an Australian marine protected area. *Tourism Management* 18(5), 259–271.

Davis, D., Banks, S., Birtles, A., Valentine, P. and Cuthill, M. (2000) Whale sharks in Ningaloo marine park: managing tourism in an Australian marine protected area. In: Ryan, C. and Page, S. (eds) *Tourism Management: Towards the New Millennium*. Pergamon, Oxford, pp. 281–299.

Dearden, P. (2006) Effects of participating in wildlife tourism: a case study from diving. In: *Conference Program: The 2nd Australian Wildlife Tourism Conference, 13–15 August 2006*. Promaco Conventions, Western Australia, p. 39.

Dobson, J. (2006) Sharks, wildlife tourism, and state regulation. *Tourism in Marine Environments* 3(1), 15–23.

Duffus, D.A. and Dearden, P. (1993) Recreational use, valuation and management of killer whales (*Orcinus orca*) on Canada's pacific coast. *Environmental Conservation* 20(2), 149–156.

Finkler, W. and Higham, J. (2004) The human dimensions of whale watching: an analysis based on viewing platforms. *Human Dimensions of Wildlife* 9(2), 103–117.

Forestell, P.H. (1993) If leviathan has a face, does Gaia have a soul? Incorporating environmental education in marine eco-tourism programs. *Ocean and Coastal Management* 20, 267–282.

Forestell, P.H. and Kaufman, G.D. (1990) The history of whalewatching in Hawaii and its role in enhancing visitor appreciation for endangered species. In: Miller, M.L. and Auyong, J. (eds) *Proceedings of the 1990 Congress on Coastal and Marine Tourism*, Vol. 11. National Coastal Resources Research and Development Institute, Newport, Oregon, pp. 399–407.

Foxlee, J. (2001) Whale watching at Hervey Bay. *Parks and Leisure Australia* 4(3), 17–18.

Frohoff, T.G., and Packard, J.M. (1995) Human interactions with free-ranging and captive bottlenose dolphins. *Anthrozoos* 8(1), 44–57.

Groff, A., Lockhart, D., Ogden, J. and Dierking, L.D. (2005) An exploratory study of the effect of working in an environmentally themed facility on the conservation-related knowledge, attitudes and behavior of staff. *Environmental Education Research* 11(3), 371–387.

Higginbottom, K. and Tribe, A. (2004) Contributions of wildlife tourism to conservation. In: Higginbottom, K. (ed.) *Wildlife Tourism: Impacts, Management and Planning*. Common Ground/Sustainable Tourism CRC, Altona, Victoria, pp. 99–123.

Higham, J.E.S. (1998) Tourists and albatrosses: the dynamics of tourism at the Northern Royal albatross colony, Taiaroa Head, New Zealand. *Tourism Management* 19(6), 521–531.

Higham, J.E.S. (2001) Managing ecotourism at Taiaroa Head Royal Albatross colony. In: Shackley, M. (ed.) *Flagship Species: Case Studies in Wildlife Tourism Management*. The International Ecotourism Society, Burlington, Vermont, pp. 17–31.

Higham, J.E. and Carr, A.M. (2003) Sustainable wildlife tourism in New Zealand: an analysis of visitor experiences. *Human Dimensions of Wildlife* 8, 25–36.

Higham, J. and Lusseau, D. (2004) Ecological impacts and management of tourist engagements with marine mammals. In: Buckley, R. (ed.) *Environmental Impacts of Ecotourism*. CAB International, Wallingford, UK, pp. 171–186.

Higham, J.E.S. and Lusseau, D. (2007) Urgent need for empirical research into whaling and whale watching. *Conservation Biology* 21(2), 554–558.

Howard, J. (2000) Research in progress: does environmental interpretation influence behaviour through knowledge or affect? *Australian Journal of Environmental Education* 15/16, 153–156.

Hoyt, E. (2003) Towards a new ethic for experiencing dolphins and whales. In: Frohoff, T. and Peterson, B. (eds) *Between Species: Celebrating the Dolphin–Human Bond*. Sierra Club Books, San Francisco, California, pp. 168–177.

Hoyt, E. (2005) Sustainable ecotourism on Atlantic islands, with special reference to whale watching, marine protected areas and sanctuaries for cetaceans. *Biology and Environment: Proceedings of the Royal Irish Academy* 105B(3), 141–154.

Hoyt, E. and Hvenegaard, G.T. (2002) A review of whale-watching and whaling with applications for the Caribbean. *Coastal Management* 30(4), 381–399.

Hughes, M. and Saunders, A.M. (2005) Interpretation, activity participation, and environmental attitudes of visitors to Penguin Island, Western Australia. *Society and Natural Resources* 18, 611–624.

Hughes, P. (2001) Animals, values and tourism – structural shifts in UK dolphin tourism provision. *Tourism Management* 22(4), 321–329.

Kellert, S.R. (1999) *American Perceptions of Marine Mammals and their Management.* Humane Society of the United States, Washington, DC.

Kirkwood, R., Boren, L., Shaughnessy, P., Szteren, D., Mawson, P., Huckstadt, L., Hofmeyr, G., Ossthuizen, H., Schiavini, A., Campagagna, C. and Berris, M. (2003) Pinniped-focused tourism in the southern hemisphere: a review of the industry. In: Gales, N., Hindell, M. and Kirkwood, R. (eds) *Marine Mammals: Fisheries, Tourism and Management Issues.* CSIRO Publishing, Collingwood, Victoria, pp. 257–272.

Lewis, A. and Newsome, D. (2003) Planning for stingray tourism at Hamelin bay, Western Australia: the importance of stakeholder perspectives. *International Journal of Tourism Research* 5, 331–346.

Lück, M. (2003) Education on marine mammal tours as agent for conservation – but do tourists want to be educated? *Ocean and Coastal Management* 46(9), 943–956.

Macgregor, K. (2006) Commerce, community and conservation: a collaborative approach to managing a sustainable marine turtle tourism industry. In: *Conference Program: The 2nd Australian Wildlife Tourism Conference.* Promaco Conventions, Western Australia, pp. 139–148.

Manfredo, M.J. and Driver, B.L. (2002) Benefits: the basis for action. In: Manfredo, M.J. (ed.) *Wildlife Viewing: A Management Handbook.* Oregon State University Press, Corvallis, Oregon, pp. 70–92.

Masters, D. (1998) *Marine Wildlife Tourism: Developing a Quality Approach in the Highlands and Islands.* A report for the tourism and environment initiative and Scottish natural heritage, May 1998. Visit Scotland, Sustainable Tourism Unit. Available at: http://www.greentourism.org.uk/publications.html

Mayes, G., Dyer, P. and Richins, H. (2004) Dolphin–human interaction: pro-environmental attitudes, beliefs and intended behaviours and actions of participants in interpretation programs: a pilot study. *Annals of Leisure Research* 7(1), 34–53.

McIntyre, M. (2006) Can whale watching save the whale? In: *Conference Program: The 2nd Australian Wildlife Tourism Conference, 13–15 August 2006.* Promaco Conventions, Western Australia, p. 28.

McKegg, S., Probert, K., Baird, K. and Bell, J. (1998) Marine tourism in New Zealand: a profile. In: Miller, M.L. and Auyong, A. (eds) *Proceedings of the 1996 World Congress on Coastal and Marine Tourism.* Oregon State University, Corvallis, Oregon, pp. 154–159.

Moscardo, G. (2000) Understanding wildlife tourism market segments: an Australian marine study. *Human Dimensions of Wildlife* 5(2), 36–53.

Moscardo, G. (2001) *Understanding Visitor–Wildlife Interactions: Factors Contributing to Satisfaction.* CRC Reef Research Centre, Townsville, Queensland, Australia.

Moscardo, G. and Saltzer, R. (2004) Understanding wildlife tourism markets. In: Higginbottom, K. (ed.) *Wildlife Tourism: Impacts, Management and Planning.* Common Ground/Sustainable Tourism CRC, Altona, Victoria, pp. 167–185.

Moscardo, G., Woods, B. and Greenwood, T. (2001) *Understanding Visitor Perspectives on Wildlife Tourism.* Wildlife Tourism Research Report Series No. 2. CRC for Sustainable Tourism, Gold Coast, Queensland, Australia.

Moscardo, G., Woods, B. and Saltzer, R. (2004) The role of interpretation in wildlife tourism. In: Higginbottom, K. (ed.) *Wildlife Tourism: Impacts, Management and Planning.* Common Ground/Sustainable Tourism CRC, Altona, Victoria, pp. 231–251.

Muloin, S. (1997) Wildlife tourism (book review). *Journal of Sustainable Tourism* 5(2), 175–178.

Muloin, S. (1998) Wildlife tourism: the psychological benefits of whale watching. *Pacific Tourism Review* 2, 199–212.

Muloin, S. (2000) The psychological benefits experienced from tourist–wildlife encounters. PhD thesis, Tourism Program, School of Business, James Cook University, Queensland, Australia.

Murphy, L. and Norris, A. (2005) Understanding great barrier reef visitors: segmentation according to reef trip benefits. *Tourism in Marine Environments* 1(2), 71–87.

Neil, D.T., Orams, M.B. and Baglioni, A.J. (1996) Effect of previous whale watching experience on participants' knowledge of, and response to, whales and whale watching. In: Colgan, K., Prasser, S. and Jeffrey, A. (eds) *Encounters with Whales 1995 Proceedings*. Hervey Bay, Queensland, 26–30 July 2005. Australian Nature Conservation Agency, Canberra, Australia, pp. 182–188.

Newsome, D., Dowling, R.K. and Moore, S.A. (2005) Managing potential impacts. In: *Wildlife Tourism*. Channel View, Clevedon, UK, pp. 190–195.

O'Neill, F., Barnard, S. and Lee, D. (2004) *Best Practice and Interpretation in Tourist/Wildlife Encounters: A Wild Dolphin Swim Tour Example*. Wildlife Tourism Research Report Series No. 25. CRC Sustainable Tourism, Gold Coast, Queensland, Australia.

Orams, M. (1999) *Marine Tourism: Development, Impacts and Management*. Routledge, London.

Orams, M. (2001) From whale hunting to whale watching in Tonga: a sustainable future? *Journal of Sustainable Tourism* 9(2), 128–146.

Orams, M. (2005) Dolphins, whales and ecotourism in New Zealand: what are the impacts and how should the industry be managed? In: Hall, C.M. and Boyd, S. (eds) *Nature-based Tourism in Peripheral Areas: Development or Disaster?* Channel View, Clevedon, UK, pp. 231–245.

Orams, M.B. (1994) Creating effective interpretation for managing interaction between tourists and wildlife. *Australian Journal of Environmental Education* 10(2), 21–34.

Orams, M.B. (1995a) Towards a more desirable form of ecotourism. *Tourism Management* 16, 3–9

Orams, M.B. (1995b) A conceptual model of tourist–wildlife interaction: the case for education as a management strategy. *Australian Geographer* 27(1), 39–51.

Orams, M.B. (1995c) Development and management of a feeding program for wild bottlenose dolphins at Tangalooma, Australia. *Aquatic Mammals* 21(2), 137–147.

Orams, M.B. (1995d) Using interpretation to manage nature-based tourism. *Journal of Sustainable Tourism* 4(2), 81–94.

Orams, M.B. (1996) Cetacean education: can we turn tourists into 'greenies'. In: Colgan, K., Prasser, S. and Jeffrey, A. (eds) *Encounters with Whales 1995 Proceedings*. Hervey Bay, Queensland, July 26–30. Australian Nature Conservation Agency, Canberra, Australia, pp. 167–178.

Orams, M.B. (1997a) Historical accounts of human–dolphin interaction and recent developments in wild dolphin based tourism in Australasia. *Tourism Management* 18, 317–326.

Orams, M.B. (1997b) The effectiveness of environmental education: can we turn tourists into 'greenies'? *Progress in Tourism and Hospitality Research* 3, 295–306.

Orams, M.B. (2000) Tourists getting close to whales, is it what whale watching is all about? *Tourism Management* 21, 561–569.

Orams, M.B. (2002) Feeding wildlife as a tourism attraction: a review of issues and impacts. *Tourism Management* 23, 281–293.

Orams, M.B. (2003) Marine ecotourism in New Zealand: an overview of the industry and its management. In: Garrod, B. and Wilson, J.C. (eds) *Marine Ecotourism: Issues and Experiences*. Channel View, Clevedon, UK, pp. 233–248.

Orams, M.B. and Hill, G.J.E. (1998) Controlling the ecotourist in a wild dolphin feeding program: is education the answer? *Journal of Environmental Education* 29, 33–38.

Orsini, J.P. and Newsome, D. (2005) Human perceptions of hauled out Australian sea lions (*Neophoca cinerea*) and implications for management: a case study from Carnac Island, Western Australia. *Tourism in Marine Environments* 2(1), 23–37.

Packer, J. (2006) Learning for fun: the unique contribution of educational leisure experiences. *Curator: The Museum Journal* 49(3), 329–344.

Parsons, E.C.M., Warburton, C.A., Woods-Ballard, A., Hughes, A., Johnston, P., Bates, H. and Lück, M. (2003) Whale-watching tourists in West Scotland. *Journal of Ecotourism* 2(2), 93–113.

Parsons, E.M.C., Lewandowski, J. and Lück, M. (2006) Recent advances in whale-watching research: 2004–2005. *Tourism in Marine Environments* 2(2), 119–132.

Pearce, D.G. and Wilson, P.M. (1995) Wildlife-viewing tourists in New Zealand. *Journal of Travel Research* 34(2), 19–26.

Rawles, C.J.G. and Parsons, E.C.M. (2005) Environmental motivation of whale-watching tourists in Scotland. *Tourism in Marine Environments* 1(2), 129–132.

Reef Futures (2003) Marine wildlife. Available at: http://www.reeffutures.org/topics/monitoring/marinewildlife.cfm

Reynolds, P.C. and Braithwaite, D. (2001) Towards a conceptual framework for wildlife tourism. *Tourism Management* 22, 31–42.

Russell, C.L. and Hodson, D. (2002) Whalewatching as critical science education? *Canadian Journal of Science, Mathematics and Technology Education* 7, 54–66.

Saltzer, R. (2001) *Understanding Visitor Wildlife Interactions: Seaworld Visitor Survey*. Data Summary Report December 2001. CRC for Sustainable Tourism. Available at: http://www.crctourism.com.au/CRCBookshop/Documents/WT_Seaworld.pdf

Saltzer, R. (2003a) *Understanding Visitor–Wildlife Interactions: A Case Study of the Penguin Place, Otago Peninsula, New Zealand*. Data summary report March 2003. Sustainable Tourism CRC. Available at: http://www.crctourism.com.au/CRCBookshop/Documents/WT_PenguinPlace-NZ.pdf

Saltzer, R. (2003b) *Understanding Visitor–Wildlife Interactions: A Case Study of the Royal Albatross Centre, Otago Peninsula, New Zealand*. Data summary report March 2003. Sustainable Tourism. Available at: http://www.crctourism.com.au/CRCBookshop/Documents/WT_AlbatrossCentre-NZ.pdf

Saltzer, R. (2003c) *Understanding Visitor–Wildlife Interactions: A Case Study of Monarch Wildlife Cruises, Otago Peninsula, New Zealand*. Data summary report March 2003. Sustainable Tourism. Available at: http://www.crctourism.com.au/CRCBookshop/Documents/WT_MonarchCruises-NZ.pdf

Samuels, A., Bejder, L. Constantine, R. and Heinrich, S. (2003) Swimming with wild cetaceans, with a special focus on the southern hemisphere. In: Gales, N., Hindell, M. and Kirkwood, R. (eds) *Marine Mammals: Fisheries, Tourism and Management Issues*. CSIRO, Collingwood, UK, pp. 277–303.

Scarpaci, C., Nugegoda, D. and Corkeron, P.J. (2005) Tourists swimming with Australian fur seals (*Arctocephalus pusillus*) in Port Philip Bay, Victoria, Australia: are tourists at risk? *Tourism in Marine Environments* 1(2), 89–95.

Schäenzel, H. (1998) The effectiveness of environmental interpretation: understanding the values gained from wildlife viewing tourism experiences. *Environmental Perspectives* 21, 10–13.

Schänzel, H.A. (2004) Educational entertainment: the emotive and personal context of environmental interpretation. In: Smith, K.A. and Scott, C. (eds) *Proceedings of the New Zealand Tourism and Hospitality Research Conference*, Wellington, 8–10 December. Victoria University of Wellington, Wellington, UK, pp. 348–356.

Schänzel, H.A. and McIntosh, A.J. (2000) An insight into the personal and emotive context of wildlife viewing at the Penguin Place, Otago Peninsula, New Zealand. *Journal of Sustainable Tourism* 8(1), 36–52.

Shackley, M. (1998) 'Stingray City' – managing the impacts of underwater tourism in the Cayman Islands. *Journal of Sustainable Tourism* 6(4), 328–338.

Smith, A., Newsome, D., Lee, D. and Stoeckl, N. (2006) *The Role of Wildlife Icons as Major Tourist Attractions – Case Studies: Monkey Mia Dolphins and Hervey Bay Whale Watching*. CRC for Sustainable Tourism, Gold Coast, Queensland, Australia.

Smith, L. (2006) Evaluating the effectiveness of the Jurabi turtle experience. In: *Conference Program: The 2nd Australian Wildlife Tourism Conference, 13–15 August 2006*. Promaco Conventions, Canning Bridge, Western Australia, pp. 165–174.

Sorice, M.G., Shafer, C.S. and Ditton, R.B. (2006) Managing endangered species within the use-preservation paradox: the Florida manatee (*Trichechus manatus latirostris*) as a tourism attraction. *Environmental Management* 37(1), 69–83.

Stokes, T., Dobbs, K. and Recchia, C. (2002) Management of marine mammal tours on the Great Barrier Reef. *Australian Mammalogy* 24, 39–49.

Takei, A. (1998) A cross-cultural assessment of the effectiveness of the interpretive program at the dolphin provisioning program, Tangalooma, Australia. Diploma in Marine Science thesis, University of Queensland. Available at: http://www.tangalooma.com/dolphinweb/research.asp

Tisdell, C. and Wilson, C. (2001a) Wildlife-based tourism and increased support for nature conservation financially and otherwise: evidence from sea turtle ecotourism at Mon Repos. *Tourism Economics* 7(3), 233–249.

Tisdell, C. and Wilson, C. (2001b) Tourism and the conservation of sea turtles: an Australian case study. In. Tisdell, C. (ed.) *Tourism Economics, the Environment and Development: Analysis and Policy*. Edward Elgar, Cheltenham, UK, pp. 356–368.

Tisdell, C. and Wilson, C. (2002) *Economic, Educational and Conservation Benefits of Sea Turtle Based Ecotourism: A Study Focused on Mon Repos*. Wildlife Tourism Research Report Series No. 20. CRC for Sustainable Tourism, Gold Coast, Queensland, Australia

Tisdell, C. and Wilson, C. (2005) Perceived impacts of ecotourism on environmental learning and conservation: turtle watching as a case study. *Environment, Development and Sustainability* 7(3), 291–302.

Topelko, K.N. and Dearden, P. (2005) The shark watching industry and its potential contribution to shark conservation. *Journal of Ecotourism* 4(2), 108–128.

Townsend, C. (2003) Marine ecotourism through education: a case study of divers in the British Virgin Islands. In: Garrod, B. and Wilson, J.C. (eds) *Marine Ecotourism: Issues and Experiences*. Channel View, Clevedon, UK, pp. 138–154.

Tremblay, P. (2002) Tourism wildlife icons: attractions or marketing symbols. *Journal of Hospitality and Tourism Management* 9(2), 164–180.

Valentine, P.S., Birtles, A., Curnock, M., Arnold, P. and Dunstan, A. (2004) Getting closer to whales – passenger expectations and experiences, and the management of swim with dwarf minke whale interactions in the Great Barrier Reef. *Tourism Management* 25, 647–655.

Waayers, D., Newsome, D. and Lee, D. (2006) Observations of non-compliance by tourists to a voluntary code of conduct: a pilot study of turtle tourism in the Exmouth region, Western Australia. *Journal of Ecotourism* 5(3), 211–222.

Warburton, C.A., Parsons, E.C.M. and Goodwin, H. (2000) Marine wildlife tourism and whale-watching on the Isle of Mull, Scotland. Paper presented to the Scientific Committee at the 52nd meeting of the International Whaling Commission, 11–28 June 2000, Australia. Available at: http://www.whaledolphintrust.co.uk/research/documents/iwc_ecotourism.pdf

Whale and Dolphin Conservation Society (WDCS) (2005) The benefits of high quality whale watching. WDCS Australasia. Available at: http://www.wdcs.org.au/info_details.php?select=1133325744

Whales Alive (2005) Whale watching. Whales alive. Available at: http://www.whalesalive.org.au/whalewatching.html

Wildlife Extra (2006) Marine wildlife. Available at: http://www.wildlifeextra.com/hm-marine.html

Wilson, C. and Tisdell, C. (2001) Sea turtles as a non-consumptive tourism resource especially in Australia. *Tourism Management* 22, 279–288.

Wilson, C. and Tisdell, C. (2003) Conservation and economic benefits of wildlife-based marine tourism: sea turtles and whales as case studies. *Human Dimensions of Wildlife* 8(1), 49–58.

Woods, B. and Moscardo, G. (2003) Enhancing wildlife education through mindfulness. *Australian Journal of Environmental Education* 19, 97–108.

Würsig, B. (1996) Swim with dolphin activities in nature: weighing the pros and cons. *Whale Watcher: The Journal of the American Cetacean Society* 30(2), 11–15.

3 Shark! A New Frontier in Tourist Demand for Marine Wildlife

J. Dobson

Introduction

Shark! A word that can strike fear into the hearts of many people is synonymous with unseen monsters from the deep, crushing jaws and violent death. Yet it is also a word that is increasingly becoming associated with the marine wildlife tourism industry. Research into the development, impact and management of shark-based tourism has been limited and is certainly not as well developed as that of research focusing on cetacean watching. This chapter will attempt to outline some of the key features of shark-based tourism, especially in terms of its potential management challenges, and areas in need of further research.

The Shark

In order to understand the utilization of sharks by the wildlife tourism industry it is first necessary to briefly outline the conservation status of sharks and the reason they are able to act as attractions. Sharks belong to the Chondrichthyan taxonomic class, which is further divided into two subclasses: the Holocephali (the chimaeras) and the Elasmobrachii (sharks, skates and rays). This chapter will primarily focus on the utilization of sharks as tourist attractions, but will refer to skates and rays where relevant. Although popular in terms of Hollywood notoriety, little is actually known about the biology of sharks and other chondrichthyan fishes compared to other marine fauna (Camhi et al., 1998). Compagno et al. (2005) identify 416 species of shark (although it is difficult to generate an exact figure as new species are being discovered or reclassified). Of these, 45 sharks (and a further 65 rays) are listed as being globally threatened (those species considered as being critically endangered, endangered or vulnerable) by the IUCN Red List of Threatened Species (IUCN SSG, 2006).

©CAB International 2008. *Marine Wildlife and Tourism Management:*
Insights from the Natural and Social Sciences (eds J.E.S. Higham and M. Lück)

Sharks have few natural predators; however, they face a number of threats from human activities including targeted fisheries, incidental by-catch, increases in oceanic pollution and the destruction of nursery grounds by coastal developments. Targeted shark fisheries are a particular problem, cartilage (used in traditional medicines), jaws and teeth (for the souvenir trade) and shark fins (used to make shark fin soup, a particular delicacy in the Far East) are all valuable commodities (Cunningham-Day, 2001; Philpott, 2002). The tourism industry can also have a negative impact on shark populations. Coastal tourism developments can result in the removal of important mangrove habitat which act as nursery grounds for young sharks (e.g. large-scale hotel development on Bimini in the Bahamas has destroyed important nursery grounds for Lemon sharks (*Negaprion brevirostris*; The Shark Trust, no date, a) and shark nets used to protect tourist beaches indiscriminately kill many sharks and other forms of marine wildlife each year (Cunningham-Day, 2001; Compagno *et al.*, 2005). It is estimated that approximately 100 million sharks are killed each year and that over the last 20–30 years, shark populations have decreased between 70% and 90% worldwide (Compagno *et al.*, 2005). Most shark fisheries worldwide go unmanaged and the origins of shark products entering international trade go virtually unrecorded (Camhi *et al.*, 1998) resulting in many shark populations being put at risk of total collapse. As will be explained below, sharks do not possess the levels of affection held by humans for terrestrial top predators such as lions and tigers; the marine environment is also considered alien to humans and therefore conservation efforts tend to lag behind those for terrestrial environments. Unfortunately, shark eradication programmes are often better funded than shark conservation and management schemes (Compagno *et al.*, 2005).

However, sharks are slowly gaining protected status. The great white shark (*Carcharaodon carcharias*), basking shark (*Cetorhinus maximus*) and whale shark (*Rhincodon typus*) are listed on Appendix II of the Convention on International Trade in Endangered Species (CITES), requiring any trade in these species to be licensed and monitored (CITES, 2006). Some sharks receive protection when they are present in the territorial waters of certain countries. The great white shark is protected in several countries, including South Africa, Malta and Australia (Dobson *et al.*, 2005) and the basking shark is protected in the UK waters under Schedule 5 of the Wildlife and Countryside Act, 1981 (JNCC, 2002). The utilization of sharks and rays by the tourism industry does offer the potential to contribute towards their conservation (Anderson and Waheed, 2001; Graham, 2004; Dobson *et al.*, 2005; Topelko and Dearden, 2005) and will be discussed later in this chapter.

Sharks as Wildlife Tourism Attractions

Authors, such as Tremblay (2002) and Smith *et al.* (2006), suggest that certain species are popular with tourists due to their status as 'wildlife icons'. These icons tend to possess certain 'charismatic' properties, such as cuteness or approachability making them appealing to humans. Caughley (1985) produced a taxonomy of human perceptions of animals ranging from 'lovelies' (animals we like, revere or honour) through to 'nasties' (animals we fear or loathe). Charismatic marine mega fauna, such as whales and dolphins, can be considered 'lovelies' (Corkeron, 2006), indeed Bulbeck (2005, p. 82) suggests that one reason for the popularity of dolphins as tourist attractions is that humans 'project onto dolphins our own dreams and desires'.

Apart from the basking and whale sharks (the two largest fish in the sea), it may be difficult to see (if charisma is a significant factor) how and why other shark species can act as wildlife attractions. Sharks, especially in Western societies, tend to possess negative charisma and are perceived as non-human, dangerous and man-eaters. Under Caughley's (1985) typology they may be considered 'nasties'. Bart's 1972 study of the popularity of animals ranked the shark a lowly 26th out of 30 listed species, beating only the spider, snake, rat and scorpion in popularity (cited in Shackley, 1996, p. 20).

In Western culture, the shark has a long history of demonization. Early paintings, such as John Singleton Copley's 1778 picture *Watson and the Shark* and Winslow Homer's 1899 picture *The Gulf Stream*, represented sharks as significant hazards of the sea (Morey, 2002). This theme has continued through into popular culture with Peter Benchley's 1974 book 'Jaws' and subsequent 1975 Steven Spielberg film enshrining the shark as a 'monster' in the public's consciousness. Shark attacks on humans, although rare (only 94 people have died as a result of shark attacks worldwide since 1990; ISAF, 2006) make headlines around the world. It is important to note that the term 'shark attack' is an anthropocentric interpretation of this form of human–shark encounter. There is growing evidence that many apparent 'attacks' are often cases of mistaken identity (sharks will test bite objects to see if they are edible) or that the shark is acting in self-defence after feeling threatened by the presence of humans in the water (Peschak and Scholl 2006).

Despite this, the media frequently interpret these encounters as 'attacks' and tend to provide high levels of sensationalist coverage which generates headlines such as 'I Fought Off Jaws With My Bare Hands' (Byrne, 2005), 'Brit's Shark Terror at Sea' (Armstrong, 2005) and 'Just When You Thought It Was Safe To Go Back In The Water . . . Jaws Attacks' (Perrie, 2005). This helps further entrench the sharks' supposedly anthropophagous nature in human consciousness. Negative coverage can also help establish what Cohen (2002) describes as a 'moral panic'. A moral panic can be generated by a wide range of social phenomena where society overreacts to perceived threats (Goode and Nachman, 2003). Although predominately used to describe threats from deviant social behaviour, the term is also relevant to the public reaction that can be generated by the horror of shark attacks and the sense of outrage generated in response to the growth of shark-based tourism (Dobson, 2006).

Paradoxically, it is these negative connotations held primarily in Western society that hold the key as to why sharks can act as wildlife attractions. As with the growing appeal of saltwater crocodiles in Australia (Ryan, 1998; Ryan and Harvey, 2000), the shark embodies aspects of primitive nature, wild and untamed. This along with its perceived dangerousness has made shark encounters especially appealing to the growing eco-adventure market (Dobson *et al.*, 2005). The next part of this chapter will go on to explore the different ways in which tourists can interact with sharks.

Tourism/Shark Encounters

The Spectrum of Tourist–Wildlife Interaction Opportunities model identifies that tourist wildlife encounters can take place in captive, semi-captive and wild environments (Orams, 2002) and encounters can then be consumptive or non-consumptive in nature (Tremblay, 2001). Encounters with sharks can take place in both captive and wild locations; however, some encounters in the wild rely on attracting sharks to tour boats to increase the likelihood of sightings. These contrived experiences sit between semi-captive and wild experiences (Orams, 2002).

Encounters in Captive Environments

Sharks are often a key feature in many aquaria and usually form part of the central exhibit (e.g. the I & J Predator Exhibit at Cape Town's Two Oceans Aquarium). In many aquaria, tourists are able to walk through acrylic viewing tunnels as sharks swim around and above them, or observe them through large viewing windows. Skates and rays are also popular attractions and can often be found, along with smaller species of sharks, in touch pools where visitors can directly interact with them (Fig. 3.1). The 2006 American Elasmobranch Society International Captive Elasmobranch Census (ICES) identified that 86 individual species of sharks and 103 species of skates and rays were held in 112 institutions around the world.

Fig. 3.1. A visitor interacting with small sharks and rays in a touch pool at Underwater World, Singapore. (Photograph John Dobson.)

The most popular shark species held in aquaria tend to be sedentary/coastal species, such as the nurse shark (*Ginglymostoma cirratum*) or the raggedtooth shark (*Carcharias taurus*) (AES, 2006). However, improvements in aquaria technology and shark husbandry have seen larger pelagic species being kept in captivity. The Monterey Bay Aquarium in California had some success in keeping a great white shark for 198 days between September 2004 and March 2005. Whale sharks (the largest fish in the sea) are exhibited at both the Georgia Aquarium in the USA and the Okinawa Churaumi Aquarium in Japan. Although keeping such large pelagic species in captivity has drawn criticism concerning general health and life expectancy issues, these species prove to be popular attractions for visitors. Visitor attendance at the Monterey Bay Aquarium in California reached record levels when a second great white shark was put on display in August 2006 (Grayson, 2006) and over 3 million people have visited the Georgia Aquarium to see four whale sharks on display there since it opened in November 2005 (Gross, 2006).

Some aquaria provide tourists with the only opportunity to have a guaranteed underwater encounter with sharks by allowing divers to swim in exhibit tanks (Jackson, 2000). Deep Sea World in Edinburgh, Scotland, Underwater World in Singapore and Sea World on the Gold Coast in Australia are amongst many that now provide opportunities to dive with their sharks.

Encounters in the Natural Environment

Carwardine and Watterson (2002) identify 267 individual locations in 43 different countries around the world where tourists can have a shark encounter at some point during the year and it is estimated that over 500,000 people pay to dive with sharks in the wild every year (Topelko and Dearden, 2005). Not all encounters with sharks take place on organized shark-viewing trips as they may be encountered as part of the resident fauna on general dives.

Contrived encounters (feeding/baited encounters)

Sharks, by their very nature as apex predators, tend to be much harder to view in the wild than other species that are utilized by the marine wildlife tourism industry. Outside of natural congregations there is often a need to use some method for attracting sharks to dive sites. Operators can use bait (which may or may not be fed to the sharks), chum (a mixture of blood and fish parts, which, as it disperses across the water, is sensed by the sharks that follow the slick to tour boats) and decoys to lure sharks close to boats or dive sites to ensure that visitors have a close experience. Perhaps the best known baited shark encounters take place in the Bahamas where a number of operators offer shark-diving experiences with a wide range of shark species.

Cage diving to view potentially dangerous sharks often involves baiting and sometimes feeding activities. Cage diving with great white sharks has become a popular adventure activity and can be found in four areas globally: the Western Cape of South Africa, the Farallon Islands off the California coast, Ille de Guadalupe, Mexico and Port Lincoln, South Australia. Cage diving also takes place with other species of sharks; tourists can dive with Galapagos sharks (*Carcharhinus galapagensis*) and sandbar sharks (*Carcharhinus plumbeus*) off Hawaii's north shore and blue sharks (*Prionace glauca*) in the UK. In very few locations tourists can also view sharks from the shoreline. Bull sharks (*Carcharhinus leucas*) have been attracted to scraps being thrown into the water by fishermen at Walkers Cay resort in the Bahamas and have congregated close to the shoreline, which provided tourists with a good view of the sharks (Carwardine and Watterson, 2002). Baiting and feeding sharks is a highly controversial activity raising concerns for both shark and human welfare and will be explored later in this chapter.

Wild encounters

Natural aggregations of sharks do occur at reef drop offs, 'cleaning' stations (areas where sharks and other larger fish congregate to have parasites removed by smaller fish) and seasonal feeding and breeding sites making them important dive locations (Jackson, 2000). Whale sharks seasonally congregate at various locations around the world to feed and (possibly) breed. They often feed at or near the surface allowing tourists to view them from boats or more frequently swim with them without the need to be scuba trained. This rise in popularity has enabled whale shark tourism to develop in 18 locations around the world (IWC, 2005), including Ningaloo Reef in Australia, Donsal in the Philippines and Gladen Spit in Belize. Other aggregation sites which have facilitated the development of shark-based tourism include Aliwal Shoals and Protea Banks in South Africa where raggedtooth sharks, bull sharks and scalloped hammerheads (*Sphyrna lewini*) congregate at certain times each year. However, it is the Cocos Islands in the Pacific which have reputedly the greatest congregation of sharks in the world, including various species of hammerhead and reef sharks (Jackson, 2000).

Sports fishing

Fishing sharks for sport is a popular although highly controversial activity. Interest in shark fishing witnessed a dramatic increase after the release of Spielberg's film 'Jaws'. During this time shark fishing was predominately consumptive and was perceived as a 'macho' activity. Trophy photographs of fishermen standing next to their catch on the dock were prize souvenirs for participants. Shark-fishing tournaments are still popular activities especially in the USA with several taking place each year along the eastern seaboard. One of the best known is the Oak Bluffs 'Monster Shark Tournament' in Martha's Vineyard. These tournaments are popular not only with competitors (The Humane Society of the United States (HSUS) estimated that approximately

240 boats took part in the Oak Bluffs tournament in 2005 (HSUS, 2006)) but also with tourists who come to see the boats landing their catch at the end of each day. The HSUS has been actively campaigning against the Oak Bluff tournament raising concerns that their emphasis on catch size encourages the extraction of large mature species that are critical to the long-term survival of shark populations. Unfortunately, these are the sharks that are of most value to tournament competitors as they tend to be large specimens that provide more chance of winning (HSUS, 2006). The tournament organizers defend the tournament as the rules have been changed in recent years towards catch and release and boats are limited to landing one shark per day. Those sharks that are landed during these tournaments are often donated to scientific organizations for research purposes (BBGFC, 2006). In recent years, decreases in the number of sharks being caught have seen wider adoption of a more conservation-orientated catch, and release ethos began to spread throughout the industry (Peirce, 2000).

Benefits of Shark-based Tourism

As noted above, approximately 100 million sharks are killed every year and key shark species are considered to be at risk of extinction. Wildlife tourism has been identified as a potential tool for conservation due to its ability to raise awareness and educate tourists, enhance local economic benefits, provide a platform for scientific research and carry out lobbying activities (Tisdell and Wilson, 2001; Higginbottom *et al.*, 2003; Tapper, 2006). Although Carwardine and Watterson (2002) have identified only 24 shark species that are regularly utilized by the dive tourism industry, they can help wider shark conservation through the flagship species concept (Dobson, 2004; IWC, 2005). Flagship species are commonly used by conservation organizations as they are iconic species capable of generating public interest in conservation and increasing funding opportunities (Walpole and Leader-Williams, 2002).

Education and attitude change

The exposure of the public to sharks is probably one of the most significant contributions that wildlife tourism can make towards the conservation of sharks. Psychological theory supports the notion of exposing individuals to stimuli that can result in the enhancement of their attitudes towards it (Zajonic, 1968; Cassidy, 1997). Previous studies by Beaumont (2001) and Gray (1985) have related this to tourism in the natural environment finding that such exposure can help engender a positive conservation ethic within tourists. A comment made by a research scientist reported by Dobson *et al.* (2005, p. 6) helps illustrate the potential of the South African cage-diving industry in challenging the Jaws stereotype that is associated with the great white shark.

[I]f it is done properly and if the information is given to the customers properly there is nothing better in the world than having people come out and see the Great White sharks . . . you see people arrive in the morning and they still come with Jaws in their heads. When they see the sharks swimming peacefully around the boat it sort of shatters all of that.

This potential for wildlife tours to influence attitudes towards sharks is an area that deserves further research.

Economic benefits

The development of shark-based tourism can have significant economic benefits for local communities and can help engender a conservation ethic through highlighting the value of live sharks (Dobson, 2004; Dobson *et al.*, 2005; Topelko and Dearden, 2005). Whale shark tourism at Ningaloo Reef in Western Australia generates a revenue of US$7.8 million in its short 2-month season (Graham, 2004) and cage diving in Gansbaai, South Africa contributes 289 million rand to the local economy (Hara *et al.*, 2003). Anderson and Waheed's (2001) study of the development of shark tourism in the Maldives highlighted the significant contribution this form of tourism can make to developing economies. Shark diving was worth approximately US$2.3 million per year with a single Grey Reef shark (*Carcharhinus amblyrhynchos*) worth in the region of US$3300 per year if used for tourism purposes (as opposed to a one-off value of US$32 if caught and sold by a fisherman). Recognition of the value of sharks as tourist attractions resulted in the government of the Maldives introducing a number of conservation initiatives aimed at protecting them from overfishing between 1995 and 1998, culminating in the banning of all shark fishing in tourist areas.

These obvious economic benefits can also create a 'poacher turned gamekeeper' approach where shark fishermen move from earning a living through consumptive fishing to non-consumptive wildlife watching. However, these benefits are predicated on ensuring that fishermen are retrained and are able to gain access to tangible benefits from the growth of the tourism industry (Graham, 2004; Topelko and Dearden, 2005).

Research and lobbying

The shark-watching industry also has an ability to help contribute towards shark research. Whale shark boats are considered ideal research platforms for gathering data (IWC, 2005). As part of the regulatory framework established to control the cage-diving industry in South Africa, operators are required to collect rudimentary data (e.g. size and gender) on the sharks they encounter. Some operators have taken this a stage further and have been involved in projects to tag great white sharks as well as collecting data on predation patterns (Dobson *et al.*, 2005).

As mentioned previously, sport fishing, especially in the UK, has recognized the need for shark conservation and the move from consumptive to non-consumptive forms of fishing. In the UK this has resulted in the establishment of two key initiatives: catch and release policies, which have enabled shark-tagging programmes to be implemented. Sports fishermen who volunteer attach tracking tags to sharks before their release; this then enables migratory routes, growth rates, population fluctuations and levels of non-natural predation to be monitored by research scientists. Sports fishermen have also formed 'Save Our Sharks', a lobby group that has been campaigning for the protection of threatened species such as Tope (*Galeorhinus galeus*) and Spurdog (*Squalus acanthias*) in British waters.

Problems with Shark-based Tourism

The impact of wildlife tourism on associated target species has been well documented (Shackley, 1996; Orams, 2002). These impacts range from behavioural changes (such as habituation to food sites) to injury and death. The impact of wildlife tourism on whale sharks is perhaps the best documented with several authors (Davis and Banks, 1997; Davis, 1998; Graham, 2004; Quiros, 2005) noting that the growth in tourism centred on swimming with whale sharks can have a direct effect on their behaviour. Quiros (2005) raises concerns that the whale shark tourism industry tends to be based around seasonal aggregations where sharks congregate to feed. Therefore, any disturbance caused by the industry (e.g. overcrowding of tour boats or swimmers) can cause the shark to divert from feeding to avoidance behaviour (such as directional changes or diving). This then negatively impacts upon the time in which the sharks can feed and store energy. The growth in the popularity of whale shark tourism at Gladden Spit in Belize has also coincided with a decrease in whale shark sightings (Graham, 2004; Quiros, 2005). In 2004, average sightings were down to just one or two sharks per trip, compared with a high of eight or nine a few years before (Quiros, 2005), which Graham (2004) attributes to potential disturbance from the increase in the number of tour boats visiting the area. Basking and whale sharks are also vulnerable to boat impact when they feed at the surface. They can be difficult to spot and collisions between boats and sharks can result in injuries to the sharks (C. Speedie, UK, 2006, personal communication). Dobson *et al.* (2005) also observed great white sharks colliding with cages and boats during trips in South Africa.

As stated earlier, some sharks and rays are popular attractions in touch pools in aquaria. This practice has been criticized by the Born Free Foundation (no date) as causing unnecessary stress and disturbance to the animals. Both Shackley (1998) and Newsome *et al.* (2004) identified that touching, attempting to ride and manhandling stingrays at Stingray City in the Cayman Islands and Hamelin Bay in Australia were a problem. These actions can cause stress to the animal and result in physical problems, such as skin lesions, and increased risk of infection. Touching or attempting to touch great white sharks by tourists is not an uncommon occurrence on cage-diving trips in South Africa (personal observation), although this raises concerns for the welfare of tourists and the industry rather than the shark. One cage-dive tour operator commented 'touching the shark is a big 'no'. We come down hard on any tourist who we observe trying to do it, it only takes one accident and that would be the end of the industry' (White shark cage-dive operator, South Africa, 2006, personal communication).

Baiting and feeding

The feeding of wildlife in order to attract them to areas where they are accessible for tourists to view is a common aspect of wildlife tourism today although this artificial feeding can have significant consequences for the target species, including growing dependency on and habituation to food sites, increased aggression (both inter- and intraspecies), and the potential for disease transmission (Orams, 2002; Newsome *et al.*, 2004). These concerns are especially pertinent towards shark-based tourism. Burgess (1998) raises particular concerns about the ecological disruption caused by the feeding of sharks. He notes that the large concentration of sharks at popular dive sites, caused by feeding, is unnatural and that at some dive sites in the Bahamas sharks have become so habituated to the sound of tour boat motors that it elicits a Pavlovian-type response. The large numbers of sharks that congregate at feeding sites can render them vulnerable to opportunistic fishing leading to localized extinction. This unnatural congregation of sharks may also denude other areas of their apex predator causing wider ecological problems. Stingrays fed in the Cayman Islands have also exhibited signs of learned behaviour which could adversely affect their natural feeding patterns should feeding stop (Shackley, 1998).

The underlying philosophy of feeding wildlife is to increase the likelihood that tourists will be able to view and in many cases photograph the targeted species. This can then lead to animals being manipulated to ensure tourists gain the best or most memorable view. The Bahamas became a particularly well-known shark dive destination due to shark wranglers manipulating sharks for tourist pictures and even enticing sharks to bite divers who are protected by chain mail suits (Jackson, 2000). Great white sharks are often lured past cages with bait to try to ensure tourists have the best possible view (Fig. 3.2). This manipulation of sharks has been taken a stage further with some cage-dive operators holding sharks on bait lines to make them thrash around and even using a technique to open the mouths of white sharks (Fig. 3.3). Far from being educational these actions tend to reinforce the stereotypical image of *Jaws* in the minds of tourists (Dobson *et al.*, 2005).

Baiting and feeding sharks for tourism purposes also highlights one of the more unique issues of shark-based tourism in that it can create a moral panic among other marine user groups and local residents, especially where sharks are baited or fed (Dobson, 2006). The main concern is for the potential for sharks to become habituated to the presence of humans in the water (Cunningham-Day, 2001; Environment Australia, 2002) which may lead to an increase in aggressive behaviour towards divers (Nelson *et al.*, 1986) and even an increase in shark attacks on other marine users. As highlighted above shark attacks, although rare, often generate high levels of media interest and the development of shark-based tourism can therefore generate intense negative

Fig. 3.2. Bait lines are used to lure great white sharks towards tourist boats in Gansbaai, South Africa. (Photograph John Dobson.)

Fig. 3.3. A shark wrangler opens the mouth of a great white shark, Gansbaai, South Africa. (Photograph Neil Crooks.)

reaction from local communities in some locations where it takes place. The industry can often attract negative media coverage especially when a shark attack occurs within the area operated by industry, due to speculation that sharks are conditioned to humans through chumming and feeding during dives. Headlines such as 'Cages Could Spark Attacks' (Anon. 2005), 'Feeding Frenzy' (Uglow, 2002), 'Shark-dive Vacation Endangers Tourists' (Anon., 1998) and 'Dangerous Liaisons: Teaching Great White Sharks To Link People With Food Is A Recipe For Disaster' (York, 1998) help illustrate the levels of concern over the development of the industry. This negative publicity can also have a positive side for shark-based tourism. As noted previously, shark-based tourism tends to appeal to the adventure market and so any story that increases the notion of its dangerousness can help increase the popularity of the industry. As one cage-dive operator, referring to a widely publicized story circulated in 2005 about a shark supposedly attacking a diver in a cage, commented 'It was very good for business' (White shark cage-dive operator, South Africa, 2006, personal communication).

Regulating Shark-based Tourism

The management of shark-based tourism has been low on the agenda of many state authorities but as the controversy surrounding baiting and feeding of sharks has grown and its potential economic worth recognized, so has state interest in managing the industry. Strategies range from non-intervention to prohibition. Lewis and Newsome (2003) and Newsome *et al.* (2004) illustrate the problem of non-intervention strategies in their study of the development of stingray feeding at Hamelin Bay in Australia. Without any formal regulation the uncontrolled nature of the feeding raised serious questions about the long-term sustainability of the industry.

Banning certain practices is a more extreme response to controlling the development of the industry. Public concerns about habituation of sharks to humans can lead to state authorities banning shark feeding. During the summer of 2001, a series of shark attacks took place in Florida generating intense media interest, resulting in *Time* magazine dubbing the period 'The Summer of the Shark' (McCarthy, 2001). Despite being only a very small and localized industry, shark-based tourism was considered to be a significant causal factor in the spate of attacks. The development of an organized opposition group (The Marine Safety Group who lobbied for a ban on shark feeding) and the growing media interest generated public debate about the future of shark viewing in Florida waters. This heavily influenced the decision by authorities to adopt a precautionary approach to the problem by banning the feeding of sharks (and other fish) for tourist purposes in January 2002 (sport fishing for sharks was exempt). Although the ban did not stop people viewing sharks in the wild it prohibited the attraction of sharks to tour boats, essentially making it very difficult to run a profitable business in Florida waters (Dobson, 2006). Other areas in the USA that are adopting precautionary approaches to managing shark-based tourism include Hawaii and California. In Hawaii, fish feeding (specifically for the purposes of attracting sharks to tour boats) is banned in state waters (up to 3 miles offshore) and further debate on extending the ban to cover federal waters (up to 12 miles offshore) was taking place during 2006. A total ban on the use of bait and decoys to attract white sharks to tour boats in the Farallon Marine Sanctuary (California) was being considered by the National Oceanic and Atmospheric Administration (NOAA) during late 2006. The proposal stated that boats must stay at least 50 m away from any shark and must not use any method of attracting sharks to tour boats (NOAA, 2006). These proposals will provide a more naturalistic encounter as operators will need to rely on observing natural predations. However, as with Florida, not being able to increase the chances of providing tourists with an encounter may undermine the financial viability of shark-based tourism in the area.

Sitting between non-intervention and prohibition lays the concept of licensing operators. Licensing strategies have been implemented to regulate both the whale shark industry at Ningaloo Reef in Australia and the cage-diving industry in South Africa, requiring any operator who wishes to observe sharks to hold a permit. This ensures that the number of operators can be controlled and as part of the permit condition abide by a set of rules and regulations. The value of any form of regulation does depend upon the ability of authorities to enforce regulations. The permit system developed by the South African government's Marine and Coastal Management (MCM) department to regulate the cage-diving industry in South Africa has been fraught with problems. Legal challenges by operators, objections to the industry by local marine user groups (some of whom founded 'Shark Concern', an anti-cage-dive lobby group) and an inability to enforce the permit conditions has seen the system become unworkable (Dobson, 2006). A more successful system has been implemented in the Ningaloo Reef Marine Park. Authorities introduced an additional charge of AUS$15 per tourist and although controversial it has allowed the funding of observers who are able to monitor tourist behaviour and operator practice (Davis and Tisdell, 1998).

Self-regulation and voluntary schemes are also prevalent in the management of shark-based tourism. The Wise scheme (WIldlife SafE) is a UK-based voluntary training scheme that aims to train boat operators (both private and commercial) who wish to view marine wildlife. Upon successful completion of the training, participants are awarded Wise accreditation; this scheme is being used to help train operators on how to safely approach and observe basking sharks. Codes of conduct have been developed to inform people on how to approach and swim with basking and whale sharks (The Shark Trust, 2004; The Shark Trust, no date, b).

Conclusion

Sharks are perhaps one of the most demonized creatures in human history and yet it is the shark that has the most to fear from any shark–human encounter. An estimated 100 million sharks are killed each year as a result of human activities placing many populations at risk from total collapse. Effective conservation measures need to be put in place if sharks are to have a sustainable future.

Despite the many negative connotations that surround sharks, they have a unique attractiveness to tourists, which has led to the development of a global shark-based tourism industry. Shark–tourist encounters take place in a range of environments ranging from passive encounters via viewing tunnels and windows in aquaria through to naturalistic active encounters where sharks are encountered by chance as part of the natural marine fauna.

This utilization by the marine wildlife tourism industry has the potential to help contribute towards shark conservation, especially through placing greater economic value on living rather than dead sharks, and changing attitudes towards sharks by exposing tourists to them in their natural environment. Unfortunately, shark-based tourism can also exert a range of negative impacts on sharks including disturbance, behaviour modification and injury, as well as impacting on human perception through media hyperbole. In order to maximize benefits and reduce potential problems the industry requires appropriate regulation to ensure both shark and tourist safety. At present, regulation is sporadic, ranging from licensing and permit schemes through to adopting a precautionary approach and banning certain activities used to attract sharks to boats. In order to have any value, whatever method used to regulate shark-based tourism must be enforceable.

This area of marine wildlife tourism would benefit from further academic scrutiny. Current research has tended to focus on whale shark tourism (perhaps due to its status as a flagship species); however, shark-based tourism is much more diverse than this in terms of locations, species used and impacts caused. Any future research will require a multidisciplinary approach. The natural sciences should further assess the potential impacts of the industry (especially the issue of habituation resulting from feeding and baiting) on the welfare and behaviour of sharks. Social science research should focus on issues concerning appropriate and workable management strategies (especially relating to the safeguarding of sharks and the management of any concerns from local communities where shark-based tourism occurs). Developing an understanding of the nature of the visitor experience (e.g. assessing motivations for undertaking trips whether adventure or wildlife in nature) and comparing captive and non-captive experiences of sharks would help broaden our understanding of this form of tourism. There is also an urgent need to establish the economic worth of non-consumptive shark-based tourism (particularly in developing countries) so that communities can be encouraged and assisted in moving away from shark fishing. Finally, and perhaps most importantly, researchers should examine the potential for shark-based tourism to influence peoples' perception and attitude towards sharks.

The Senegalese poet Baba Dioum stated that 'in the end, we will conserve only what we love. We will love only what we understand and we understand only what we are taught' (Soulé and Orians, 2001, p. 125). At present many humans are not educated about sharks, we certainly do not understand sharks and, for many, love is not a word they would use to describe their feelings towards sharks. Perhaps exposing tourists to sharks through shark-based tourism may go some way towards rectifying this and helping achieve a more balanced understanding of the plight of the ocean's ultimate predator.

References

AES (2006) International Captive Elasmobranch Census. Available at: www.flmnh.ufl.edu/fish/organizations/aes/census2006.htm

Anderson, C. and Waheed, A. (2001) The economics of shark and ray watching in the Maldives. *Shark News* 13 (July). Available at: www.flmnh.ufl.edu/fish/Organizations/SSG/sharknews/sn13/shark13news19.htm

Anon. (1998) Shark-dive vacation endangers tourists. *Environmental News Network*. Available at: www.enn.com

Anon. (2005) Cages could spark attacks. *The Sun*, 29 March 2005, p. 13.

Armstrong, J. (2005) Brit's shark terror at sea. *The Mirror*, 25 March, p. 4.

BBGFC (2006) 20th Monster Shark Tournament Rules. Available at: www.bbgfc.com/BBGFCMonsterSharkTournament2006web.pdf

Beaumont, N. (2001) Ecotourism and the conservation ethic: recruiting the uninitiated or preaching to the converted? *Journal of Sustainable Tourism* 9(4), 317–341.

Benchley, P. (1974) *Jaws*. Pan, London.

Born Free Foundation (no date) Do aquaria slip through the legislative net? Available at: www.bornfree.org.uk/zoocheck/aquaria

Bulbeck, C. (2005) *Facing the Wild: Ecotourism, Conservation and Animal Encounters*. Earthscan, London.

Burgess, G.H. (1998) Diving with elasmobranchs: a call for restraint. *Shark News* 11 July. Available at: www.flmnh.ufl.edu/fish/organizations/ssg/sharknews/sn11/shark11news1.htm

Byrne, P. (2005) I fought off jaws with my bare hands. *The Mirror*, 29 March, p. 7.

Camhi, M., Fowler, S., Musick, J., Bräutigam, A and Fordham, S. (1998) *Sharks and their Relatives. Ecology and Conservation*. IUCN, Gland, Switzerland.

Carwardine, M. and Watterson, K. (2002) *The Shark Watcher's Handbook*. Princeton University Press, Princeton, New Jersey.

Cassidy, T. (1997) *Environmental Psychology: Behaviour and Experience in Context*. Psychology Press, Hove, UK.

Caughley, G. (1985) Problems in wildlife management. In: Messel, H. (ed.) *The Study of Populations*. Pergamon Press, Sydney, pp. 129–135.

CITES (2006) How CITES works. Available at: www.cites.org/eng/disc/how.shtml

Cohen, S. (2002) *Folk Devils and Moral Panic*, 3rd edn. MacGibbon & Kee, London.

Compagno, L., Dando, M. and Fowler, S. (2005) *Sharks of the World*. Collins, London.

Corkeron, P.J. (2006) How shall we watch whales. In: Lavigne, D.M. (ed.) *In Pursuit of Ecological Sustainability*. IFAW, Guelph, Canada, pp. 161–170.

Cunningham-Day, R. (2001) *Sharks in Danger: Global Shark Conservation Status with Reference to Management Plans and Legislation*. Universal Publishers, Parkland, Florida.

Davis, D. (1998) Whale shark tourism in Ningaloo Marine Park, Australia. *Anthrozoos* 5(11), 5–11.

Davis, D. and Banks, S. (1997) Whale sharks in Ningaloo Marine Park: managing tourism in an Australian marine protected area. *Tourism Management* 18(5), 259–271.

Davis, D. and Tisdell, C.A. (1998) Tourist levies and willingness to pay for a whale shark experience. *Tourism Economics* 5(2), 161–174.

Dobson, J. (2004) *The Potential for Wildlife Tourism to Contribute Towards Elasmobranch Conservation: A Case Study of the South African Cage Diving Industry*. Paper presented at the European Elasmobrach Society meeting, London Zoological Society, London.

Dobson, J., Jones, E. and Botterill, D. (2005) Exploitation or conservation: can wildlife tourism help conserve vulnerable and endangered species? *Interdisciplinary Environmental Review* 7(2), 1–13.

Dobson, J. (2006) Sharks, wildlife tourism and state regulation. *Tourism in Marine Environments* 3, 25–34.

Environment Australia (2002) *White Shark* (Carcharodon carcharias) *Recovery Plan*. Department of Environment and Heritage, Australia.

Goode, E. and Nachman, B.-Y. (2003) *Moral Panics The Social Construction of Deviance*. Blackwell, Oxford.

Graham, R.T. (2004) Global whale shark tourism: a golden goose of sustainable and lucrative income. *Shark News* 16(October), 8–9.

Gray, D.B. (1985) *Ecological Beliefs and Behaviours: Assessment and Change*. Greenwood Press, Westport, Connecticut.

Grayson, B. (2006) Shark draws crowds, criticism to aquarium. Available at: www.montereyherald.com/mld/montereyherald/news/16106324.htm

Gross, D. (2006) Aquarium conducts exam on whale sharks. Available at: www.examiner.com/a-89889 Aquarium_Conducts_Exam_on_Whale_Sharks.html

Hara, M., Maharaj, I. and Pithers, L. (2003) *Marine-based Tourism in Gansbaai: A Socio-economic Study*. Department of Environmental Affairs, Cape Town.

Higginbottom, K., Tribe, A. and Booth, R. (2003) Contributions of non-consumptive wildlife tourism to conservation. In: Buckley, R., Pickering, C. and Weaver, D.B. (eds) *Nature-based Tourism, Environment and Land Management*. CAB International, Wallingford, UK, pp. 181–196.

Humane Society of the United States (HSUS) (2006) Matha's vineyard is no place for a monster shark fishing tournament. Available at: www.hsus.org/wildlife/wildlife_news/oak_bluffs_shark_hunt_ 4–06

ISAF (2006) ISAF Statistics for the Worldwide Locations with the Highest Shark Attack Activity Since 1990. Available at: www.flmnh.ufl.edu/fish/sharks/statistics/statsw.htm

IUCN SSG (2006) IUCN Shark Specialist Group Red List Summary Tables 2000–2006. Available at: http://128.227.186.212/fish/organizations/ssg/RLsummary2006.pdf

IWC (2005) Unofficial summary of international whale Shark conference held in Perth, May 2005. Available at: www.mcss.sc/iws_presentation_summary.pdf

Jackson, J. (2000) *Diving with Sharks and other Adventure Dives*. New Holland Publishers, London.

JNCC (2002) *Wildlife and Countryside Act 1981*. JNCC, London.

Lewis, A. and Newsome, D. (2003) Planning for stingray tourism at Hamelin Bay, Western Australia: the importance of stakeholder perspectives. *International Journal of Tourism Research* 5, 331–346.

McCarthy, T. (2001) Why can't we be friends? *Time Magazine*, 30 July, pp. 46–52.

Morey, S. (2002) The shark in modern culture: beauty and the beast. *Journal of Undergraduate Research* 4(1). Available at: www.clas.ufl.edu/jur/200209/papers/paper_morey.html

Nelson, D.R., Johnson, P.R., McKibben, J.N. and Pittenger, G.C. (1986) Antagonistic attacks on divers and submersibles by grey reef sharks, *Carcharhinus amblyrhynchos*: antipredatory or competitive. *Bulletin of Marine Science* 38(1), 68–88.

Newsome, D., Lewis, A. and Moncrieff, D. (2004) Impacts and risk associated with developing, but unsupervised, stingray tourism at Hamelin Bay, Western Australia. *International Journal of Tourism Research* 6, 305–323.

NOAA (2006) Gulf of the farallones national marine sanctuary regulations: proposed rules. *Federal Register* 71(194).

Orams, M.B. (2002) Feeding wildlife as a tourism attraction: a review of issues and impacts. *Tourism Management* 23, 281–293.

Peirce, R. (2000) Changing times. *Shark Focus* 21, 10–11.

Perrie, R. (2005) Just when you thought it was safe to go back in the water . . . jaws attacks. *The Sun*, 25 March, p. 5.

Peschak, T. and Scholl, M.C. (2006) *South Africa's Great White Shark*. Struik Publishers, Cape Town.

Philpott, R. (2002) Why sharks may have nothing to fear more than fear itself: an analysis of the effect on human attitudes on the conservation of the great white shark. *Colorado Journal of International Environmental Law and Policy* 13(2), 445–472.

Quiros, A. (2005) Whale shark ecotourism in the Philippines and Belize: evaluating conservation and community benefits. *Tropical Resources Bulletin* 24(Spring), 42–48.

Ryan, C. (1998) Saltwater crocodiles as tourist attractions. *Journal of Sustainable Tourism* 6(4), 314–327.

Ryan, C. and Harvey, K. (2000) Who likes saltwater crocodiles? Analysing socio-demographics of those viewing tourist wildlife attractions based on saltwater crocodiles. *Journal of Sustainable Tourism* 8(5), 426–433.

Shackley, M. (1996) *Wildlife Tourism*. International Thomson Business Press, London

Shackley, M. (1998) Stingray City – managing the impact of underwater tourism in the Cayman Islands. *Journal of Sustainable Tourism* 6(4), 328–338.

Smith, A., Newsome, D., Lee, D. and Stoeckl, N. (2006) *The Role of Wildlife Icons as Major Tourist Attractions. Case Studies: Monkey Mia dolphins and Hervey Bay whale watching*. CRC Sustainable Tourism Technical Report, Gold Coast, Queensland, Australia.

Soulé, M.E. and Orians, G.H. (2001) *Conservation Biology: Research Priorities for the Next Decade*. Island Press, Covelo, California.

Tapper, R. (2006) *Wildlife Watching and Tourism: A Study on the Benefits and Risks of a Fast Growing Tourism Activity and Its Impact on Species*. UNEP/CMS Secretariat, Bonn, Germany.

The Shark Trust (2004) *Basking Shark Code of Conduct*. The Shark Trust, Plymouth, UK.

The Shark Trust (no date a) *Bimini Mangrove Habitat Destruction*. Available at: www.sharktrust.org/content.asp?did = 26543

The Shark Trust (no date b) *The Whale Shark Project: Code of Conduct*. Available at: www.whalesharkproject.org/prospectus.asp?siteid=2&content=&=1&curpage=

Tisdell, C. and Wilson, C. (2001) Wildlife-based tourism and increased support for nature conservation financially and otherwise: evidence from sea turtle ecotourism at Mon Repos. *Tourism Economics* 7(3), 233–249.

Topelko, K.N. and Dearden, P. (2005) The shark watching industry and its potential contribution to shark conservation. *Journal of Ecotourism* 4(2), 108–128.

Tremblay, P. (2001) Wildlife tourism consumption: consumptive or non-consumptive. *International Journal of Tourism Research* 3(1), 81–86.

Tremblay, P. (2002) Tourism wildlife icons: attractions or marketing symbols? *Journal of Hospitality and Tourism Management* 9, 164–181.

Uglow, D. (2002) Feeding frenzy. *Geographical* 74(8), 84–89.

Walpole, M.J. and Leader-Williams, N. (2002) Tourism and flagship species in conservation. *Biodiversity and Conservation* 11(3), 543–547.

York, A. (1998) Dangerous liaisons: teaching great white sharks to link people with food is a recipe for disaster. *New Scientist*, 24 October, p. 4.

Zajonic, R.B. (1968) Attitudinal effects of mere exposure. *Journal of Personality and Psychology* 9, 1–29.

4 Tourist Interactions with Sharks

P. Dearden, K.N. Topelko and J. Ziegler

Introduction

The small town of Parksville on Vancouver Island in British Columbia once touted itself as the 'shark fishing mecca of the Pacific Northwest' (Wallace and Gisborne, no date). The main quarry was the world's second biggest fish, the basking shark (*Cetorhinus maximus*). The sharks destroyed too many salmon fishing nets, and in 1949, extermination programmes were initiated. A large steel-cutting ram was attached to the bow of a fisheries protection vessel and the harmless baskers were sliced into oblivion. A caption under a basking shark photograph in a local newspaper captured the mood of the day: 'This is a basking shark, basking and leering. But the smirk will soon be wiped off its ugly face by the fisheries department, which is cutting numerous sharks down to size' (Wallace and Gisborne, no date, p. 51).

The actions of fishermen 50 years ago have precluded Vancouver Island from participating in the now burgeoning global shark-watching industry. The actions also illustrate the importance of the historical context in understanding the development of non-consumptive wildlife oriented recreation (NCWOR). Duffus and Dearden (1990) pointed out the importance of history in their wildlife tourism model (Fig. 4.1), which sought to clarify the elements of the non-consumptive experience and draw attention to the need for appropriate management actions that would embrace both natural and social science.

Beginning with a historical review of attitudes towards sharks, this paper uses Duffus and Dearden's (1990) wildlife tourism model to illustrate the development of shark watching as NCWOR. Sharks are a particularly appropriate group for consideration because attitudes towards them have changed dramatically over the last 30 years – once routinely exterminated they are now the basis of a flourishing shark-watching industry in many parts of the world (Topelko and Dearden, 2005). Growth of the shark-watching industry is of particular interest to conservationists since sharks worldwide are facing

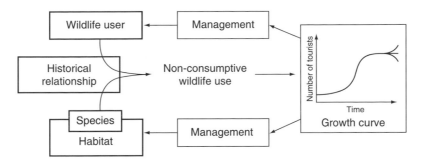

Fig. 4.1. The core components of non-consumptive wildlife use. (Adapted from Duffus and Dearden, 1990.)

unprecedented pressure from fisheries, largely due to the value of their fins (Magnussen *et al.*, 2007). The economic value of shark watching has provided an incentive to protect sharks and/or their habitats in several countries, including the Maldives, the Philippines, and the USA (Topelko and Dearden, 2005). But can the shark-watching industry grow too quickly for its own good? What are some of the different manifestations of shark watching, and what are some of the management interventions that might be applied to avoid some of the negative impacts associated with uncontrolled growth in tourism? Answers to these questions require an understanding of the biophysical and socio-economic contexts of the activity, and are explored with reference to the growth of whale shark watching.

The Wildlife User

History influences demand for wildlife contact through cultural conditioning of human perceptions towards species (Duffus and Dearden, 1990). With respect to sharks, this conditioning has been strong, as is the case with many predators that have the capability to eat humans. Of particular interest here are the perceptions of divers towards sharks. In the next section, changes in divers' perceptions over time are examined using editorials, letters, columns and advertisements appearing in *Skin Diver* magazine from January 1974 to December 2002. *Skin Diver* was selected for analysis because it has been reporting on trends in the scuba-diving industry for over 50 years, and with over 200,000 subscribers worldwide, it is one of the world's most popular dive magazines. Figure 4.2 shows the main periods of evolution of attitudes towards sharks.

1974–1984: Humans need protection – from sharks

Prior to the release of Peter Benchley's *Jaws* in 1975, sharks demanded little attention from divers. Only two articles featuring sharks appeared in *Skin Diver* from January 1974 to November 1975. Both articles sought to educate

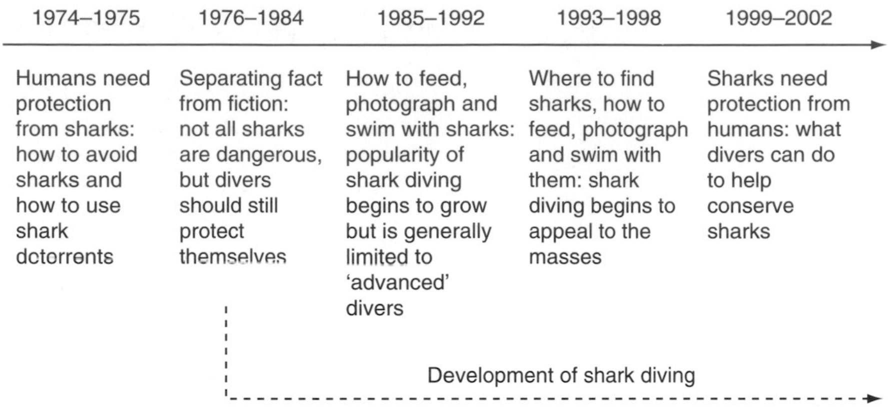

1974–1975	1976–1984	1985–1992	1993–1998	1999–2002
Humans need protection from sharks: how to avoid sharks and how to use shark deterrents	Separating fact from fiction: not all sharks are dangerous, but divers should still protect themselves	How to feed, photograph and swim with sharks: popularity of shark diving begins to grow but is generally limited to 'advanced' divers	Where to find sharks, how to feed, photograph and swim with them: shark diving begins to appeal to the masses	Sharks need protection from humans: what divers can do to help conserve sharks

Development of shark diving

Fig. 4.2. *Skin Diver*'s coverage of sharks from January 1974 to December 2002.

readers about sharks and their behaviour (the *real* threat of sharks to divers, why sharks attack, how to avoid shark attacks), but readers were also given tools that would enable them to protect themselves against sharks (how to recognize 'threat displays', what to do when attacked and how to use defensive weapons properly) (e.g. Barada, 1974; McNair, 1975). Although an emphasis was placed on shark avoidance, *Skin Diver* also tried to provide readers with the mental tools (e.g. stay calm and still), as well as the physical tools to enable divers to dive with peace of mind. Advertisements for underwater safety featured weapons that could be used against sharks, and these weapons were designed to kill (e.g. contact pressure explosives, spears). Figure 4.3 is an advertisement for a book that describes 'how sharks behave and how man might control them', which appeared in *Skin Diver*'s June 1976 issue.

In 1975, *Jaws* was released and 'divers rushed to cancel their scuba lessons and people stayed away from the dive shops in droves, while equipment manufacturers suffered one of their worst years in a long, long time' (McNair, 1976, p. 45). The July 1976 issue of *Skin Diver* featured a great white shark on the cover, with the caption: 'Are sharks really dangerous to divers? *Skin Diver* compares fact to fiction.' The author of this article asked readers not to fall victim to the paranoia of shark mania (McNair, 1976). In fact, from 1976 to 1988, most of the articles featuring sharks attempted to separate fact from fiction by reporting the 'real' danger that sharks posed to divers, although there were some mixed messages. Several articles sought to organize sharks into categories – sharks that bite (bad sharks) and sharks that do not (good sharks) (e.g. Hauser, 1975a; McNair, 1975; Cooluris, 1977; Roessler, 1980). These authors downplayed the danger of diving in 'shark-infested' waters, by presenting a balanced view of the danger of shark attack to divers (e.g. there is a greater risk of being chewed up by propellers than getting chewed up by a shark). Authors like Hauser (1975b), McNair (1976) and Hall (1981) tried to assure divers that sharks were not deserving of their reputation, as not all sharks are dangerous and not all dangerous sharks are dangerous all the time.

Fig. 4.3. Advertisement appearing in *Skin Diver* July 1975.

Other articles may have fed the shark hysteria. For example, in an article titled 'Anatomy of a Shark Bite' the author stressed caution while diving with blue sharks: 'If you do see a blue shark approaching, it may well bite....You must constantly be turning and checking the sharks around you. To relax for a moment could be fatal' (Hall, 1981, p. 18). Another author describes his experience while filming a movie:

> [T]hree mako sharks arrived on the scene at the same time. They moved in and out of visual range continuously and there was no way to anticipate from which direction they would reappear. One of the divers decided to 'dispatch' one of

them so that he could visually cope with the other two. He 'did the job neatly' with the McNair Bangstick [an equalizer] which *all safety divers carried for use in time of imminent attack* [emphasis added].

(Waterman, 1979, p. 12)

The last line is particularly telling – divers are not safe, unless they carry weapons to protect themselves from 'imminent attack'. But divers were not the only ones affected by the impact of *Jaws*. As the movie approached the US$150 million mark less than 6 months after its release, shark teeth sold around the country for as much as US$100, shark-hunting clubs rose in popularity, and shark meat made a big dent in the fish business:

> One fish wholesaler on the Eastern seaboard sold 60,000 pounds of shark in 1973. In 1975 he sold more than 300,000 pounds. Baked Mako was served in Atlanta's American Motor Hotel with a card on the table announcing: 'Jaws: for a jaw-ful revenge.'
>
> (Hauser, 1975b, p. 100)

1985–1992: Shark diving popularized

Around 1985, there was a shift in emphasis from shooting sharks with lethal devices to shooting sharks with underwater cameras. Divers began to accept sharks as part of the reef community and the popularity of shark feeding and photography grew, although shark diving as a sport was limited to experienced divers who were willing to accept the risks. In 1986, *Skin Diver* printed an article that featured cage diving with blue sharks off the California coast (Walker, 1986), and in 1988, *Skin Diver* began to advertise whale shark expeditions that employed private airplanes, helicopters, chase boats and local fishers to spot the sharks for divers and underwater photographers (Wagner, 1988).

In 1991, the popularity of shark feeding intensified. Under normal conditions, sharks are difficult to find and photograph. In response to the growing demand to experience the adrenaline rush associated with shark diving, dive masters in the Caribbean developed shark-feeding programmes that secured regular interactions with reef sharks (Frink, 1991). *Skin Diver* began running articles to assure readers that shark-feeding dive sites were safe (e.g. Frink, 1991; Gleason, 1991; Lawrence, 1992).

Divers also seemed to be developing a healthier respect for sharks, as evidenced by the movement towards use of non-lethal shark deterrent devices, such as anti-shark wet suits (e.g. chain mail suits), billy clubs, ski poles, chemical repellants and eventually, hands:

> Many groups that swim with sharks carry shark billies. In the 1970s, these were poles with contact pressure explosives. More recently, divers have used ski poles or polespears as shark clubs. Several groups, however, report observing that a shark's skin marks when rendered off with a pole. At the end of a dive certain animals could be observed with multiple marks. Dive masters have begun pushing away – by hand – sharks that come too close.
>
> (Sleeper, 1991, p. 129)

There are other indications that attitudes towards sharks were improving. In 1988, *Skin Diver* featured an article titled 'Man Bites Shark', which was the magazine's first attempt to educate its readers about shark fisheries and the impact on shark populations:

> In 1977, two years after the movie *JAWS* came out, thresher shark became king of the dinner plate. . . . By 1982, the thresher shark fishery in Santa Barbara was showing signs of decline. . . . Fishers turned to the angel shark. . . . Santa Barbara became the major port for angel sharks, with local fishers bringing them in by the boatload. After only three years, the angel shark fishery was showing signs of decline. There have been a number of shark fisheries developed worldwide, and they've all failed.
>
> (Hauser, 1988, pp. 58–59)

In 1989, *Skin Diver* printed another article, 'Sharks and Shipwrecks', which mentioned the diminishing number of Atlantic sand tiger sharks due to over-harvesting (Farb, 1989). However, even though there was greater concern for the welfare of sharks, the hysteria created by *Jaws* continued to retard growth in the shark-diving industry, and also slowed the adoption of more sympathetic attitudes. Shark conservation would not become a hot topic until after the explosion of the shark-diving industry.

1993–1998: Explosion of the shark-diving industry

> Is this a dream come true or what? I'm swimming in the clear blue waters of the Bahamas, surrounded by sharks! Dozens and dozens of them!
>
> (Cardone *et al.*, 1999, p. 78)

Twenty years earlier, this experience would have been a nightmare but more divers were trying shark diving and the activity exploded in popularity around 1993. Once a sport, appealing only to professional photographers, shark feeding and photography began to appeal to the masses of the dive community. Carl Roessler, a regular contributor to *Skin Diver*, summarizes the development of dive travel:

> In 1972, divers sought quiet coral reefs on which they would (they hoped) see a shark rarely, if ever. I can remember constantly reassuring callers that their dive vacations were surely safe from sharks. Of course, there was a tiny, rather lunatic fringe of divers who actually hoped they could photograph sharks underwater. Before them, only professional filmmakers braved putting themselves in the water with fearsome creatures such as sharks . . . . Two decades later, some divers won't even consider a destination unless they are nearly guaranteed encounters with some kind of shark.
>
> (Roessler, 1993, p. 72)

Shark dive sites in Belize, the Bahamas and the Cayman Islands popularized the sport of shark feeding, allowing divers the chance to watch professionals touch, feed and placate sharks. Advertisements for dive resorts around the world featured divers petting, holding and feeding sharks (Fig. 4.4). In the

Fig. 4.4. Advertisement featuring a professional diver holding a shark in *Skin Diver*'s February 1999 issue.

1970s and 1980s, resorts seldom put a photograph of a shark in an advertisement, aware of the need to assure recreationists that their beaches were safe. But the 1990s ushered in a new era in the dive industry, and resorts were eager to use sharks as an attraction. Dive sites were increasingly rated according to the number and variety of sharks that divers could expect to find, and the proximity of the sharks to the divers – the closer the better (Murphy, 1993a,b; Frink, 1996; Harrigan, 1998).

Skin Diver also capitalized on the trend. When the July 1976 cover featuring a great white shark hit the news-stands, readers reacted negatively, and the magazine did not run another shark cover for almost 18 years, when 'thrill seeking adventurers made diving with sharks one of the sport's most sought after experiences' (Collins, 1999, p. 129). Between January 1993 and December 2002 (inclusive), sharks were featured on the cover 17 times.

1999–2002: Sharks need protection – from humans

Beginning around 1999, another trend developed in the shark-dive industry – concern for the survival of sharks. Although a significant amount of print space was still being devoted to dispelling the myth that sharks regularly consume humans, a sympathetic view of sharks was entering the mainstream, and *Skin*

Diver increasingly ran reports on the decline in shark populations worldwide (e.g. Cousteau, 1999a,b, 2001; Wyland, 2001; Bird, 2002). Recognizing the threat to sharks (and the threat to the shark-diving industry), the magazine began to call on divers to apply pressure on legislators to protect sharks (Cousteau, 2002).

One of the greatest challenges of shark conservation is reversing the traditional image of sharks as 'human eating machines'. To address this challenge, shark-dive resorts featured in *Skin Diver* incorporated shark education into the dive experience, and education components were increasingly used as a selling feature. For example, before participating in shark dives at Shark Rodeo, Bahamas, divers are required to attend a presentation on sharks, wherein participants learn about the threats facing many shark populations around the globe (Harrigan, 1998). Similarly, while diving at Shark Alley in the Cayman Islands, divers participate in a shark awareness course, which teaches shark identification, dispels myths and teaches facts about shark behaviour and stresses the need for divers to coexist with sharks (Carwardine and Watterson, 2002).

2003 and beyond

As a popular dive magazine, *Skin Diver*'s positive portrayal of sharks has had a tremendous impact on the evolution of the shark-dive industry, demonstrated by the development of a small, but growing faction of divers who now need to be convinced that sharks *can* be dangerous. A similar evolution is taking place with shark feeding, which was the catalyst for the explosion of the dive industry. In its early development, divers watched professionals feed sharks, but as the sport continues to evolve, divers are no longer content watching. Professionals made shark feeding look safe, as 'local dive masters [were] able to pet and placate these predators to the extent that they rest in the feeder's arms like domestic lapdogs' (Frink, 1996, p. 98). Convinced that the risk of harm is negligible, the demand to experience shark feeding first hand is escalating, and dive masters are beginning to offer training courses designed to give divers the skills necessary to feed sharks on their own.

In terms of the framework in Fig. 4.1, the last 40 years has seen an abrupt change in divers' (the 'wildlife users') perceptions of sharks. The other necessary component for NCWOR is a predictable occurrence of the target species and the ability to access concentrations of target species, and this is discussed in the next section.

Target Species

Duffus and Dearden (1990) suggest that history influences the demand for wildlife contact in that the abundance and distribution of almost all wild species have been dramatically affected by human activities. The example given in the introduction regarding basking sharks is a good illustration of this. In many parts of the world the extraction of sharks on a large scale still continues. However, there are still many opportunities for shark watching, and these have been summarized in publications such as *The Shark Watcher's Handbook* (Carwardine and Watterson, 2002).

To provide focus to the rest of the discussion, a more detailed consideration will be given to an overview of whale shark watching as a case study. The whale shark is one of the most watched shark species, and perhaps not coincidentally, the first shark to be listed under the Convention of Migratory Species and the Convention on Trade in Endangered Species. Whale shark watching is atypical in that the sharks are harmless to humans and, since they are regularly found at the surface, they can be observed easily by divers. Unfortunately, whale sharks can also be easily spotted by poachers, and as the basking shark story in the introduction emphasizes, historically, humans have not distinguished between predatory and non-predatory sharks.

NCWOR is dependent upon predictable occurrences of the target species within an accessible location. Such areas often coincide with special life history requirements of the species. This is the case with whale sharks.

Whale shark ecology and distribution

The whale shark (*Rhincodon typus*), which is the largest living fish in the world attaining lengths of up to 20 m and a weight of greater than 30 t, has a widespread distribution (Fig. 4.5) (Stevens, 2007). Its diet consists of planktonic and nektonic prey including small crustaceans and fishes, cephalopods and invertebrate spawn (DEH, 2005). Although capable of passive filter feeding, the whale shark is primarily a suction filter feeder restricting its ability to concentrate dif-

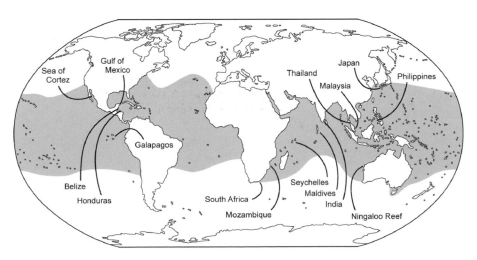

Fig. 4.5. Whale shark distribution and sites of some main whale shark-watching opportunities.

fuse planktonic food, thereby making it dependent on areas with dense plankton blooms (Heyman *et al.*, 2001). Normally a solitary species, the whale shark can occasionally be found in seasonal aggregations in areas of high primary productivity (Colman, 1997). The nature and length of duration of these aggregations has a marked influence on whale shark-watching opportunities and the total amount of income that can be generated. A short season at greater feeding depth only accessible by scuba divers will yield less return than a longer season with dominantly surface feeding.

Whale shark watching

Whale sharks have been called the 'Ambassador of Sharks' (Gibson, 2006). Their large size, docile nature, planktivorous diet, surface feeding tendencies and slow movement make them attractive from a tourism standpoint, and they do not require baiting or feeding for viewing like many other shark species (e.g. great white sharks, bull sharks). Their tendency to aggregate in predictable locations has enabled many countries (see Fig. 4.5) to establish a whale shark-watching tourism industry that is worth approximately US$66 million worldwide (Graham, 2005), the majority of which is captured by developing countries (Table 4.1) (Graham, 2004).

Whale shark ecotourism is an important source of sustainable revenue for developing countries, while simultaneously giving live whale sharks economic value. For example, fresh whale shark meat in the Taiwanese market retails for US$4.91–17.16/kg or US$12,948 for a 2800 kg individual (Chen and Phipps, 2002), while a live whale shark in Belize is estimated to be worth at least US$34,906 a year or approximately US$2 million over its lifetime, assuming a minimum life expectancy of 60 years (Graham, 2004). Countries with established whale shark-watching industries like Australia and Belize have reported annual economic returns of US$24 million and US$1.35 million, respectively (Graham, 2004; MRCM, 2006). This financial incentive has helped convert several South-east Asian fishing villages targeting whale sharks as prey into community-based ecotourism ventures supporting their protection (WWF Philippines, 2005).

Figure 4.1 shows the wildlife user and target species coming together to create NCWOR, but the activity is far from homogenous. Different manifestations develop reflecting a country's state of development, number of tourists, the length of time watching has been in operation, type of infrastructure provided, social and ecological impacts, possible management interventions and other influences. Duffus and Dearden (1990) represented this in a model (Figs 4.1 and 4.6) that showed the growth in number of tourists over time at a wildlife attraction. As the numbers increase there is a change in clientele from one dominated by adventurous and more specialized watchers to one dominated by a generalist tourist market. As this shift occurs, there is a concomitant change in infrastructure to cater to larger numbers of generalists (e.g. larger boats with less in-depth interpretation), and this further displaces more specialized watchers. Duffus and Dearden (1990) suggest that in the absence of management interventions this progression will occur, and may proceed to a point where

Table 4.1. Summary of main characteristics of some whale shark-watching sites.

	No. of tour operators	Length of operations	Regulatory environment	Visitation	Economic returns (US$)	Mode of interaction	Seasonality	Research involvement
SE Asia								
Donsol, Philippines	60	1998	– Protected since 1998 – Encounter controls 1992	7,100	623,000	Snorkel	January–June	WWF Philippines, The Shark Trust, CCC
Phuket, Thailand	85		– Protected since March 2000	10,000	>3–6 million	Snorkel, diving, viewing	November–April	
MA Reef								
Gladden Spit, Belize	>30	1997	– Protected since 2003 – WS guides require licence – Encounter controls	1,299	3.7 million	Snorkel, diving, viewing	March–June; 10 days around full moon	USAID, The Shark Trust, Project AWARE
Utila, Honduras	11	1998	– Protected since 1999 – Suggested guidelines 2005, developing national guidelines as of 2007	N/A	N/A	Snorkel, diving (research purposes only)	February–June	Utila Whale Shark Research Project, SRI, Whale Shark and Oceanic Research Centre

Holbox, Mexico	130	2002	– Protected since 2000 – Encounter controls and licences (2003) – Management plan since 2006	14,500	>1 million	Snorkel, viewing	June–September	
Australia Ningaloo Reef	15	1993	– Tours require WS licence – Encounter controls	5,000–7,000	10–24 million	Snorkel (with spotter planes)	March–July	CSIRO, ECOCEAN, Earthwatch Institute

ecological impacts and crowding are so severe that visitor numbers start to fall as target species avoid the area and crowds are too distracting even for generalized visitors. In other words, in the absence of management interventions the NCWOR activity is unsustainable.

Management interventions are based upon assessments of Limits of Acceptable Change (LAC) in both environmental and social realms (Fig. 4.6). Managers must determine the appropriate limits necessary to meet their objectives and establish indicators, standards and monitoring programmes to ensure

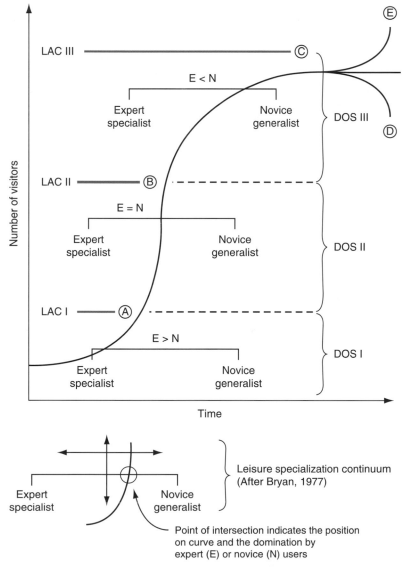

Fig. 4.6. NCWOR growth over time showing change in clientele and the need to determine Limits of Acceptable Change (LAC). (Adapted from Duffus and Dearden, 1990.)

that objectives are met. Different standards can be set in different areas through zoning to meet various objectives (Dearden *et al.*, 2006).

In the case of whale shark watching, there needs to be considerably more investigation to assess the relative position of operations in different locations (e.g. Australia, Belize) on the model, the management interventions being applied and the potential sustainability of the industry. The following section makes some observations on a few key global sites. The sites will be discussed in order of their establishment of commercialized whale shark-watching activities.

Phuket, Thailand

Phuket has been one of the main destinations worldwide for whale shark diving since the sharks were first observed in the area in 1990 during their usual migration from October to May. Although numbers vary, a marked reduction in observed whale sharks did occur in the late 1990s. There was a lack of scientific evidence to document this change but subsequent analysis of dive company logs by researchers enabled past numbers to be approximated (Theberge and Dearden, 2006). The drop in shark numbers was of particular concern to Phuket, since research indicated that viewing sharks was one of the main motivations for divers to come to Phuket and low numbers of shark sightings were failing to satisfy divers (Bennett *et al.*, 2003).

Phuket differs from most other sites in that whale sharks are almost totally observed underwater and diving is the primary activity, not whale shark watching. Advertising for the US$150 million a year diving industry has concentrated very heavily on images of whale sharks and the possibility of whale shark watching (Fig. 4.7) (Bennett *et al.*, 2003), but it is difficult to determine a precise value for the opportunity to view whale sharks. One study estimated a whale shark in Thai waters to be worth about US$5 million annually (WildAid, 2002). Although current population estimates are lacking, Theberge and Dearden (2006) reported an average of 1.3 whale shark sightings per season in the Phuket area between the 1998/99 and 2000/01 seasons. Assuming Thailand has at least these many sightings in subsequent years, it results in a minimum of US$6.5 million per annum in economic returns.

In terms of the model (Fig. 4.6) there are some trend data to help assess the relative location of Phuket. For example, Bennett *et al.* (2003) reported 85 dive companies in operation in 2002; by 2005, this had dropped to 65 (Fig. 4.8) (Main and Dearden, 2007). Dearden *et al.* (2006) shed some light on this decline by documenting the lower levels of satisfaction and lower likelihood of returning to Phuket amongst more specialized divers, who place a significantly higher value on the presence of whale sharks as a motivation to visit Phuket than those who are less specialized. It seems highly likely that Phuket has run the full course of the evolution suggested in Fig. 4.6, and low whale shark sightings are strong contributors to this decline. The fact that there are virtually no management interventions for whale shark watching in Phuket, in contrast to all other sites discussed below, is worthy of mention.

Fig. 4.7. Whale shark advertising in Thailand.

Ningaloo Reef, Australia

Ningaloo Reef in Western Australia has been a main global site for whale shark ecotourism since 1993 (Fig. 4.9) (Colman, 1997), and is the leader in undertaking research and in regulation. Whale sharks congregate in the area March to May each year, coinciding with coral spawning and other fish aggregations (Wilson and Newbound, 2001). Whale shark watching at Ningaloo is largely a land-based operation offering day trip snorkelling or viewing tours using spotter planes. Ningaloo Reef has the greatest estimated economic returns of all the sites, approximately US$10 million (Todd-Miller, 2007), due not only to the long history, but also to the relatively long season (2–6 months) (DEH, 2005), high visitation numbers (7000) (Todd-Miller, 2007) and high tour fees (about US$160–285). Although there is a greater demand for whale shark interaction licences than the 15 available, the number has not increased since 1994 due to the uncertainty surrounding the impacts of this industry on the whale shark population (Graham, 2004).

In relation to the curve (Fig. 4.6), Ningaloo has seen substantial growth with Catlin and Jones (2006) reporting a fivefold increase in visitor numbers between 1993 and 2005, to 5000 visitors (Fig. 4.8). Numbers, however, are only one indicator of progression along the growth curve. Dearden *et al.* (2006), for example, suggest that monitoring visitor specialization over time also provides an indicator of progression and ultimately of sustainability. Catlin and Jones' (2006) work suggests that the clientele has changed at Ningaloo with more local tourists, a greater focus on the service elements of the experience, and higher tolerance to crowding, all of which would be indicators of a less specialized market as shark watching enters the mainstream of tourist activities.

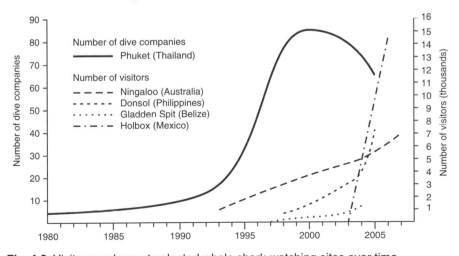

Fig. 4.8. Visitor numbers at selected whale shark-watching sites over time.

Fig. 4.9. Whale shark advertising in Australia.

Gladden Spit, Belize

The potential for whale shark tourism at Gladden Spit was realized in 1997, when researchers determined that sharks aggregated to feed on the spawn of cubera and dog snappers (Heyman *et al.*, 2001). This led to the establishment of the Gladden Spit and Silk Cayes Marine Reserve (GSSCMR) in 2000. Similar to Thailand, the industry in Belize offers scuba-diving for advanced divers, along with snorkelling and viewing within reserve boundaries. The industry is worth an estimated US$1.35 million locally, and US$3.7 million nationally (Graham, 2004). Like Australia, tour operators must be licensed to provide whale shark tours, thereby limiting the overall number of operators. Belize is further restricted in that its whale shark season lasts approximately 6 weeks in the March–June period, corresponding to the spawning activities of snappers. The shorter season results in lower visitation and, therefore, lower economic returns compared to Australia, even though tour fees are comparable (US$90–210). Access to the GSSCMR is restricted to six boats each with a maximum of 14 divers during any of the four 2 h daily time slots (Graham, 2004). This translates into a potential 84 divers in the water at the same time, possibly resulting in crowding and disruption of the aggregated spawning snappers and, therefore, whale shark predictability (Graham, 2004).

Donsol, Philippines

In 1998, a report of whale sharks off Donsol led to the killing of six sharks. The subsequent media uproar resulted in the banning of killing and trading whale sharks in the Philippines, and Donsol declared its municipal waters a whale shark sanctuary (Experience Donsol, 2006a). According to the WWF Philippines (2006), the whale shark-watching tourism industry established in Donsol led to the creation of 300 jobs within the community and an estimated economic return of about US$623,000 in 2005. It is a land-based industry offering snorkelling day trips in boats with a maximum capacity of seven people (Quiros, 2005).

Despite the fact that Donsol has a long season and amongst the greatest number of tourists (Table 4.1), it has low economic returns for the local community as there are a greater number of tour operators and lower fees compared to elsewhere (Quiros, 2005), with only 20% of the money being retained within the community in terms of shared benefits (WWF Philippines, 2006). Hiring a whale shark tour boat in Donsol costs about US$70 and holds seven people (Experience Donsol, 2006b), while a similar tour in Belize or Australia costs more than US$200 per person and can take up to 20 tourists. Donsol has a management system in which tourists must register with the Donsol Municipal Tourism Council (DMTC) and pay a fee of US$6 before being assigned to a boat (Chiu, 1998). Each boat is required to have one Butanding Interaction Officer (BIO), a spotter and a skipper. As of 2005, Donsol had 26 qualified BIOs and 60 members of the Boat Operator's Association (BOA) who work on a rotational basis to provide these tours (Quiros, 2005). As such, there is no opportunity for the boats to obtain a greater economic return as it is the DMTC that decides which boat gets tourists.

Impacts

Several studies have examined the potential for visitors to have a negative impact on whale sharks (e.g. see work at Ningaloo summarized by Mau, 2006). Possible long-term effects are summarized in Table 4.2. There is evidence for

Table 4.2. Possible negative long-term impacts of watchers on whale sharks. (From Mau, 2006).

- Disruption of feeding behaviour
- Displacement from important feeding areas
- Disruption of mating, reproductive and other social behaviour
- Abandonment of preferred breeding sites
- Changes to regular migratory pathways to avoid human interaction zones
- Stress
- Injury
- Mortality

some of these impacts, for example, propeller scars and high boat avoidance by scarred whale sharks. Cordenos-Torres *et al.* (2007) report 50% of whale sharks with fresh propeller scars in the Sea of Cortes, Mexico, but there is little evidence to tie this directly to tourism. Due to very elementary knowledge of normal long-term behaviour of whale sharks (Martin, 2007) establishing deviations that can be attributed to tourist activity is not easy. However, several researchers have documented short-term impacts. Norman (1999) recorded eye-rolling after flash photography and frequent banking when tourists swam under the head of the shark, or in the presence of scuba divers. In the Philippines, Quiros (2007) found that touch, swimmer dives and flash photography all had a significant impact on sharks.

The main challenge is whether possible long-term effects and documented short-term impacts can be attributed to the growth of tourism. For example, in Belize, the industry has expanded from two whale shark operators in 1998 to 30 in 2005 (Graham and Roberts, 2007). Meanwhile, the probability of shark sightings fell from over 80% in 1999 to less than 20% in 2004, and the maximum number of sharks fell from 13 in 1997 when shark watching started to only six in 2004 (Carne, 2005). Theberge and Dearden (2006) report similar declines in Thailand in the late 1990s as the industry grew, and in Australia whale shark sightings have declined by 50% over the last 10 years (Meekan *et al.*, 2006). However, there is no unequivocal evidence to link tourism growth with declining whale shark numbers as of yet.

Other impacts relate to the quality of the whale shark experience for participants. Is industry growth leading to more people participating but having lower satisfaction levels? Unfortunately, there is little trend data on these aspects of social carrying capacity. Catlin and Jones (2006) undertook one such study at Ningaloo and found changes between the clientele of 1995 and 2005, with a shift to a more generalist visitor emerging. Crowding was ranked third in both surveys as a detracting element. Overcrowding is also an issue at Donsol on days with low sightings where several boats crowd the only whale shark present (Quiros, 2005). There may also be too great a demand for shark tours than is possible to accommodate due to licensing restrictions. Consequently, more than 20 unregulated guides may provide tours on busy days (Quiros, 2005).

Management

The framework (Fig. 4.1) is completed by specification of management activities designed to meet the management objectives for the site. Although these activities are shown providing feedback to both animal and human management, in the case of marine species there is very little that can be undertaken in the way of animal management. Management of human activities has taken two main forms. First, regulation of possible conflicting activities (particularly fishing) and second, regulation of the mode of tourist interaction with whale sharks.

Bans on fishing have been implemented in most jurisdictions where shark-watching is taking place, and most bans have been implemented since whale shark watching was established (Table 4.1). This is important as it emphasizes the conservation potential of NCWOR. Many communities, including Donsol, have discovered that the economic returns from keeping whale sharks alive exceed those that can be made by killing them.

Most countries have developed interaction guidelines based on those observed at Ningaloo Reef (Fig. 4.10) but have allowed for local conditions. In the plankton-rich waters of Donsol, for example, it would be difficult to abide to the minimum 3 m distance that must be observed in Australia, as participants would be unable to see the whale sharks (Quiros, 2005). As such, BIOs encourage swimmers to approach within 1 m of the sharks (Quiros, 2005). Also, the 90 min limit observed in areas like Ningaloo Reef and Gladden Spit cannot apply in places like Utila, Honduras, where most encounters last less than a minute (Graham and Bustamante, 2007).

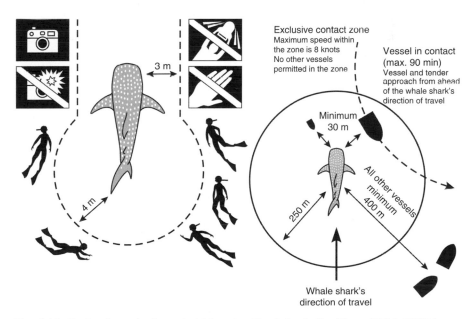

Fig. 4.10. Code of conduct used at Ningaloo Reef, Australia. (From DEH, 2005.)

Many areas require operators to be licensed in order to provide whale shark tours. In Donsol, interested members of the community must take a training course or be a member of the BOA before being able to offer tours (Quiros, 2005). Operators at Gladden Spit must also take a course (Graham and Bustamante, 2007). These types of courses may create conflict in that certified operators may be unwilling to increase membership as it would decrease individual incomes (Quiros, 2005). Increased demand has also caused management in some areas to implement a lottery system for qualified applicants. This method is used to assign licences at Ningaloo Reef (Colman, 1997), as well as to allocate access to the Whale Shark Zone within the GSSCMR prior to the start of the season at Gladden Spit (Placencia Tourism Center, 2007).

In Central America, Belize is working with Mexico and Honduras to create a common set of encounter guidelines to be included in a required whale shark tour briefing in order to promote the sustainability of the whale shark-watching industry along the Mesoamerican Reef (Graham and Bustamante, 2007). Such briefings are already included at Ningaloo Reef, Donsol, Gladden Spit and Holbox in order to ensure that tourists fully understand what is and is not permissible with respect to swimming with whale sharks.

However, the presence of regulations does not guarantee compliance, especially when tourists receive conflicting messages. For example, although tourists in Donsol must watch a video that describes the rules for interaction (Experience Donsol, 2006a), some tourists still touch the whale shark and obstruct its path – behaviours that are known to elicit a violent shudder response in the shark (Quiros, 2007). In the Philippines, BIOs are the primary enforcers of the regulations and yet there is no accountability forcing them to follow the regulations, let alone enforce them (Quiros, 2007). In fact, BIOs rarely reprimand tourists for breaking the rules, with some BIOs encouraging tourists to touch or ride the sharks (Quiros, 2005). In order to create a sustainable whale shark industry it is important to monitor compliance with regulations. For example, rangers are deployed within the GSSCMR in Belize to ensure the correct numbers of boats are present to combat overcrowding (Graham and Bustamante, 2007). It is also possible that a code of conduct instigated by a community, such as that described by Cordenos-Torres *et al*. (2007) in Mexico, will have a greater chance of long-term success than the one introduced from outside.

Conclusion

The last 20 years have witnessed a remarkable change in attitudes towards sharks amongst those that are the most exposed to them, divers. From a conservation viewpoint, this has been very beneficial in that many locations around the world have now enacted shark conservation regulations in order to protect the subjects of the shark-watching industry. This rapid growth is well demonstrated with the whale shark, where watching now occurs in some 18, mostly under-developed, countries where the income is sorely needed. However, in terms of management, the challenges are just beginning. The growth of whale shark watching has been explosive, with most of it occurring over the last 10 years. Some of the sites that have experienced this growth have concurrently experienced reductions in whale shark numbers. As of yet there is no evidence that ties the reductions directly to interference by watchers. None the less some sites, such as Gladden Spit in Belize, have reduced boat numbers as a precautionary measure. Others, such as Thailand, have yet to introduce any watching regulations, let alone monitoring systems. Unfortunately, we know so little about whale sharks that it is impossible to know whether whale sharks, once displaced from a feeding source by watchers, might return in the future or simply permanently move to other feeding sites. In view of this it would be very wise to adopt the precautionary approach displayed by Belize in implementing and enforcing regulations and invest in further research in shark reactions to watchers.

There are also many variations in the type of shark watching that have developed. In the absence of management interventions it is likely that the industry will grow to maximize numbers. Holbox Island in Mexico, for example, has expanded from virtually no watchers in 2003 to a 2006 total of 14,500 as the tourists of Cancun have discovered the attraction (Holbox Tours and Travel, 2007). The World Wildlife Fund (WWF) reports that tourists pay up to US$250 a trip, which compares handsomely to the US$25 a day that fishers used to earn from fishing (WWF Central America, 2003). The financial incentives to expand are obviously strong, but with such a rapid expansion the potential for severely violating ecological and social carrying capacities before they are even determined is also very high.

Other locales have focused on attracting lower numbers of people who stay for a longer time, pay more and often take part in helping generate knowledge about whale sharks. Examples include DNA sampling and photo-identification studies in the Maldives (MRCM, 2006), La Paz, Mexico (SRI, 2007) and Utila, Honduras (Utila Whale Shark Research, no date), as well as photo-identification projects in Sogod Bay, Philippines (CCC, 2006) and Ningaloo Reef (Earthwatch, 2007).

Overall, the development of shark watching is a good thing for both sharks and people. It provides an incentive for conservation and an opportunity to increase awareness amongst people about the ocean environment. But can it expand too rapidly? Is there a danger of creating negative impacts before we realize they are happening? Unmanaged NCWOR is often unsustainable and should not be driven solely by the market. Ultimately managers and stakeholders must decide what kind of shark watching best meets the objectives for each area and then manage to meet those objectives.

References

Barada, B. (1974) Shark research: fact or fiction? *Skin Diver* 23(10), 28–32.

Bennett, M., Dearden, P. and Rollins, R. (2003) The sustainability of dive tourism in Phuket, Thailand. In: Landsdown, H., Dearden, P. and Neilson, W. (eds) *Communities in SE Asia: Challenges and Responses.* University of Victoria, Centre for Asia Pacific Initiatives, Victoria, pp. 97–106.

Bird, J. (2002) Tagging whale sharks. *Skin Diver* 51(4), 22–23.

Cardone, B., Stearns, W., Murphy, G., Drafahl, J., Drafahl, S., Mahaney, C., Hornsby, A., Roessler, C. and Strickland, M. (1999) Sharks of the Pacific and Indian Oceans. *Skin Diver* 48(2), 78–79, 82–87, 90–92, 94–96.

Carne, L. (2005) *Unofficial Proceedings of the First International Whale Shark Conference.* Perth, Australia, May 2005. Available at: http://www.mcss.sc/iws_presentation_summary. pdf

Carwardine, M. and Watterson, K. (2002) *The Shark Watcher's Handbook.* Princeton University Press, Princeton, New Jersey.

Catlin, J. and Jones, R. (2006) Evolution of the whale shark tourism market. *Proceedings Getting Real about Wildlife Tourism: The 2nd Australian Wildlife Tourism Conference.* Freemantle, Western Australia, pp. 85–95.

Chen, V.Y. and Phipps, M.J. (2002) *Management and Trade of Whale Shark in Taiwan.* A Traffic East Asia Report, Traffic East Asia, Taipei.

Chiu, S. (1998) Donsol's whale sharks draw tourists. Available at: http://www.oneocean.org/ overseas/may98/coastal_alert.html#20b_sharks

Collins, T. (1999) The great white flight. *Skin Diver* 48(5), 129.

Colman, J. (1997) *Whale Shark Interaction Management, with Particular Reference to Ningaloo Marine Park 1997–2007.* Western Australian Wildlife Management Program, No 27.

Cooluris, J. (1977) It's an angel. *Skin Diver* 26(4), 62–63.

Coral Cay Conservation (CCC) (2006) Whale shark photo-identification project. Sogod bay, Philippines. Available at: http://www.coralcay.org/science/publications/philippines_2006_ whaleshark_report.pdf

Cordenos-Torres, N., Enríquez-Andrade, R. and Rodríguez-Dowdell, N. (2007) Community-based management through ecotourism in Bahia de los Angeles, Mexico. *Fisheries Research* 84, 114–118.

Cousteau, J.M. (1999a) Sharks – the jaws of life. *Skin Diver* 48(5), 42–43.

Cousteau, J.M. (1999b) Hope for sharks – and people. *Skin Diver* 48(9), 40–41.

Cousteau, J.M. (2001) Beyond the cage. *Skin Diver* 50(1), 28–29.

Cousteau, J.M. (2002) Sharks need more than words. *Skin Diver* 51(8), 27–28.

Dearden, P., Bennett, M. and Rollins, R. (2006) Dive specialization in Phuket: implications for reef conservation. *Environmental Conservation* 33(4), 353–363.

Department of the Environment and Heritage (DEH) (2005) Whale shark (*Rhincodon typus*) recovery plan issues paper. Available at: http://www.environment.gov.au/biodiversity/ threatened/publications/recovery/r-typus-issues/pubs/r-typus-issues-paper.pdf

Duffus, D.A. and Dearden, P. (1990) Non-consumptive wildlife-oriented recreation: a conceptual framework. *Biological Conservation* 53(3), 213–231.

Earthwatch (2007) Available at: http://www.earthwatch.org/site/pp2.asp?c = dsJSK6PFJnH&b = 1170789

Experience Donsol (2006a) Ecotourism development. Available at: http://www.ecotourdonsol. com/ecotourism.php)w

Experience Donsol (2006b) Rates. Available at: http://ecotourdonsol.com/rates.php

Farb, R. (1989) Sharks and shipwrecks. *Skin Diver* 38(8), 42–43.

Frink, S. (1991) Grand Bahamas: an island overview. *Skin Diver* 40(2), 132–135, 166–168.

Frink, S. (1996) UNEXSO and Bahamas princess resort and Casino. *Skin Diver* 45(8), 56–59, 71.

Gibson, M. (2006) A shark like no other. *Bay Islands Voice* October 2006, 4(10). Available at: http://www.bayislandsvoice.com/issue-v4-10.htm

Gleason, B. (1991) Shark dive in Fiji. *Skin Diver* 40(5), 88, 90, 114–116, 118, 128, 130–133.

Graham, R.T. (2004) Global whale shark tourism: a 'golden goose' of sustainable and lucrative income. *Shark News* 16, 8–9.

Graham, R.T. (2005) *Unofficial Proceedings of the First International Whale Shark Conference*. Perth, Australia, May. Available at: http://www.mcss.sc/iws_presentation_ summary.pdf

Graham, R.T. and Bustamante, G. (2007) Whale shark tourism management: exchanging information, networking and developing guidelines for best practices in the Mesoamerican reef region. *Proceedings of the Whale Shark Workshop in Placencia, Belize*, 25–27 September 2006.

Graham, R.T. and Roberts, C.M. (2007) Assessing the size, growth rate and structure of a seasonal population of whale sharks (*Rhincodon typus*, Smith, 1828) using conventional tagging and photo identification. *Fisheries Research* 84(1), S71–S80.

Hall, H. (1981) Anatomy of a shark bite. *Skin Diver* 30(4), 18–19.

Harrigan, B. (1998) Sharks, sharks, and more sharks! *Skin Diver* 47(6), 64–65, 102–104.

Hauser, H. (1975a) Dangerous sharks studied. *Skin Diver* 24(1), 34.

Hauser, H. (1975b) Eat them before they eat you. *Skin Diver* 24(12), 47.

Hauser, H. (1988) Man bites shark. Sharkburgers to go? *Skin Diver* 37(3), 58–61.

Heyman, W.D., Graham, R.T., Kjerfve, B. and Johannes, R.E. (2001) Whale sharks *Rhincodon typus* aggregate to feed on fish spawn in Belize. *Marine Ecology Progress Series* 215, 275–282.

Holbox Tours and Travel (2007) Swim with whale sharks! Available at: http://www.travelyucatan. com/whale_shark_holbox/whale_shark_holbox.php

Lawrence, M. (1992) Nassau Scuba Centre: The dive you will never forget! *Skin Diver* 41(5), 100–102, 142–147.

Magnussen, J.E., Pikitch, E.K., Clarke, S.C., Nicholson, C., Hoelzel, A.R. and Shivji, M.S. (2007) Genetic tracking of basking shark products in international trade. *Animal Conservation (Online Early Articles)*, 1469–1795.

Main, M. and Dearden, P. (2007) Tsunami impacts on Phuket's diving industry: Geographical implications for marine conservation. *Coastal Management*, 35(4), 1–15.

Marine Research Centre Maldives (MRCM) (2006) South Ari Atoll, Republic of Maldives: site fidelity of *Rhincodon typus* to the Republic of Maldives. Available at: www.whaleshark-project.org/do_download.asp?did = 26782

Martin, R.A. (2007) A review of behavioural ecology of whale sharks (*Rhincodon typus*). *Fisheries Research* 84, 10–16.

Mau, R. (2006) Managing for conservation and recreation: the Ningaloo whale shark experience. *Proceedings from Getting Real about Wildlife Tourism: The 2nd Australian Wildlife Tourism Conference*. Freemantle, Western Australia, pp. 127–137.

McNair, R. (1975) Sharks I have known: basic knowledge of local species behaviour key to safety. *Skin Diver* 25(7), 53–57.

McNair, R. (1976) The other side of the coin. *Skin Diver* 25(7), 44–49.

Meekan, M.G., Bradshaw, C.J.A., Press, M., McLean, C., Richards, A., Quasnichka, S. and Taylor, J.G. (2006) Population size and structure of whale sharks *Rhincodon typus* at Ningaloo Reef, Western Australia. *Marine Ecology Progress Series* 319, 275–285.

Murphy, G. (1993a) UNEXSO's shark dive! 30 minutes of high-voltage action! *Skin Diver* 42(1), 48–54.

Murphy, G. (1993b) 26 great shark dives around the world. *Skin Diver* 42(3), 120–123.

Norman, B.M. (1999) *Aspects of the Biology and Ecotourism Industry of the Whale Shark in North-Western Australia*. MA thesis, Murdoch University, Western Australia.

Placencia Tourism Center (2007) Whale shark tourism interaction guidelines 2007. Available at: http://www.placencia.com/WhaleSharkGuidelines2007.htm

Quiros, A.L. (2005) Whale shark ecotourism in the Philippines and Belize: evaluating conservation and community beliefs. *Yale University Bulletins*. Available at: www.yale.edu/tri/pdfs/bulletin2005/042Bull05-Quiros.pdf

Quiros, A.L. (2007) Tourist compliance to a code of conduct and the resulting effects on whale shark (*Rhincodon typus*) behavior in Donsol, Philippines. *Fisheries Research* 84(1), S102–S108.

Roessler, C. (1980) The oceanic whitetip shark (*Carcharhinus longimanus*). *Skin Diver* 29(10), 26–27.

Roessler, C. (1993) Shark action! *Skin Diver* 42(7), 72, 141–143.

Shark Research Institute (SRI) (2007) La Paz whale shark expedition. Available at: http://www.sharks.org/expedition_pdfs/LaPaz-2007.pdf

Sleeper, J.B. (1991) Advanced diving, shark diving. *Skin Diver* 40(5), 16, 114.

Stevens, J.D. (2007) Whale shark (*Rhincodon typus*) biology and ecology: a review of the primary literature. *Fisheries Research* 84, 4–9.

Theberge, M. and Dearden, P. (2006) Using ecotourist data for detecting a decline in whale shark (*Rhincodon typus*) sightings in the Andaman Sea, Thailand. *Oryx* 40, 337–342.

Todd-Miller, C. (2007) One marine biologist having a whale of a time. Bangkok Post, Thailand, March 26, 2007. Available at: www.flmnh.ufl.edu/fish/sharks/InNews/whale2007.html

Topelko, K.N. and Dearden, P. (2005) The shark watching industry and its potential contribution to shark conservation. *Journal of Ecotourism* 4, 108–128.

Utila Whale Shark Research (no date) Available at: http://www.utilawhalesharkresearch.com/

Wagner, M. (1988) Whale shark encounter (*Rhincodon typus*). *Skin Diver* 37(11), 48–49, 51.

Walker, J. (1986) Blue water, blue sharks: high voltage action – just off the California Coast. *Skin Diver* 35(3), 92, 94, 96.

Wallace, S. and Gisborne, B. (no date) *Basking Sharks: The Slaughter of BCs Gentle Giants*. New Star Books, Vancouver, British Columbia.

Waterman, S. (1979) Focus on sharks. *Skin Diver* 28(3), 8–13.

WildAid (2002) Wildaid's position on Mercury, tourism and extinction. *Thai Diver*. Available at: http://www.diver.com.ph/thaidiver/TD2-1/TD-2.1Environment.htm

Wilson, S.G. and Newbound, D.R. (2001) Two whale shark faecal samples from Ningaloo reef, Western Australia. *Bulletin of Marine Science* 68, 361–362.

World Wildlife Fund (WWF) Central America (2003) Playing dominos in the sea. Press release. May. San Jose, Costa Rica. Available at: http://www.panda.org/about_wwf/where_we_work/latin_america_and_caribbean/country/honduras/news/index.cfm?uNewsID = 7608

World Wildlife Fund (WWF) Philippines (2005) *The Donsol Experience Retold in Mandarin*. In Biota Filipina, April 2005, pp. 6–7. Available at: http://www.wwf.org.ph/downloads/biota/April05.pdf

World Wildlife Fund (WWF) Philippines (2006) *Whale Shark Ecotourism Contributes to Philippine Economy*. In Biota Filipina. January 2006, pp. 6–7. Available at: http://www.wwf.org.ph/downloads/biota/Jan2006.pdf

Wyland (2001) Fear of sharks. *Skin Diver* 50(10), 54–55.

5

Human–Polar Bear Interactions in Churchill, Manitoba: The Socio-ecological Perspective

R.H. LEMELIN

Introduction

For over four decades Churchill, Manitoba (Canada), has been known as 'the polar bear capital of the world', receiving thousands of bear-viewing tourists annually, frequent international media exposure and scientific attention from around the world (Lemelin, 2005). The Canadian Wildlife Service's polar bear research programme, established in Churchill in 1966 (Stirling, 1998), is one of the world's most intensive, long-term studies of any large mammal. Surprisingly, despite this profile and the abundance of scientific research conducted in this region, the human dimensions of polar bear tourism has until recently (Eckhart, 2000; Dyck and Baydack, 2004; Lemelin, 2005) been relatively ignored. Considering the role of various stakeholders in the development of polar bear tourism in Churchill, this is a serious omission, an oversight that we hope to address in this chapter.

This chapter will examine polar bear–human interactions in the Churchill area, while paying close attention to the growing wildlife tourism industry (i.e. polar bear tourism) at the end of the 20th century, and the role of the community of Churchill, if any, in the management of polar bears. The chapter begins by providing a literature review of wildlife tourism, followed by an historical overview of the Churchill region of Canada. Next is a detailed examination of four eras of polar bear–human interactions in the region. An overview of legislation and regulations is then provided, followed by a discussion and conclusion.

Wildlife–Human Interactions

Impacts from wildlife tourism have been well documented (see Higginbottom, 2005; Newsome et al., 2005), yet the local socio-cultural context of human–wildlife relationships has often been overlooked. This is somewhat surprising since the value that a host community places on a particular wildlife species can and does affect host perceptions of, and enthusiasm for, a tourism venture. For example, when wildlife is used for subsistence (e.g. beluga whales hunted in the Canadian western Arctic), then the locally assigned value may be high (see Dressler *et al.*, 2001). Where wildlife is perceived as disruptive, such as crop raiding by elephants (*Loxodonta africana*) (Treves-Naughton and Treves, 2005) and jaguars (*Panthera onca*) attacking livestock (Rabinowitz, 2005), wildlife–human interactions may be negative. Complicating these relationships is the fact that many predators kill species harvested by humans for consumption or recreation (e.g. wolf (*Canis lupus*) predation on elk (*Alces alces*) – Kunkel and Pletscher, 2000), and occasionally may even kill people (Thirgood *et al.*, 2000; Treves and Karanth, 2003). In addition, human–wildlife relationships are often complex interactions of benefits and costs. One example is elk–human interactions near Riding Mountain National Park, Manitoba, Canada. On the one hand, outfitters benefit from elk hunts, while on the other, cattle herds can become infected by disease transmitted from the elk (R. Brook, 2005, Churchill, personal communication).

Missing from this examination of host community–wildlife interactions is the role of other stakeholders, including managers, scientists, tour operators, ENGOs, and of course, the attraction: the wildlife. The role of these stakeholders and their subsequent interactions (or lack thereof) with the host community will also affect how wildlife is perceived. From a social perspective, the success of wildlife tourism, or even its existence, therefore, will depend on changes to values placed on the attraction, as well as the actual and perceived impacts of wildlife tourism (Burns and Barrie, 2005). From an ecological perspective, the success of wildlife tourism will depend on the adaptability and resilience of the wildlife and its environment (M. Ramsay, 1999, Churchill, personal communication). In Churchill, many of these stakeholders affected how polar bear management strategies were implemented and enforced. These relationships will be examined in the next section.

Polar bears (*Ursus maritimus*) and human beings have an uneasy ecological relationship in the Arctic and subarctic regions: both species compete for food; each, according to Inuit lore, believes it is the dominant force in the Arctic, and each has been known to prey on the other (Herrero and Fleck, 1990; Honderich, 1991). Polar bear–human interactions in the Churchill region have ranged from uneasy mutualistic relationships during the paleo-Eskimo era, to active harvesting during the pre-colonization and colonization phases, to confrontational and contradictory management approaches during the military era, back to an uneasy somewhat mutualistic relationship during the scientific tourism era of the late 20th and early 21st centuries (Honderich, 1991). Remarkably, the number of humans killed or injured by polar bears in this area has been low considering the proximity that large numbers of bears and humans share a good portion of the year (Stirling, 1998).

Wildlife Tourism

Wildlife tourism (i.e. birding, whale watching, bear viewing) is 'undertaken to view and/or encounter wildlife. It can take place in a range of settings, from captive, semi-captive, to in the wild, and it encompasses a variety of interactions from passive observation to feeding and/or touching the species involved' (Newsome *et al.*, 2005, pp. 18–19). In addition to the increasing awareness of wildlife issues, Higginbottom (2005) argues that wildlife tourism should also foster conservation and sustainable development. To implement sustainable development strategies properly, wildlife tourism management should also identify just who exactly the stakeholders are (Burns, 2005). In the Churchill context, a stakeholder is defined as any person, group, organization or community that affects, or is affected, by polar bear tourism (Newsome *et al.*, 2005). Although often considered as a stakeholder, the host community from a wildlife tourism perspective is defined as 'those who live in the vicinity of the tourist attraction and are directly or indirectly involved with, and/or affected by, the wildfire tourism activities' (Newsome *et al.*, 2005, p. 118). These types of activities may include a range of involvement including employment, lease agreements, concessions and partnerships, as well as active participation in management strategies (Newsome *et al.*, 2005).

Despite being much smaller than the demand for birding, African wildlife safaris or whale watching, the demand for bear viewing across the world is increasing (Brown, 2006). Some of the most popular bear congregations viewing areas in North America include Anan Creek, Brooks Falls, Hyder, McNeil River Falls, Pack Creek (Alaska); Bella Coola, Knight Inlet (British Columbia); Churchill (Manitoba); as well as Svalbard in Norway, Kamchatka and Wrangel Island in Russia. In Alaska, brown bears are viewed from vehicles in Denali National Park and photographed from gravel platforms along the banks of the McNeil River, while black bears and the occasional Kermode bears (i.e. spirit bears) are viewed from boats in the Queen Charlotte Islands of British Columbia. In polar regions, bears can be viewed from cruise ships in the archipelago of Svalbard, Norway and from helicopters and/or tundra vehicles in Churchill, Canada. Although these are only a few examples of how bears can be viewed by human beings, they do illustrate the diversity within the bear-viewing industry as a component of wildlife tourism.

Valuable research on the biological impacts of bear viewing in areas of bear congregations has been undertaken in recent years, studies especially focused on aspects of the bears' behaviour (Aumiller and Matt, 1994; Fagen and Fagen, 1994; Eckhart, 2000; Dyck and Baydack, 2004) and their management (Dalle-Molle and Van Horn, 1989). A study conducted on denning black, brown and polar bears in North America and Europe demonstrated that recreational vehicles such as snowmobiles can interfere with the animals' hibernating patterns, and result in their permanent abandonment of dens and increased cub mortality (Linnell *et al.*, 2000). In contrast, behavioural studies conducted on Churchill's polar bear-viewing industry found that while vigilance among subadult bears does appear to increase in the presence of humans, the increase is somewhat modest (Watts and Ratson, 1989; Dyck and Baydack, 2004). Further, the relative absence of defensive kills and injuries to both bears and humans in these areas of large bear aggregations indicates that bear–human management strategies have been successful (Aumiller and Matt, 1994).

In contrast, a more modest number of studies have examined the human dimensions of bear–human interactions. Most of the socio-political research conducted in areas of bear aggregations (National Park Services, 1995; Herrero and Herrero, 1997, 1999) was designed to assess current anthropogenic uses and relate these to current management strategies. With respect to examples of socio-economic studies conducted in areas of bear aggregations (Creed and Mendelson, 1993; Lemelin et al., 2006), such research has revealed that as long as user fees are directed towards local initiatives at the sites (i.e. conservation, management, research), visitors are generally not opposed to extra charges. Finally, social psychological studies (Bath, 1994; Whittaker, 1997) and socio-environmental studies (Lemelin, 2006a; Lemelin and Smale, 2006) revealed that the motives and values attracting human beings to bears are quite diverse. The impacts of wildlife tourism on local communities are examined in the next section.

Local Communities and Wildlife

Literature focusing on wildlife tourism and host communities can be divided into three broad categories: Traditional uses and wildlife tourism, protected areas and wildlife tourism and critical analysis and wildlife tourism. Traditional uses, especially when animals are harvested, can conflict with wildlife tourism. Examples where management strategies have attempted to minimize these interactions or create positive dialogues between various stakeholders include Muvla's (2001) report on Zambia National Parks, and Freeman's (2003) analysis of polar bear trophy hunts by Inuit communities. Today, a number of public and private initiatives are jointly managed by local and indigenous peoples. One example is the Uluru-Kata Tjuta National Park, another is the Karanambu Ranch in Guyana (Shackley, 1998). Critical examinations of wildlife tourism and host communities include Burns' (2004) assessment of the 1990 Solomon Islands tourism plan and Belsky's (2000) examination of community-based discourses and practice in Gales Point, Belize.

In many of the cases listed above, wildlife tourism generated a range of economic benefits for local communities including revenue creation, employment, entrepreneurship, economic diversification and infrastructure improvements. In addition, community involvement in wildlife tourism can result in increased pride in, and recognition of, the cultural and natural assets of the area through the development of cultural centres, traditional cultures and crafts, as well as interpretation strategies (Higginbottom, 2005; Newsome et al., 2005). Tourism may also lead to the renewal of interests in traditional knowledge systems (Huntly et al., 2005). Profits from wildlife tourism can provide health care and education, assist conservation efforts and raise the living standards of host communities (Ashley and Roe, 1998; Muvla, 2001). Profits can also be directed to compensation strategies, offsetting the cost of wildlife-incurred damage to livestock and crops (Adams and Infield, 2003; Treves-Naughton and Treves, 2005).

Although wildlife tourism can help preserve wildlife, natural landscapes and cultures on the one hand, it can also transform landscapes and habitats, impede or restrict certain traditional practices and even displace or remove indigenous populations on the other (Chapin, 2004; Dowie, 2005). Further, the displacement of indigenous peoples for the creation of protected areas (Royal Chitwan National Park, Yellowstone National Park) can create resentment, and may be perceived as catering to non-local needs (Dowie, 2005). Last, economic leakage, corruption and seasonal or low paying jobs have also been reported (Ashley and Roe, 1998; Newsome *et al.*, 2005). Stakeholder involvement is the key to all these perspectives.

Stakeholder involvement, especially as it pertains to the host community's engagement in wildlife tourism can entail passive roles in the industry (employment), to active engagement in decision-making processes regarding establishing and managing tourism in a natural area. Stakeholder involvement in planning and training, where local communities are acknowledged as key group, is widely accepted in the developed world, whereas such involvement is a somewhat newer concept elsewhere (Timothy, 1999). The exception is stakeholder involvement in adaptive management systems (Berkes, 2004). Adaptive governance of social–ecological systems provides a framework whereby social and ecological science and traditional and local understanding are recognized, and wherever possible, incorporated into management approaches. These systems can, in combination, provide a key to reducing community vulnerability and enhancing adaptability and resilience (Berkes *et al.*, 2003; Folke *et al.*, 2005).

Increasingly, strategies promoting community-based participatory approaches or adaptive management to resource management (Knight and Meffe, 1997; Wondolleck and Yaffee, 2000), and re-examination of resource systems based on local and traditional ecological knowledge (Agrawal, 1995; Pimbert and Gujia, 1997) are being incorporated into resource management in an 'attempt to manage conflicts between competing users, negotiate through, and out of, social and ecological crisis situations and avoid or pre-empt future conflicts and crises' (Blann *et al.*, 2002, p. 212).

The involvement of local people and their knowledge systems, for example, Traditional Ecological Knowledge (TEK) and/or Local Ecological Knowledge (LEK) (i.e. the sum knowledge derived by long-term, applied experiences and observations by indigenous and local peoples), have received widespread support (Pimbert and Gujja, 1997; Olsson and Folke, 2001; Berkes *et al.*, 2003), yet little research exists on examining how this can be accomplished in tourism (Huntly *et al.*, 2005).

To summarize, wildlife tourism conservation activities have greater potential for success if local people are allowed to take part in formulating and implementing policies and programmes that incorporate safeguards against abuses and place strong emphasis on sustainability, equity and social justice (Belsky, 2000). The following section illustrates how community members have been involved in some discussions relating to polar bear management in the Churchill region, and excluded from others.

The Environmental Context

Churchill is located on the estuary of the Churchill River where it flows into Hudson Bay (Fig. 5.1). It has a subarctic climate and is located in the Hudson

Bay Lowlands terrestrial eco-zones: characterized by flat terrain underlain by continuous permafrost, poor drainage and a transition from boreal forest to Arctic tundra vegetation (Brook and Kenkel, 2002). Hudson Bay is a relatively shallow inland sea and is usually covered by ice (at least near shore) between late November/early December and late June–early July, though it appears that the length of the ice-free period is increasing.

The polar bears using this area comprise the Western Hudson Bay population, estimated at 1200 ± 250 bears (Lunn *et al.*, 2002). They spend most of their time on the ice hunting seals (primarily ringed seals (*Phoca hispida*), but also bearded seals [*Erignathus barbatus*] and probably harbour seals (*Phoca vitulina*)), which make up over 95% of their diet (Stirling, 1998). When they are forced ashore in summer by melting ice, they may feed opportunistically (Lunn *et al.*, 2002) but largely they fast, losing considerable body weight. On shore,

Fig. 5.1. Churchill, Manitoba. (From Chapin, 2006.)

they segregate by age and sex class, with males remaining on the coast and females moving on average 80 km inland – largely to avoid the sometimes cannibalistic males (Dyck and Daley, 2002). Pregnant females use earth dens (and later in the season, snow dens) inland, and remain there after the sea ice on the Hudson and James Bays freezes up, giving birth to 2.2 cubs on average, and returning to the sea ice with their cubs in March (Ramsay and Stirling, 1990). Most dens are located south-east of Churchill (i.e. Wapusk National Park), which has apparently been used by polar bears for centuries (Scott and Stirling, 2002).

Polar Bear and Human Interactions

The Churchill region has a long and diverse history of human occupation. Paleo-Eskimo (pre-Dorset and Dorset cultures) sites have been found near the town (1700 BC), indicating long use of the coast for harvesting marine mammals (Brandson, 2005). Inuit, Cree and Dene people all used the area prior to European contact in the 17th century, but their presence increased with the establishment of permanent fur trading posts and settlements by the Hudson's Bay Company at York Factory (approximately 250 km south-east of Churchill on the Hayes River) in 1684, and at Fort Churchill in 1717 (Brandson, 2005). The present town of Churchill dates from the 1920s, when the port facilities were constructed, linking southern and northern Manitoba by the Hudson Bay Railway in 1929 (Brandson, 2005).

Originally conceptualized as part of the Crimson Staging Route to ferry wounded personnel from overseas during World War II, Fort Churchill (1946) and later the Churchill Research Range (1958) contributed to the growth and development of the town from the mid-1940s until the late 1960s (Brandson, 2005). The sites were abandoned in the 1970s, yet the community of Churchill persisted. Today, the community's economy is well diversified ranging from transportation services (Port of Churchill, Northern Transport Canada Limited, Omnitrax and Calm Air) to health services provided by the Regional Health Authority and the Northern First Nation Transient Centre.

The abundance of lakes, rivers, forests and tundra, coupled with the long-standing tradition of wilderness outfitters, lodges and other leisure facilities has provided a firm foundation for Churchill's tourism industry. By the 1960s, Churchill was becoming a popular birding destination, while acquiring a reputation for live beluga whale captures. Beluga whales were captured and sent to Canadian and international zoos and aquariums (Brandson *et al.*, 2002). However, it was not until a decade later that some small-scale polar bear outings were offered in the region (Lemelin, 2005). Through the help of various existing industries, such as hunting, fishing, birding, whale watching, aurora borealis gazing and polar bear-viewing activities, Churchill's tourism industry continued to grow and diversify throughout the late 20th century. The economic impact of nature tourism in 2002 was estimated at well over US$3 million dollars (Lemelin, 2005). One of the most important components of Churchill's wildlife tourism industry is polar bear viewing.

Churchill became an international tourism destination because polar bears congregate along the shores of the Hudson Bay to await the formation of sea ice in early to mid-November. It is this unique natural phenomenon that attracts thousands of wildlife tourists each year to this region. The ability of tundra vehicles to traverse the subarctic environment provides the ideal mode of transportation to see and photograph this large, attractive and somewhat predictable predator in relative safety and comfort. The viewing of polar bears in this area, however, is further facilitated by the habituation and tolerance of these animals to human presence, their curiosity and their propensity to entertain wildlife viewers.

A majority of the bear viewing in Churchill actually occurs in two protected areas (i.e. Churchill Wildlife Management Area (CWMA) and Wapusk National Park (WNP)) located 21 km east of the community. Even with the creation of WNP in 1997, and the transfer of 1.14 million hectares of the CWMA from Manitoba to Canada, the CWMA remains, at 848,813 ha, the largest and most northerly Wildlife Management Area (WMA) in Manitoba (Manitoba Conservation, 1999, unpublished data). A detailed examination of polar bear–human interactions in Churchill, Manitoba from the 19th to the 21st centuries is provided next.

Polar Bear Management

In order to understand polar bear–human interactions in the Churchill region, four eras are discussed. These eras are: harvesting (pre-1940s), military (1940–1960s), research/management (1960s–present) and tourism (1970–present) eras. It is important to note that these eras are ideal types and, therefore, some overlap does occur.

In the first era (prior to the 1940s), polar bears were actively harvested by whalers, explorers, adventurers and indigenous peoples for meat, pelts and as entertainment (i.e. cubs were often sent to zoos). Regulation of polar bear hunting began only after the province of Manitoba was created in 1912, and the Natural Resources Agreement Act of 1930 established Aboriginal hunting rights for polar bears. In 1949, polar bear hunting was restricted to Aboriginal people, and in 1954, hunting polar bears for trade was banned altogether (Stirling, 1998). The establishment of registered traplines in the 1950s changed trapping and land-use patterns in the region by defining who could trap where.

Prior to the establishment of Fort Churchill and the Churchill Research Range, traditional and illegal hunting had removed and dispersed bears, approaching the village of Churchill. However, the construction of Fort Churchill and the Churchill Research Range profoundly transformed polar bear–human interactions in the area. Although somewhat recent, the military era of Churchill is poorly documented. However, discussions with stakeholders (residents and former employees of the Churchill Research Range) have indicated that various and contradictory management philosophies were used to 'manage' polar bears. While it appears that no polar bears were supposedly permitted near Fort Churchill or the Churchill Research Range, unfenced dumps provided polar bears with opportunities to forage for food (Bukowsky, 2002). In addition, frequent reconnaissance training in the field may have led to food conditioning of some subadult bears (Lemelin, 2005). Thus, it appears that on the surface, the military's management approach to polar bears was aggressive. Yet there is also evidence of contradictory, food-conditioning approaches by some military personnel.

Increasing polar bear–human encounters throughout the 1960s and 1970s was directly related to the abandonment of several coastal communities and the withdrawal of up to 4000 armed forces personnel from Fort Churchill. 'When the garbage dump at the Fort was no longer there, bears headed for the several smaller dumps at Churchill, where by-laws for garbage pickup and disposal were not strictly enforced' (Bukowsky, 2002, p. 151). Potential hazards associated with people–bear confrontations were not new to the people of Churchill, and following the military's withdrawal and increasing polar bear–human encounters, residents had learned to cope with polar bears. However, two human deaths in the 1970s prompted the Manitoba Department of Natural Resources and Transportation Services to study polar bear–human conflicts (Stirling, 1998).

Already operating in the area since the late 1960s, the Canadian Wildlife Service (CWS) of Environment Canada was commissioned to investigate polar bear–human interactions in Churchill. Among other findings, research efforts led to the creation of the Polar Bear Alert Programme and the delineation of the Western Hudson Bay polar bear population. In 1969, provincial officers were stationed in Churchill and a polar bear patrol was put into action in the fall. The goal was to ensure the safety of people and protection of property from damage by polar bears, and to ensure that bears are not unnecessarily harassed or killed (Bukowsky, 2002). This was accomplished by shooting problem bears, or clearing the area of bears by relocation (i.e. trapping and transporting). This latter process was facilitated by airlifts, which were subsidized by the International Fund for Animal Welfare (IFAW). Although IFAW is no longer involved in the Polar Bear Alert Programme, airlifts through the help of helicopters continue, as do the roles of ENGOs in polar bear management and research (Lemelin, 2005). The adoption of the Western Hudson's Bay management zone in 1974 recognized that the polar bear population was shared between Manitoba and Nunavut (Lunn *et al.*, 1998). A quota was established and Manitoba agreed to lend 15 of its tags to Nunavut communities who harvested bears from that population, retaining the rest for control kills. The quota has since been adjusted, and is now set at 27. As many as 19 are on loan to Nunavut, and 8 are retained for the Manitoba Polar Bear Alert Programme for destroyed problem bears (Lunn *et al.*, 2002).

In the 1970s, two international treaties were established. The Convention on International Trade in Endangered Species (CITES) outlined that shipments of polar bears or parts thereof must be done under permit and that governments must keep statistics on all legally exported or imported polar bear parts (Lunn *et al.*, 1998). The International Agreement on the Conservation of Polar Bears (IACPB) signed in Oslo, Norway by Canada, Denmark, Norway, the USA and the former USSR, addressed international polar bear research, hunting quotas and protection strategies pertaining to polar bears (Lunn *et al.*, 1998).

Despite all these efforts, increasing bear sightings (200 in 1976 up from 76 in 1967) and increasing numbers of bears killed by wildlife personnel continued. It soon became apparent that a polar bear programme could succeed only with the cooperation of residents and local and provincial governments. Hence, a local Churchill polar bear committee was established, consisting of residents, a council member and staff of the wildlife branch (Bukowsky, 2002).

Following extensive dialogues with stakeholders and the host community, the committee submitted 14 recommendations, including the production of education material for bear safety (posters, pamphlets and classrooms presentations) and the acquisition of Building D-20 at Fort Churchill as a temporary holding place for polar bears. The facility was designed to house 16 individuals and four family groups (Lunn et al., 1998). The importance of the 'polar bear jail' as it became known cannot be overstated, since the facility complemented the Polar Bear Alert Programme and ongoing scientific studies, and saved the lives of polar bears each year (Bukowsky, 2002).

In 1991, the province of Manitoba enhanced its polar bear management strategies by expanding on the national listing of the polar bear as globally abundant and secure by the Committee on the Status of Endangered Wildlife in Canada (COSEWIC), while provincially rare and perhaps vulnerable to extirpation. These revisions to the Manitoba Wildlife Act in 1991 changed the status of polar bears from protected species to big game species, thereby enhancing the province's ability to manage polar bears, and effectively, prohibiting all hunting of polar bears, except in special cases decreed by the Minister of Conservation (Manitoba Government, 2003).

Two additional Acts, the Polar Bear Protection Act and the Resource Tourism Operators Act, both assented in 2002 in the Manitoba legislature, further increased polar bear protection in the province of Manitoba. In addition to policies and legislations, protected areas strategies as explained earlier were also established to protect polar bears' staging and denning areas. The CWMA was created in 1978. With a network of 35 km of all-weather road, 125 km of gravel trails and a number of permanent (the Churchill Northern Studies Centre, formerly the Churchill Research Range) and semi-permanent infrastructures (tundra vehicle departing platforms and tundra hotels), the CWMA is relatively accessible to both local and non-local human users. Anthropogenic activities in the CWMA include hunting, trapping, wood harvesting, off-road vehicle recreation (snowmobiles, ATVs), scientific research and wildlife observation (birdwatchers, polar bear observers) (Teillet, 1988). These activities are controlled through a series of guidelines. Those pertaining specifically to tourism are addressed next.

1. A permit is required for tour operators; this applies to all commercial tourism operations in the CCWMA.
2. All tour operators are required to keep their tundra vehicles on designated trails.
3. Operators must avoid pursuing or harassing polar bears.
4. Polar bear viewers are not permitted to yell, harass or stick arms, hats or food out of the tundra vehicle windows.

5. There is restricted access to prime polar bear staging areas around the Gordon Point zone.

6. Tour operators are prohibited from feeding or baiting polar bears.

7. Limited, temporary overnight facilities at designated locations for extended tours are allowed. Facilities must be removed after the completion of each season.

8. Grey water and solid waste must be removed daily from all tundra vehicle locations (including hotels, tundra vehicles and departure points).

9. Helicopters must hover at no less than 200 ft over wildlife; landings are only permitted in designated areas (Manitoba Conservation, 1999, unpublished data).

These guidelines are enforced by Manitoba Conservation (formerly the Manitoba Department of Natural Resources). However, the CWMA Management Guidelines do not impose restrictions on the number of tourists allowed in the CWMA, or charge user fees. In fact, since a majority of management efforts are dedicated to the Polar Bear Alert Programme, little if any monitoring of the polar bear tourism activities in the CWMA occurs. The results have been that, throughout the last two decades, some of the guidelines on bear observation have not been respected or enforced. Further, in the spring of 1999, Manitoba Conservation initiated public consultation with local user groups in Churchill aimed at revising the 1988 CWMA Management Guidelines. Although a number of meetings have been held, no modifications to date have been made (W. Roberts, 2000, Churchill, personal communication). The Churchill Land and Resources Use Steering Committee expressed various concerns regarding the harassment of polar bears (Calvert *et al.*, 1995). Many of these concerns were addressed through cooperation between the management agencies and the operators. In 1996, WNP was established, incorporating much of the CWMA. The park was created to fulfill the National Park Systems Plan object-ive of representing all of Canada's natural regions within national parks. A management strategy, addressing anthropogenic uses in the park and tour-ism management approaches have recently been completed by the Wapusk Park Management Board and are currently under review.

To resume, the closure of York Factory and the military's withdrawal from the Churchill area in the mid to late 20th century decreased human activity along the west coast of the Hudson Bay, resulting in declining human disturb-ance of polar bears in the area, and indirectly contributing to the growth of the polar bear population in the region (Lemelin, 2005). As was the case in other locations (e.g. Yellowstone National Park), bear population growth often resulted in increased wildlife–human encounters, which in turn stimulated wild-life management needs (Schullery, 1992). Wildlife management in the region that had been previously overseen by the military, at times with dreadful impacts on the polar bear population, was taken over by the Manitoba Conservation and assisted by the Canadian Wildlife Service. Ironically, the effort of these two agencies soon garnered attention from such environmental organizations as IFAW. This attention would subsequently lead to the emergence of a new wild-life tourism industry – polar bear viewing (Lemelin, 2005).

Discussion

From active harvesting to total protection, polar bear–human interactions in the Churchill area have been dynamic and dialectical. Yet research, management, legislation and the creation of protected areas in the late 20th and early 21st centuries demonstrate how adaptive management strategies based on the ecological boundaries have successfully protected polar bears in this region.

Although the range of interested stakeholders is large, Scheyvens (2002) and Newsome *et al.* (2005, p. 115) argue that unless local communities 'gain some benefits from the conservation of wildlife they will have little incentive to sustainably manage these resources'. Community members in Churchill were largely excluded from the management of polar bears until the 1970s. Since then, they have been actively involved in some form or other, in various policies, legislation and protected areas strategies. The latest inclusion is representation on the board of directors for WNP, and the recognition of traditional knowledge in WNP. Therefore, involving the host community, which often bears the brunt of the impacts from wildlife tourism, must continue if future polar bear management strategies in this area are to be successful.

Although polar bear protection has increased and the town may have benefited from the presence of polar bears (i.e. employment, entrepreneurship), a number of residents voiced their concerns with the globalization and consolidation of the industry (i.e. two tundra vehicle owners), and the subsequent commodification of the polar bear industry (Lemelin, 2006a). The industry does not strongly promote the cultural dimensions of the community. It is only interested in promoting bears, usually large numbers of bears (Lemelin, 2006b). These developments increase the potential for economic leakage and/or increased inequity from the uneven distribution of the benefits of the industry. Ironically, many local citizens cannot afford to see polar bears from the tundra vehicles or helicopters, and while some complementary excursions are offered to Churchill residents, most Churchillians' polar bear-viewing experiences are relegated to watching them at the city dump. Safety concerns pertaining to polar bears were also expressed, and while the Polar Bear Alert Programme reduced the number of problem bear kills throughout the late 20th century (see Calvert *et al.*, 1995), the seasonality of the programme and the marginalization of people who live outside of the polar bear management zone established around the community of Churchill (i.e. Goose Creek) were highlighted as limitations of the programme.

Despite concerns over dated wildlife management guidelines in the CWMA, polar bear tourism was at one time mostly concentrated in a few protected areas. Today, there are a number of opportunities to see polar bears in various settings, from night-time tundra vehicle tours to helicopter tours to 'abandoned' den sites, photographic excursions near polar bear dens in the spring, and viewing supposedly natural interactions between husky dogs and polar bears just outside of the CWMA. These opportunities illustrate the 'creepage' of the industry outside of managed areas (also referred to as sacrifice areas) to new areas, where few, if any, management guidelines exist. Concerns over the growing wildlife photography tours offered in the denning areas of the CWMA and WNP were also expressed by some researchers, 'although this activity is not regulated or monitored, the subjective impressions of park staff, pilots and biologists are that the number of parties and the distance traveled into the denning area have increased since 1997. The effect of this on the behaviour, condition or survival of bears is not known' (Lunn *et al.*, 2002, p. 50).

Some residents noted their concerns vis-à-vis the role of non-local stakeholders (i.e. scientists, ENGOs), asserting that many of these individuals have no 'sense of belonging' to the region and provided little, if any, benefit to the community and/or the polar bears. Others questioned the research approach by the CWS, asserting that polar bears in the region are overly studied and harassed by biologists. Some interviewees also voiced their concerns regarding ENGOs in the region. As indicated earlier, IFAW played a pivotal role at the onset of polar bear management. Other ENGOs which have since then entered the polar bear arena include, Greenpeace, Born Free, Polar Bears International (formerly Polar Bears Alive) and the World Wide Fund for Nature (WWF). Often praised by researchers (Mason *et al.*, 2000; Stewart *et al.*, 2005), the WWF's Arctic Tourism Project never took hold in the community, since a visit by WWF representatives was met with a lukewarm response in 1998. By 1999, the WWF had basically pulled out of Churchill. Discussions with local residents indicated that much of the concern with the WWF and other ENGOs pertained to the paternalistic and condescending approaches employed by a number of these organizations. This pattern is not uncommon (see Brulle, 2000), as residents also questioned the accountability and transparency of these 'stakeholders', claiming that no one really knows who they go back and report to. Despite the community's resistance to ENGOs, Polar Bears International, founded by photographer Dan Guravich (one of the 'founding fathers' of the polar bear tourism industry in Churchill), revamped its image in the early 21st century. They hosted several town-hall meetings, and implemented a process much more transparent and accountable to local stakeholders. The results have been research funding, greater media awareness, the promotion of an educational mandate and local members on advisory committees.

Many authors have noted the positive economic and socio-cultural impacts from wildlife tourism (Higginbottom, 2005; Newsome *et al.*, 2005), yet few have discussed the sense of civic pride or spatial attachment that can be produced by wildlife tourism. Such is this case in Churchill, Manitoba. While it is true that many residents value the economic prosperity brought to the community by polar bear tourism, others unaffiliated with the industry also take pride in the polar bear. The polar bear is everywhere in the community – on the town's promotional material, on the welcome sign, on the jerseys of the local ice hockey team. Indeed, the positive impact from this polar icon reverberates deep within the social fabric of the community.

One stakeholder rarely mentioned in this article and the literature is perhaps the most important of all – the polar bear. The polar bears are the attractions, and without them, there would be no industry. Indeed, as noted by biologists (M. Dyck, 1998, Churchill, Manitoba, personal communication; M, Ramsay, 1999, Churchill, personal communication) and some local residents, it is largely because of the polar bear's tolerance that so few humans have been injured or killed. If arguments can be made for the protection of polar bears, it behooves us to remember that under the present polar bear management systems, the attraction has no voice. Until that can be provided, interpreters, managers and operators can remind the visitors that they 'need to behave in a fashion accept-able to *polar bears*, not the other way around' (Schackley, 1996, p. 36, italics added by the author).

Conclusion

Management strategies in the Churchill region have been largely driven by a species-selective approach to wildlife protection. This process is at best mutualistic and aimed at fulfilling anthropocentric needs. For example, while polar bears have been studied for over four decades, and they are now revered by thousands of wildlife tourists, other species such as wolves[1] remain unpro-tected and harvested. Researchers and tourists have noted that if the wolf, hunts were eliminated in this area, wolves could also become an attraction of the wildlife-observation industry. Perhaps there is hope for the wolf, and all other species, in this region. According to the late biologist Dr Malcolm Ramsay, polar bears in Churchill in the late 1980s were seen as a nuisance that should be actively hunted and destroyed. Today, the outlook on polar bears is much different as locals participate with the Polar Bear Alert Programme, while others see polar bear tourism as the only viable and sustainable industry in the region (Lemelin, 2005). That said, challenges do exist, from climate change, biomagnifications, polar bear–human conflicts in protected areas (i.e. a polar bear attack on a researcher within Wapusk National Park, 18 November 2004), the proposed removal of the Churchill dump and upgrading the current polar bear status to endangered in Canada. However, the latter is also seen as an opportunity by some local residents, for greater marketing and awareness of polar bear tourism.

It would be easy to critique wildlife management in this region. However, these stakeholders should be commended for their dedication and foresight to the polar bears. The incorporation of provincial, national and international legislation, acts and wildlife management strategies (i.e. the Polar Bear Alert Programme), in addition with the inclusion of local stakeholders in decision-making processes (i.e. the Wapusk National Park Management Board), have promoted the protection of polar bears, and subsequent coexistence with polar bears.

Despite a history of harvesting beluga whales, and later capturing and ship-ping beluga whales to various aquariums and zoos across North America (Brandson *et al.*, 2002) in 2000, the town of Churchill refused to extract a number of beluga whales from the Churchill River to be sent to a popular

---

[1] In 2003, two separate wolf–polar bear interactions were photographed by tourists and researchers. Tour-ists mentioned that although they were unaware that wolves were found in this region, they did indicate their interest in learning about this species.

aquarium in southern Ontario. The rationale was explained in the following way – the community promotes itself as a premier, wildlife tourism destination, where people from around the world come to see birds, belugas and bears. The attraction of this wildlife tourism industry is that the animals are free and wild, they should always remain so. What is more remarkable is that the decision was not imposed on the community from governmental departments, ENGOs or tourists. In fact, locals were the ones who reached that decision on their own. That is the power, or as some would say, the benefits of wildlife tourism in Churchill, which is why the viewing of polar bears in natural environments is so important. Perhaps when wolves and all the inhabitants of the CWMA are viewed like beluga whales and polar bears – animals worth protecting, worth preserving – then the province of Manitoba, the country of Canada and all of humanity will be one step closer to implementing a biospheric management approach for all inhabitants.

References

Adams, W.M. and Infield, M. (2003) Who is on the gorilla's payroll? Claims on tourist revenue from a Ugandan national park. *World Development* 31(1), 177 190.

Agrawal, A. (1995) Dismantling the divide between indigenous and scientific knowledge. *Development and Change* 26, 413–439.

Ashley, C. and Roe, D. (1998) *Enhancing Community Involvement in Wildlife Tourism: Issues and Challenges.* IIED Wildlife and Development Series 11. International Institute for Environment and Development, London.

Aumiller, L. and Matt, C.A. (1994) Management of McNeil river state game sanctuary for viewing of brown bears. *Proceedings from the International Conference – Bears Resources and Management* 9, 51–61.

Bath, A. (1994) Public attitudes toward polar bears: an application of human dimensions in wildlife resource research. *Proceedings from the International Union of Game Biologists XXI Congress* 1, 168–174.

Belsky, J.M. (2000) The meaning of the manatee: community-based ecotourism discourse and practice in gales point, belize. In: Zerner, C. (ed.) *Plants, People and Justice: Conservation and Resource Extraction in Tropical Developing Countries.* Columbia University Press, New York.

Berkes, F. (2004) Rethinking community-based conservation. *Conservation Biology* 18, 621–630.

Berkes, F., Colding, J. and Folke, C. (eds) (2003) *Navigating Social–Ecological Systems: Building Resiliency for Complexity and Change.* Cambridge University Press, Cambridge.

Blann, K., Light, S. and Musumeci, J.A. (2002) Facing the adaptive challenge: practitioners' insights from negotiating resource crises in Minnesota. In: Berkes, F., Colding, J. and Folke, C. (eds) *Social–Ecological Systems: Building Resilience for Complexity and Change.* Cambridge University Press, New York, pp. 210–240.

Brandson, L. (2005) Churchill. In: Nuttall, M. (ed.) *Encyclopedia of the Arctic.* Routledge. New York, pp. 352–254.

Brandson, L., Henry, D. and Hickes, J. (2002) The days of the whalers. In: Churchill Ladies Club (eds) *Through the Years: Churchill, North of 58°.* Friesens Corporation, Altona, Manitoba, Canada, pp. 21–25.

Brook, R.K. and Kenkel, N.C. (2002) A multivariate approach to vegetation mapping of Manitoba's Hudson Bay lowlands. *International Journal of Remote Sensing* 23(21), 4761–4776.

Brown, C.E. (2006) An overview of agency-managed bear viewing sites in Alaska: a diversity of agencies, opportunities, and visitor management strategies. *Proceedings of the International Symposium for Society and Resource Management. Global Challenges, Local Responses.* ISSRM, Vancouver, British Columbia.

Brulle, R.J. (2000) *Agency, Democracy, and Nature: The U.S. Environmental Movement from a Critical Theory Perspective.* MIT Press, Cambridge, Massachusetts.

Bukowsky, R. (2002) Manitoba Department of Resources: public safety and the polar bear jail. In: Churchill Ladies Club (eds) *Through the Years: Churchill, North of 58°.* Friesens Corporation, Altona, Manitoba, Canada, pp. 150–152.

Burns, G.L. (2005) The host community and wildlife tourism. In: Higginbottom, K. (ed.) *Wildlife Tourism: Impacts, Management and Planning.* Sustainable Tourism Cooperative Research Centre (STCRC), Gold Coast, Queensland, Australia, pp. 125–143.

Burns, P. (2004) The 1990 Solomon islands tourism plan: a critical discourse analysis. *Tourism and Hospitality: Planning and Development* 1(1), 57–78.

Burns, P.M. and Barrie, S. (2005) Race, space and 'our own piece of Africa': doing good in Luphisi village? *Journal of Sustainable Tourism* 13(5), 468–485.

Calvert, W., Taylor, M., Stirling, I., Kolenosky, G.B., Kearney, S., Crete, M. and Luttich, S. (1995) Polar bear management in Canada 1988–92. In: Wiig, Ø., Born, E.W. and Garner, G.W. (eds) *Polar Bears: Proceedings of the Eleventh Working Meeting of the IUCN/SSC Polar Bear Specialist Group.* IUCN – The World Conservation Union, Gland, Switzerland, pp. 61–79.

Chapin, M. (2004) A challenge to conservationists. *World Watch* (November/December) 17–30.

Creed, C. and Mendelson, R. (1993) The value of watchable wildlife: a case study of McNeil River. *Journal of Environmental Management* 39, 101–106.

Dalle-Molle, J.L. and Van Horn, J.C. (1989) Bear–people conflict management in Denali national park, Alaska. *Bear–People Conflict – Proceedings of a Symposium on Management Strategies.* Yellowknife, North-west Territories, Canada, pp. 121–127.

Dowie, M. (2005) Conservation refugees. *Orion* (November/December), 16–27.

Dressler, W., Berkes, F. and Mathias, J. (2001) Beluga hunters in a mixed economy: managing the impacts of nature-based tourism in the Canadian western Arctic. *Polar Record* 37, 35–48.

Dyck, M.G. and Baydack, K.R. (2004) Vigilance behaviour of polar bears (*Ursus maritimus*) in the context of wildlife-viewing activities at Churchill, Manitoba, Canada. *Biological Conservation* 116, 343–350.

Dyck, M.G. and Daley, K.J. (2002) Cannibalism of a yearling polar bear (*Ursus maritimus*) at Churchill, Canada. *Arctic* 55(2), 190–192.

Eckhart, G. (2000) The effects of ecotourism on polar bear behavior. MSc thesis, Department of Biology in the College of Arts and Science at the University of Central Florida, Orlando, Florida.

Fagen, M.J. and Fagen, R. (1994) Interactions between wildlife viewers and habituated brown bears, 1987–1992. *Natural Areas Journal* 14, 159–164.

Folke, C., Hahn, T., Olsson, P. and Norberg, J. (2005) Adaptive governance of social ecological systems. *Annual Review Environmental Resources* 30, 441–473.

Freeman, M. (2003) The growth and significance of polar bear trophy hunting. Community-based sustainable development in the Canadian Arctic. *3rd International Wildlife Management Congress*, 1–5 December, Christchurch, New Zealand.

Herrero, S. and Fleck, S. (1990) Injury to people inflicted by black, grizzly, or polar bears: recent trends and new insights. *The International Conference on Bear Resources and Management* 8, 25–32.

Herrero, J. and Herrero, S. (1997) Visitor safety in polar bear viewing activities in the Churchill region of Manitoba, Canada. *Report for Manitoba Natural Resources and Parks Canada*, Calgary.

Herrero, J. and Herrero, S. (1999) Visitors and polar bears in Wapusk National Park: Planning for safety. *Report for Parks Canada – Wapusk National Park*, Calgary.

Higginbottom, K. (ed.) (2005) *Wildlife Tourism: Impacts, Management and Planning*. Sustainable Tourism Cooperative Research Centre (STCRC), Gold Coast, Queensland, Australia.

Honderich, J.E. (1991) Wildlife as a hazardous resource: An analysis of the historical interaction of humans and polar bears in the Canadian Arctic 2000 BC to AD 1935. MA thesis, University of Waterloo, Waterloo, Ontario.

Huntly, P.M., Van Noort, S. and Hamer, M. (2005) Giving increased value to invertebrates though ecotourism. *South African Journal of Wildlife Research* 35(1), 53–62.

Knight, R.L. and Meffe, G.K. (1997) Ecosystem management: agency liberation from command and control. *Wildlife Society Bulletin* 25, 676–678.

Kunkel, K.E. and Pletscher, D.H. (2000) Habitat factors affecting vulnerability of moose to predation by wolves in southeastern British Columbia. *Canadian Journal of Zoology* 78, 150–157.

Lemelin, R.H. (2005) Wildlife tourism at the edge of chaos: complex interactions between humans and polar bears in Churchill, Manitoba. In: Berkes, F., Huebert, R., Fast, H., Manseau, M. and Diduck, A. (eds) *Breaking Ice: Renewable Resource and Ocean Management in the Canadian North*. University of Calgary Press, Calgary, Alberta, Canada, pp. 183–202.

Lemelin, R.H. (2006a) The Gawk, the glance, and the gaze: occular consumption and polar bear tourism in Churchill, Manitoba Canada. *Current Issues in Tourism* 9(6), 516–534.

Lemelin, R.H. (2006b) Polar bear tourism in Churchill, Manitoba. In: Oakes, J. and Riewe, R. (eds) *Climate Change: Linking Traditional and Scientific Knowledge*. Aboriginal Issues Press, Winnipeg, Manitoba, pp. 235–244.

Lemelin, R.H. and Smale, B.J.A. (2006) Effects of environmental context on the experience of polar bear viewers in Churchill, Manitoba. *Journal of Ecotourism* 5(3), 176–191.

Lemelin, R.H., McCarville, R. and Smale, B.J.A. (2006) The effects of context on reports of fair price for wildlife viewing opportunities. *Journal of Park and Recreation Administration* 24(3), 50–71.

Linnell, J.D.C., Swenson, J.E., Andersen, R. and Barnes, B. (2000) How vulnerable are denning bears to disturbances? *Wildlife Society Bulletin* 28(2), 400–413.

Lunn, N.J., Atkinson, S., Branigan, M., Calvert, Clark, D., Doidge, B., Elliot, C., Nagy, J., Obbart, M., Otto, R., Stirling, I., Taylor, M., Vandal, D. and Wheatley, P. (2002) Polar bear management in Canada 1997–2000. In: Lunn, N.J., Schliebe, S. and Born, E.W. (eds) *Polar Bears: Proceedings of the 13th Working Meeting of the IUCN/SSC Polar Bear Specialist Group, 23–28 June 2001, Nuuk, Greenland*. Occasional Paper of the IUCN Species Survival Commission No. 26. IUCN – The World Conservation Union, Gland, Switzerland, pp. 41–52.

Lunn, N.J., Taylor, M., Calvert, P., Stirling, I., Obbard, M., Elliot, C., Lamontagne, G., Schaeffer, J., Atkinson, S., Clark, D., Bowden, E. and Doidge, B. (1998) Polar bear management in Canada 1993–1996. In: Derocher, A.E., Garner, G.W., Lynn, N.J. and Wiig, O. (eds) *Polar Bears: Proceedings of the 12th Working Meeting of the IUCN/SSC Polar Bear Specialist Group. 3–7 February, Oslo, Norway*. Occasional Papers of the IUCN Species Survival Commission, Eland, SW, pp. 51–66.

Manitoba Government. Legislative Electronic Publications (2003) Proposed legislation would protect Manitoba polar bears. Available at: http://www.govmb.ca/chc/press/top/2002/07/2002–07–03–02.html

Mason, P., Johnston, M. and Twynam, D. (2000) The World Wide Fund for Nature Arctic Tourism project. *Journal of Sustainable Tourism* 8(4), 305–323.

Muvla, C.D. (2001) Fair trade in tourism to protected areas: a micro case study of wildlife tourism to South Luangwa National Park, Zambia. *International Journal of Tourism Research* 3, 393–405.

National Park Services (1995) Final development concept plan environmental impact statement for Brooks River Area: Katmai National Park and Preserve, Alaska. United States Department of the Interior, Denver, Colorado.

Newsome, D., Dowling, R.K. and Moore, S.A. (eds) (2005) *Wildlife Tourism*. Multilingual Matters Ltd, Clevedon, UK.

Olsson, P. and Folke, C. (2001) Local ecological knowledge and institutional dynamics for ecosystem management: a study of Lake Racken watershed, Sweden. *Ecosystems* 4, 85–104.

Pimbert, M.P. and Gujja, B. (1997) Village voices challenging wetland management policies: experiences in participatory rural appraisal from India and Pakistan. *Nature and Resources* 33, 34–42.

Rabinowitz, A. (2005) Jaguars and livestock: living the world's third largest cat. In Woodroffe, R., Thirgood, S. and Rabinowitz, A. (eds) *People and Wildlife, Conflict or Co-existence?* Cambridge University Press, Cambridge, pp. 278–286.

Ramsay, M. and Stirling, I. (1990) Fidelity of female polar bears to winter-den sites. *Journal of Mammalogy* 71(2), 233–236.

Scheyvens, R. (2002) *Tourism for Development: Empowering Communities*. Prentice-Hall, London.

Schullery, P. (1992) *The Bears of Yellowstone*. High Plains Publishing, Worland, Wyoming.

Scott, P.A. and Stirling, I. (2002) Chronology of terrestrial den use by polar bears in western Hudson Bay as indicated by tree growth anomalies. *Arctic* 55(2), 151–166.

Shackley, M. (1996) *Wildlife Tourism*. International Thomson Business Press, Boston, Massachusetts.

Shackley, M. (1998) Designating a protected area at Karanambu ranch, Rupuni savannah, Guyana: resource management and indigenous communities. *Ambio* 27, 207–210.

Stewart, E.J., Draper, D. and Johnston, M.E. (2005) A review of tourism research in the polar regions. *Arctic* 58(4), 383–394.

Stirling, I. (1998) *Polar Bears*. Fitzhenry & Whiteside, Toronto.

Teillet, D.J. (1988) The Churchill Wildlife Management Area: Management Guidelines. Report produced for the Manitoba Department of Natural Resources.

Thirgood, S., Redpath, S., Newton, I. and Hudson, P. (2000) Raptors and red grouse: conservation conflicts and management solutions. *Conservation Biology* 14, 95–104.

Timothy, D.J. (1999) Participatory planning: a view of tourism in Indonesia. *Annals of Tourism Research* 26(2), 371–391.

Treves, A. and Karanth, K. (2003) Human–carnivore conflict and perspectives on carnivore management worldwide. *Conservation Biology* 17, 1491–99.

Treves-Naughton, L. and Treves, A. (2005) Socio-ecological factors shaping local support for wildlife crop raiding by elephant and other wildlife in Africa. In: Woodroffe, R., Thirgood, S., and Rabinowitz, A. (eds) *People and Wildlife, Conflict or Co-existence?* Cambridge University Press, Cambridge, pp. 252–277.

Watts, P.D. and Ratson, S.P. (1989) Tour operators avoidance of deterrent use and harassment of polar bears. *Proceeding of a Symposium on Bear–People Conflicts Management Strategies* 189–191.

Whittaker, D. (1997) Capacity norms on bear viewing platforms. *Human Dimensions of Wildlife* 2(2), 37–49.

Wondolleck, J.M. and Yaffee, S.L. (2000) *Making Collaboration Work: Lessons From Innovation in Natural Resource Management*. Island Press, Washington, DC.

6 Specialization of Whale Watchers in British Columbia Waters

C. Malcolm and D. Duffus

Introduction

In the early 1990s, authors like Duffus and Dearden (1990, 1992) and Forestell (1993) advocated the importance of combining social with ecological research in the development of whale-watching management, based on wildlife management principles espoused by Decker and Goff (1987). This call for interdisciplinary research has been echoed in the promotion of conceptual frameworks for general wildlife-viewing management that incorporate ecological and social science understanding (e.g. Hvenegaard, 1994; Orams, 1996; Davis et al., 1997; Reynolds and Braithwaite, 2001).

With respect to marine mammal viewing, neither the ecological nor social research, either alone or integrated, has provided significant input into management practice (Constantine, 1999). More to the point, little of the research has demonstrated the lasting social or conservation value that is espoused as the overarching *raison d'etre* for using wild marine mammals as ecotourism subjects (e.g. Forestell, 1993; Carlson, 1996; IFAW, 1997). Much of the natural science aspect of this equation has been burdened by poor research design, a serious lack of data and inconclusive results (Duffus and Dearden, 1993; Duffus and Baird, 1995; Trites and Bain, 2000), and only recently have some conclusive, and constructive studies surfaced (e.g. Williams et al., 2002, 2006; Lusseau, 2003a,b; Lusseau and Higham, 2004); however, little of this recent work has been adopted by management bodies as of yet (although the results of Williams et al. (2002) have been incorporated into whale-watching guidelines by whale-watching industries in British Columbia).

The social science aspects of the equation have not been as problematical. Recent human dimensions research in marine mammal viewing (primarily whale watching) includes: (i) general data collection on motivation, satisfaction and demographics; (ii) exploration of the psychological domain of what whales and whale watching mean to humans; and (iii) the use of education as a management tool (e.g. Amante-Helwig, 1996; Muloin, 1998; Orams and Hill, 1998; Giroul et al., 2000; Orams, 2000; Russell and Hodson, 2002; Parsons et al., 2003; Finkler and Higham, 2004). However, it is only recently that empirical research in this area has incorporated specific, management-oriented design and therefore little is known of the whale watchers themselves on a scale that is relevant for managers (Constantine, 1999; Orams, 2000).

Orams (1996) proposes education strategies to control interaction with marine wildlife through increasing contextual understanding that fosters behavioural change and achieves voluntary compliance with guidelines or regulations. To do so, an understanding of the wildlife viewers is required; this includes knowledge of participants' previous experiences, attitudes towards the wildlife resource and expectations for their viewing experience. Much of this requirement can be drawn together under the rubric of specialization.

Although there is considerable research pertaining to recreation specialization (primarily bird watching), the concept has not yet been applied to whale watchers. In this chapter we present the case for using specialization as a tool for the management of whale watching, particularly as part of an educational basis. We devise a specialization index specific to whale watchers and apply it to a data set derived from three well-developed centres of whale watching on Vancouver Island, British Columbia, Canada.

Specialization in Ecotourism

Bryan (1977) first defined recreation specialization as 'a continuum of behavior from the general to the particular, reflected by equipment and skills used in the sport, and activity setting preferences' (p. 175). Ecotourism specialization is therefore based on the idea that ecotourists are a heterogeneous assemblage and that subgroups, or segments, of participants may require distinct management techniques (Butler and Fenton, 1987; Duffus and Dearden, 1990; Hvenegaard, 2002), although few studies of specialization include the management consequent.

Specialization indexes measure participants on a scale constructed from variables such as prior experience, level of education and interest, type of equipment used, time and economic commitments, travel patterns, including tourism infrastructure desires, membership in organizations and centrality to the participants' lifestyles. Methods of creating specialization indexes vary. Techniques such as z-scores, cluster analysis, factor analysis or original designs, such as summed scoring for different levels of experience or education, have all been attempted (Wellman et al., 1982; Schreyer et al., 1984; Donnelly et al., 1986; Watson et al., 1991; Ditton et al., 1992; McFarlane, 1994, 1996; Cole and Scott, 1999; Hvenegaard, 2002; Scott and Thigpen, 2003; Lee and Scott, 2004; Scott et al., 2005). Most indexes are composed of a maximum of four groups.

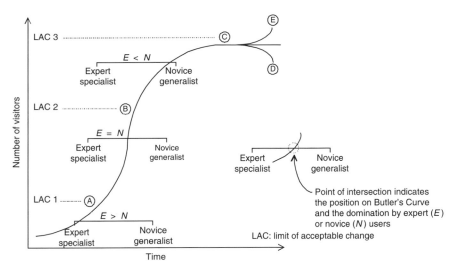

Fig. 6.1. User specialization and site evolution (Duffus and Dearden, 1990).

Duffus and Dearden (1990) suggest that specialization in wildlife viewing will dictate an ecotourist's site choice. They adapt Butler's (1980) tourism site life cycle to illustrate the evolution of the user alongside the site (Fig. 6.1). At the beginning, when few visitors utilize the site, the infrastructure is underdeveloped. At this point (Fig. 6.1: 'A'), the user group is dominated by the 'expert-specialist', who has prior experience in the activity, is knowledgeable, has realistic expectations and does not demand extensive infrastructure. As more tourists begin to use the site, additional infrastructure is established and the site eventually becomes dominated by the 'novice-generalist' (as the curve approaches 'C'). The novice-generalist is inexperienced, has little prior knowledge, general expectations and demands a high level of infrastructure. Limits of acceptable change (LACs) (Stankey *et al.*, 1985; Cole and Stankey, 1998) represent thresholds where more specialized users are no longer attracted to the site due to characteristics such as increased infrastructure development, perceived crowding with less-specialized users or decreased wildlife viewing potential, and search for experiences elsewhere.

This model also implies an increased impact upon resources as the site evolves through increased numbers of users and infrastructure development. If impact on the resource base causes its deterioration (e.g. wildlife populations move away, or die out), then the site may attract fewer users and struggle to maintain the level of infrastructure developed for many generalized users (point 'D'). Active management of the resource and the users is then needed to maintain the site at point 'C' or further increase the number of visitors ('E'). Education of the users will play a pivotal role in management of these problems, but the application of effective educational programming requires detailed knowledge of the special (or lack thereof) interests of the users.

Whale-watching Sites in British Columbia

The three principal whale-watching centres in British Columbia are at Johnstone Strait, Clayoquot and Barkley Sounds, and on the waters off Southern Vancouver Island, near Victoria (including the San Juan Islands, Washington State) (Fig. 6.2). Differences in geographical location (accessibility, travel time), tourism infrastructure and the cetacean species viewed result in three distinct whale-watching situations in British Columbia.

Whale watching in Johnstone Strait is focused on the 'northern resident' killer whales (*Orcinus orca*) that inhabit the waters surrounding the northern half of Vancouver Island to the southern Gulf of Alaska (Ford *et al.*, 2000; Baird, 2006). Travel to the area from larger population centres (which also act as tourist collector centres) requires a 450 km drive north from Victoria, or a

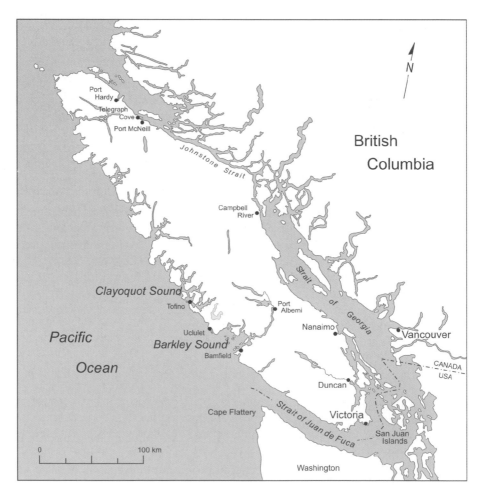

Fig. 6.2. Vancouver Island, British Columbia, including whale-watching centres.

1.5 h ferry trip and 350 km drive from Vancouver. There are few other tourist attractions (except limited sport fishing) in this area of Vancouver Island and consequently little tourist infrastructure.

Whale watching in Clayoquot and Barkley Sounds is focused on gray whales (*Eschrichtius robustus*), and more recently on humpback whales (*Megaptera novaeangliae*). During the summer months fluctuating numbers of gray whales (from 1 to 30) forage in the area (Duffus, 1996). Whale watching in this area originates in the towns of Tofino and Ucluelet, both a 300 km drive from Victoria, or a 1.5 h ferry trip and 200 km drive from Vancouver. In addition to whale watching, Pacific Rim National Park Reserve attracts visitors for camping, hiking, beach-combing and surfing. Clayoquot Sound is also a popular area for sport fishing and sea kayaking. Tourism infrastructure, based on the nature-oriented attractions of the area, includes some limited shopping, primarily focused on local British Columbia artwork, fine dining and various types of accommodations (resort hotels, motels, bed and breakfasts and camping).

Whale watching on the cross-border waters of southern Vancouver Island is focused on the 'southern resident' population of killer whales that frequent the waters surrounding southern Vancouver Island, including Puget Sound, Washington and unknown areas off southern British Columbia and Washington (Ford *et al.*, 2000; Baird, 2006). These whales are the focus of increasing scrutiny as their population declined for 6 consecutive years during 1990–2001 (Baird, 2006), while whale watching, both private and commercial, increased, along with concerns about the effects of over-fishing and toxin loading (Baird, 2001; Ross, 2006; Ross *et al.*, 2006). The southern residents were listed as 'threatened' in 1999, and upgraded to 'endangered' in 2001 by the Committee on the Status of Endangered Wildlife in Canada (COSEWIC, 1999, 2001). When the Canadian Species-at-Risk Act was passed in 2003, the southern residents were listed in Schedule 1 as 'endangered'. The population currently holds no conservation status in the USA, although a petition was filed in 2001 to list the southern residents under the US Endangered Species Act.

The combined human population of this area, which includes Victoria, Vancouver and Seattle, is close to 6 million, surrounding the core habitat of the southern residents. Whale-watching vessels depart from both Canadian and US ports, although the majority of whale-watching companies and vessels operate from Victoria, only a 1.5 h ferry trip from Vancouver or Seattle. The whale-watching industry increased in size every year in the 1990s, peaking at approximately 78 vessels (44 Canadian, 34 US) in 2001. The number of vessels in 2005 was approximately 73, 50 operated by 20 companies from British Columbia ports and 23 operated by 19 companies from Washington ports (Koski and Osborne, 2005). These waters are also popular for recreational boating. Private vessels are often drawn to killer whales by clusters of commercial whale-watching vessels, and their numbers have increased each year during the 1990s as well (Otis and Osborne, 2001). This area's tourism infrastructure is highly developed, with marine tourism activities composing only one segment of the tourism industry.

Investigating Specialization in BC Whale Watchers

Data collection

We designed a questionnaire, based on Dillman (1978), to collect data required to assess specialization of whale watchers in British Columbia. We collected data on: (i) previous whale watching and learning experience; (ii) 'attitude towards whale management'; (iii) 'general attitude towards the environment', using the 'New Environmental Paradigm' (Dunlap and van Lierre, 1978); and (iv) demographics.

We collected data from 1 June to 30 September 2002 and chose whale-watching charter companies in each location that possessed a large, constant flow of whale-watching passengers during the study period. The head naturalist with each company was responsible for administration and collection of the questionnaires and all naturalists were trained in survey administration procedures. Administration of the questionnaires was determined primarily by the operational logistics of the whale-watching companies; the procedure was undertaken to minimize disruption of company operation.

Questionnaires were handed out on randomly chosen days during the study period, with every whale-watching passenger a potential participant. We did not include participants who arrived on the dock less than 15 min prior to departure or passengers who did not have a sufficient ability to read and understand English. While this may have inserted a 'cultural bias' into the data, the naturalists were not able to provide the time needed to address language issues. The main component missing in this case is Asian tour groups, which are most numerous in Victoria. Naturalists were asked to estimate refusal rate by dividing the number of total passengers by the number of returned questionnaires on each sampling day, and then calculating an average refusal rate for the study period.

Creation of the whale-watcher specialization index

We created a specialization index based on prior experience, importance of participating in whale watching and education (Table 6.1), and identified three groups along the specialization continuum (Novice, Intermediate and Advanced), based on the principle of increasing ecotourism specialization described by Duffus and Dearden (1990). We then compared specialization scores between the three whale-watching locations.

To explore whether specialization had any bearing on the conservation attitudes of whale watchers in British Columbia, we classified the responses in Section 2, 'attitude towards whale management' and Section 3, 'general attitude towards the environment' (Appendix 6.1) by specialization group. We then looked for correlation between specialization groups and conservation attitudes by creating summary scores for both sections. To test for trends between specialization groups we calculated summary scores for 'attitude towards whale management' and 'general attitude toward the environment'. The summary scores were calculated as the mean of the sum score for all questions in the section for each case.

Table 6.1. Questions and values used to construct a whale-watcher specialization index in British Columbia.

Question	Level of experience/ specialization	Value in index
1. How many times seen whales in wild?		
Never	None	1
One time	Low	2
2–5 times	Moderate	3
6–10 times	High	4
More than 10 times	Very high	5
2. How many times on commercial whale-watching trip?		
Never	None	1
Once	Low	2
Twice	Moderate	3
Three times	High	4
More than three times	Very high	5
3. Priority of whale watching:		
Unplanned activity	Low	1
One of several activities	Medium	3
Main purpose of trip	High	5
4. Previous learning[a]:		
0–1 items	Very low	1
2–3 items	Low	2
4–5 items	Medium	3
6–7 items	High	4
8–9 items	Very high	5
Minimum specialization score	–	4
Maximum specialization score	–	20
Specialization index	Specialization group	
4–8	Novice	
9–14	Intermediate	
15–20	Advanced	

[a]Options available to the participants were: 'never', 'books', 'magazines', 'internet', 'television', 'educational videos', 'aquarium', 'museum' and 'other:___'.

Appendix 6.1: Questionnaire Addressing the Specialization of Whale Watchers.

Answers for both sections below were scored: strongly disagree = −2, slightly disagree = −1, no opinion = 0, slightly agree = 1, strongly agree = 2.

Questions in both sections were composed of both biocentric- and anthropocentric-oriented conservation attitudes. For example, an answer of 'strongly agree' for 'The number of whale watching boats around whales should be limited' and

'Commercial hunting of whales should be allowed' would both score 2. Therefore, the results for Questions 1 and 8 to 11, for 'attitude towards whale management', and 9 to 12, for 'general attitude towards the environment' (the anthropocentric-oriented questions) were reversed (i.e. −2 = 2, −1 = 1) for each case in order that all results conformed towards a common orientation.

Questions for 'attitude towards whale management'.

1. Paying for a whale-watching trip should guarantee I see whales.
2. The number of whale-watching boats around whales should be limited.
3. Boats should have to stay a minimum distance away from whales.
4. There should be some time set aside when whales get a break from whale watching.
5. There should be some areas set aside where whale watching is not allowed.
6. Whale populations that are endangered should be off limits to whale-watching boats.
7. A portion of the cost to go whale watching should go directly to whale research and management.
8. Commercial hunting of whales should be allowed.
9. First Nations peoples should be allowed to hunt whales for ceremonial purposes.
10. First Nations peoples should be allowed to hunt whales for commercial purposes.
11. Whale populations should be reduced when they compete with human food resources.
12. Protecting whales for future generations is important.
13. The government has an obligation to protect whales.

Questions for 'general attitude towards the environment', using the 'New Environmental Paradigm' (Dunlap and van Liere, 1978).

1. The balance of nature is very delicate and easily upset.
2. When humans interfere with nature it often produces disastrous results.
3. Humans must live in harmony with nature in order to survive.
4. Humans are severely abusing the environment.
5. We are approaching the limit to the number of people that the earth can support.
6. The earth is like a spaceship with only limited room and resources.
7. There are limits to growth beyond which our industrialized society cannot expand.
8. To maintain a healthy economy we will have to develop a 'steady state' economy where industrial growth is controlled.
9. Humans were created to rule over the rest of nature.
10. Humans have the right to modify the natural environment to suit their needs.
11. Plants and animals exist primarily to be used by humans.
12. Humans need not adapt to the environment because they can remake it to suit their needs.

Results of the study

We collected a total of 1617 surveys, 486 in Johnstone Strait, 528 in Clayoquot and Barkley Sounds and 603 in Victoria. Whale watchers who participated in the survey were 41.1% male and 57.1% female (3.4% of the questionnaires were answered as a couple). There was no significant difference in gender between locations (χ^2 = 3.317, df = 2, p = 0.190). The majority of respondents were spread equally between the ages of 30–60 (30–39 = 22%, 40–49 = 21.6%, 50–59 = 21%), and there was no significant difference in the age of participants between locations (χ^2 = 3.980, df = 2, p = 0.137).

The highest proportion of participants was Canadian (38.1%), followed by American (26.6%) and British (18.4%); participants came from 25 different countries altogether. There was no statistically significant difference in the origin of respondents between locations (χ^2 = 3.318, df = 2, p = 0.208). With respect to the highest level of education achieved, 28.1% of the respondents had completed a university degree, 20.1% a post-graduate degree, 16.8% some college or university, 13.7% a college diploma, 13.3% high school and 7.2% grade school. There was no statistically significant difference in education level between the three locations (χ^2 = 1.441, df = 2, p = 0.487). As no statistically significant differences existed between locations for the demographic data collected, we considered the participants to be representative of the same population.

Table 6.2 presents the results of the four questions used to create the specialization index, as well as the specialization scores for each location. Participants in Johnstone Strait scored the highest and those from Victoria the lowest, for all four of the index items. Overall, whale watchers in British Columbia in 2002 had a specialization score of 9.11 out of 20. Whale watchers in Johnstone Strait were the most specialized (10.68), followed by Clayoquot and Barkley Sounds (8.68) and Victoria (8.19). The scores are statistically significant between all three locations (F = 113.48, df = 2, p < 0.000). A reliability test on the index gives an alpha value of 0.66. A discriminant function analysis, based on a hypothesis of equal representation within each specialization group, indicates a 97% agreement with our index (Novice, 100%; Intermediate, 93.3%; Advanced, 99%).

For the three locations combined, 49.4% of the participants are classified as Novice, 44.3% as Intermediate and 6.1% as Advanced (Table 6.3), and the differences between the locations are statistically significant (χ^2 = 146.580, df = 2, p < 0.000). The greatest difference is between Johnstone Strait and Victoria (χ^2 = 173.293, df = 16, p < 0.000), followed by Johnstone Strait and Clayoquot and Barkley Sounds (χ^2= 111.352, df = 16, p < 0.000) and Clayoquot and Barkley Sounds and Victoria (χ^2 = 29.020, df = 15, p = 0.016).

Attitudes towards conservation issues were related to specialization of British Columbia whale watchers (Tables 6.4 and 6.5). For 'attitude towards whale management' there was a statistically significant increase in conservation orientation with increased specialization for 11 of the 13 statements. The Advanced group exhibited statistically significant increases in four of the 11 statements compared to the Intermediate group and for all 11 with respect to the Novice group. The Intermediate group displayed statistically significant increases in eight of the 11 statements to the Novice group.

Table 6.2. Calculation of specialization index, by location and combined.

Dependant variables	Mean score				Kruskal–Wallace (χ^2)	Chi-square (χ^2)		
	JS	CBS	VIC	Locations combined		JS/CBS	JS/VIC	CBS/VIC
How many times seen whales in wild?	2.22	1.73	1.66	1.85	80.963 $p < 0.000$	48.987 $p < 0.000$	73.134 $p < 0.000$	4.761 $p = 0.313$
How many times on commercial whale-watching trip?	1.98	1.53	1.46	1.64	62.665 $p < 0.000$	49.145 $p < 0.000$	57.909 $p < 0.000$	7.384 $p = 0.117$
Priority of whale watching	3.84	3.11	2.89	3.25	186.959 $p < 0.000$	104.790 $p < 0.000$	187.569 $p < 0.000$	13.016 $p = 0.001$
Previous learning	3.74	3.10	2.84	3.20	49.467 $p < 0.000$	27.964 $p = 0.001$	52.895 $p < 0.000$	14.234 $p = 0.114$
Specialization index	10.68	8.68	8.19	9.11	113.48[a] $p < 0.000$	<0.000[b]	<0.000[b]	0.003[b]

JS = Johnstone Strait, CBS = Clayoquot and Barkley Sounds, VIC = Victoria.
[a]Results from ANOVA.
[b]Results from Tamhane's Post-hoc Test.

For 'general attitude towards the environment', statistically significant increases in conservation orientation with increased specialization existed for 11 of the 12 statements. The Advanced group exhibited statistically significant increases for six of 12 statements with respect to the Intermediate group and 11 of the 12 compared to the Novice group. The Intermediate group displayed statistically significant increases in five of the 12 statements compared to the Novice group.

Table 6.3. Percentage of participants in each specialization group by location and combined.

Location	Percentage within specialization group		
	Novice	Intermediate	Advanced
Johnstone Strait	27.9	56.8	15.3
Clayoquot and Barkley Sounds	54.3	42.8	2.9
Victoria	70.0	35.5	1.6
Combined Locations	49.4	44.3	6.1

Table 6.4. Attitude towards whale management by specialization group.

Statement	Mean response			Kruskal–Wallace χ^2	χ^2		
	N	*I*	*A*		*N* vs *I*	*N* vs *A*	*I* vs *A*
Paying for a whale-watching trip should guarantee I see whales	0.23	0.01	−0.73	101.169 $p < 0.000$	10.116 $p = 0.039$	36.778 $p < 0.000$	24.261 $p < 0.000$
The number of whale-watching boats around whales should be limited	1.40	1.49	1.66	63.514 $p < 0.000$	10.319 $p = 0.035$	15.896 $p = 0.003$	9.272 $p = 0.055$
Boats should have to stay a minimum distance away from whales	1.37	1.61	1.82	40.078 $p < 0.000$	31.313 $p < 0.000$	25.293 $p < 0.000$	7.937 $p = 0.094$
There should be some time set aside when whales get a break from whale watching	1.11	1.25	1.46	17.002 $p < 0.000$	15.412 $p = 0.004$	21.409 $p < 0.000$	9.207 $p = 0.056$
There should be some areas set aside where whale watching is not allowed	1.25	1.45	1.68	69.931 $p < 0.000$	21.305 $p < 0.000$	21.980 $p < 0.000$	9.327 $p = 0.053$
Whale populations that are endangered should be off limits to whale-watching boats	1.03	1.13	1.03	8.511 $p = 0.014$	5.192 $p = 0.268$	7.511 $p = 0.111$	9.904 $p = 0.042$
A portion of the cost to go whale watching should go directly to whale research and management	1.46	1.50	1.61	14.322 $p = 0.001$	2.513 $p = 0.642$	4.998 $p = 0.287$	2.986 $p = 0.560$
Commercial hunting of whales should be allowed	−1.62	−1.68	−1.80	9.113 $p = 0.010$	2.431 $p = 0.657$	4.310 $p = 0.366$	2.485 $p = 0.647$

Continued

Table 6.4. *Continued*

| Statement | Mean response | | | Kruskal–Wallace χ^2 | χ^2 | | |
	N	I	A		N vs I	N vs A	I vs A
First Nations peoples should be allowed to hunt whales for ceremonial purposes	−0.86	−0.96	−1.18	5.651 $p = 0.059$	11.470 $p = 0.022$	10.017 $p = 0.040$	4.749 $p = 0.314$
First Nations peoples should be allowed to hunt whales for commercial purposes	−1.49	−1.58	−1.74	5.537 $p = 0.063$	5.423 $p = 0.247$	8.168 $p = 0.086$	6.074 $p = 0.194$
Whale populations should be reduced when they compete with human food resources	−1.05	−1.24	−1.74	57.348 $p < 0.000$	11.359 $p = 0.023$	33.599 $p < 0.000$	19.591 $p = 0.001$
Protecting whales for future generations is important	1.85	1.87	1.95	12.179 $p = 0.002$	4.459 $p = 0.347$	7.914 $p = 0.048$	5.230 $p = 0.264$
The government has an obligation to protect whales	1.63	1.65	1.88	17.341 $p < 0.000$	8.997 $p = 0.061$	13.645 $p = 0.009$	8.754 $p = 0.068$

N = Novice, I = Intermediate, A = Advanced.

The 'attitude towards whale management' and 'general attitude towards the environment' summary scores also support a general trend towards increasing conservation orientation with increased specialization: statistically significant differences existed between all specialization groups for both summary scores (Table 6.6). Correlation between the summary scores and specialization groups was also statistically significant, indicating an increase in conservation attitude with increasing specialization.

Application of Specialization Understanding to Management

Whale watchers in this study are composed of specialized subgroups and the three different sites examined received whale watchers in different

Table 6.5. General attitude towards the environment (NEP) by specialization group.

Statement	Statement response mean			Kruskal–Wallace χ^2	Chi-square (χ^2)		
	N	I	A		N vs I	N vs A	I vs A
The balance of nature is very delicate and easily upset	1.34	1.49	1.67	14.890 $p = 0.001$	14.394 $p = 0.006$	20.325 $p < 0.000$	9.483 $p = 0.050$
When humans interfere with nature it often produces disastrous results	1.25	1.36	1.51	7.799 $p = 0.020$	9.618 $p = 0.047$	11.981 $p = 0.017$	9.158 $p = 0.057$
Humans must live in harmony with nature in order to survive	1.60	1.67	1.82	4.593 $p = 0.101$	6.718 $p = 0.152$	9.522 $p = 0.049$	6.890 $p = 0.142$
Humans are severely abusing the environment	1.34	1.38	1.64	10.436 $p = 0.005$	8.213 $p = 0.084$	14.795 $p = 0.005$	8.795 $p = 0.066$
We are approaching the limit to the number of people that the earth can support	0.73	0.80	1.10	11.385 $p = 0.003$	8.413 $p = 0.078$	11.214 $p = 0.024$	5.648 $p = 0.227$
The earth is like a spaceship with only limited room and resources	1.01	1.07	1.34	17.282 $p < 0.000$	3.669 $p = 0.453$	11.051 $p = 0.026$	6.808 $p = 0.146$
There are limits to growth beyond which our industrialized society cannot expand	0.93	0.98	1.34	14.409 $p = 0.001$	7.888 $p = 0.096$	17.919 $p = 0.001$	12.646 $p = 0.013$
To maintain a healthy economy we will have to develop a 'steady state' economy where industrial growth is controlled	0.95	1.02	1.16	19.581 $p < 0.000$	9.844 $p = 0.043$	6.892 $p = 0.142$	4.494 $p = 0.293$
Humans were created to rule over the rest of nature	−1.03	−1.22	−1.59	43.691 $p < 0.000$	12.835 $p = 0.012$	26.398 $p < 0.000$	14.196 $p = 0.007$

Continued

Table 6.5. *Continued*

Statement	Statement response mean			Kruskal–Wallace χ^2	Chi-square (χ^2)		
	N	I	A		N vs I	N vs A	I vs A
Humans have the right to modify the natural environment to suit their needs	−0.92	−1.07	−1.46	39.479 $p < 0.000$	8.726 $p = 0.068$	26.786 $p < 0.000$	14.405 $p = 0.006$
Plants and animals exist primarily to be used by humans	−1.24	−1.38	−1.72	45.688 $p < 0.000$	11.638 $p = 0.020$	22.805 $p < 0.000$	11.688 $p = 0.020$
Humans need not adapt to the environment because they can remake it to suit their needs	−1.26	−1.36	−1.67	25.793 $p < 0.000$	7.291 $p = 0.121$	16.819 $p = 0.002$	10.278 $p = 0.036$

N = Novice, I = Intermediate, A = Advanced.

Table 6.6. ANOVA and correlation of attitude towards whale management (ATWM) and general attitude towards the environment (NEP) scores by specialization group.

	Scores by group	ANOVA	Tamhane's T2 Post-hoc Test			Spearman's rho correlation
			N vs I	N vs A	I vs A	
Attitude towards whale management summary score (max. score = 26)	N = 15.13 I = 16.98 A = 19.34	F = 26.099 $p < 0.000$	0.000	0.000	0.002	0.189 $p < 0.000$
General attitude towards the environment (NEP) summary score (max. score = 24)	N = 12.46 I = 14.15 A = 17.16	F = 19.142 $p < 0.000$	0.000	0.000	0.001	0.152 $p < 0.000$

N = Novice, I = Intermediate, A = Advanced.

proportions from these groups. Overall, lesser-specialized, generalist participants dominate whale watching in British Columbia. Novice and Intermediate specialization groups accounted for 93.7% of British Columbia whale watchers, resulting in an overall specialization score of 9.11 out of 20. Similar to other studies, such as Muloin (1998), Neil *et al.* (1996), Tourism Queensland (1999) and Finkler (2001), over 60% of respondents in this study were first-time participants on a commercial whale-watching trip.

An understanding of specialization in British Columbia whale watchers is important in order to identify education requirements for management. The components used to construct the specialization index in this study (Tables 6.1 and 6.2) illustrate important differences: 25% more participants at Johnstone Strait had previously seen whales in the wild than at Victoria, and 20% more than at Clayoquot and Barkley Sounds. Similarly, 20% more respondents at Johnstone Strait had previously been commercial whale watching than in Victoria, and 17% more than in Clayoquot and Barkley Sounds.

Education is also important factor: while there was no significant difference between locations for highest education level achieved, there was a statistically significant difference in previous learning about whales between all three locations. Johnstone Strait participants were the most learned about whales, followed by those at Clayoquot and Barkley Sounds, and then Victoria. Finally, while 48.5% of Johnstone Strait whale watchers travelled there specifically to view whales, only 18.1% did so to Clayoquot and Barkley Sounds, and 10.8% to Victoria.

A greater number of highly specialized whale watchers are therefore attracted to Johnstone Strait, followed by Clayoquot and Barkley Sounds, then Victoria, where there are relatively few whales (Table 6.3). Duffus and Dearden's (1990) discussion of specialization in wildlife ecotourists, along with the principle of distance decay (Hägerstrand, 1957), explains this trend. The required travelling distance to Johnstone Strait, and its low level of tourism infrastructure and tourism activities, is not acceptable for many tourists. Unless whale watching is important enough to warrant a dedicated trip, most tourists will not whale watch in Johnstone Strait. Due to the dominance of the unspecialized whale watcher, for whom whale watching is unplanned or simply one of numerous planned activities, many would-be whale watchers will not expend the time or money to go whale watching in Johnstone Strait.

Duffus and Dearden's (1990) application of LAC to wildlife viewing also explains this trend. For some whale watchers the acceptable limits of industry size (crowding, i.e. number of whale-watching vessels), and proximity to urban areas with highly developed tourism infrastructure are exceeded in Victoria. More specialized whale watchers are therefore pushed away from Victoria and pulled to Clayoquot and Barkley Sounds, and particularly Johnstone Strait, where the whale-watching experience occurs in a less urban, less crowded and more wilderness-oriented environment. These push and pull factors are indicative of the behavioural model of recreation participation (Manning, 1999), which reflect a participant's motivations and expectations for engaging in an activity.

Although all three specialization groups possess conservation-oriented attitudes, more specialized whale watchers tend to hold those attitudes more strongly. This trend is evident with respect to whale-watching management issues (Table 6.4): responses to statements such as 'The number of whale watching boats around whales should be limited', 'Boats should have to stay a minimum distance away from whales', 'There should be some time set aside when whales get a break from whale watching' and 'There should be some areas set aside where whale watching is not allowed' reveal significant statistical increases in agreement along the continuum from Novice to Advanced whale watchers. Increased specialization also leads to a statistically significant decrease in agreement with the statement 'Paying for a whale watching trip should guarantee I see whales', which may indicate a more realistic understanding of the whale-watching experience, including the unpredictability of wildlife.

Correlations between specialization and the 'attitude towards whale management' and 'general attitude towards the environment' summary scores (Table 6.6) reveal that there are statistically significant differences between all specialization groups and reinforce the theory that more specialized whale watchers possess stronger environmental attitudes. Currently, whale-watcher specialization in British Columbia indicates a generalist-dominated activity that requires basic whale and marine ecology education programmes. However, specialized whale watchers who possess more conservation-oriented attitudes and greater understanding of the whale-watching experience are currently attracted in greater proportions to Clayoquot and Barkley Sounds, and particularly Johnstone Strait, than to Victoria. A programme that addresses the ecological and social aspects of whale watching and marine conservation, such as scientific uncertainty, the precautionary approach and self-regulation, may be appropriate in Johnstone Strait. In Clayoquot and Barkley Sounds, and particularly Victoria, careful attention to the development of education programmes that provide a basic ecological context upon which to build high-quality marine conservation education that can affect change in environmental attitudes, and help create more specialized whale watchers, is warranted.

If whale watching is to be a vehicle for conservation education, it is the less specialized whale watchers upon whom education should be concentrated. While more research needs to be done to determine the specific learning desires of less specialized whale watchers, the development of basic environmental education, linked to the whale-watching experience, should be examined. Other research supports this need. Lück (2003) found that whale watchers in New Zealand indicated a desire for structured on-board education programmes. Neil *et al.* (1996) reported that those who had been whale watching previously answered general cetacean ecology questions well (although with high variability), but answered questions about cetacean management very poorly. Russell and Hodson (2002) suggest that interpretation programmes aboard whale-watching vessels in Quebec may be weak, their results indicating low satisfaction with on-board education programmes. These issues should be investigated to further identify priorities for whale-watching education programmes. Satisfaction studies classified by specialization should also be considered, although controlling for the quality of education programmes aboard a variety of whale-watching vessels and at different sites is currently problematic.

The application of a monitored education programme (e.g. Orams, 1997) may serve to improve the delivery of interpretation during whale watching. Such a programme requires knowledge of what whale watchers would specifically like to know, followed by development of an education programme that applies this knowledge, and then examination of whether the information was received. To be effective, monitoring whether the information is received should be both short and long term. In the long term, the goal would be to create more specialized whale watchers.

Conclusion

Whale watching is a recreational activity that requires no special skills or equipment, no need to be in good physical condition and no prior education. Only the economic means to pay for whale-watching fares is required. It is therefore a recreational wildlife viewing activity that is available to most tourists, particularly, as it becomes more available in large coastal tourist destinations around the world.

The results of this study indicate that education programmes for whale watchers should be developed based primarily on the characteristics of the tourists, and secondarily on the species, habitat and conservation issues at a given site. Degree of specialization within the tourist-base should dictate the level at which education is delivered, whether basic ecology of cetaceans and introduction to marine conservation for generalist participants, or applied conservation concepts to local issues for highly specialized whale watchers. In order to determine where to aim education programmes, specialization research is required at each whale-watching site. For sites where a wide range of specialization is observed, whale-watching interpreters may need to develop a method of assessing each group, and possess a series of interpretive programmes that can be delivered along the specialization continuum.

The LAC concept is a useful tool for education managers to understand and observe trends in whale-watcher specialization as sites develop. The dominance of the unspecialized whale watcher observed in British Columbia may be similar in locations where whale watching is a highly developed aspect of a large tourism industry, such as southern California, Hawaii, New England, eastern Australia and New Zealand. Greater proportions of specialist whale watchers may occur in more remote locations, where ecotourism is the sole tourism attraction, such as sites in the Canadian subarctic and Arctic, the Karnali River Basin in Nepal,[1] and Midway Island, USA. Areas where whale watching remains extremely undeveloped, such as the Solomon Islands, Eastern Russia and coastal African countries, may also attract more specialized whale watchers.[2] As these areas become more developed, incorporating more varied tourist attractions and greater infrastructure, the proportion of less specialized whale watchers will increase, along with required attention to education programmes.

---

[1] The Ganges River dolphin (*Platanista gangetica*) has historically inhabited the Ganges, Meghna, Brahmputra and Karnaphuli river systems of India, Bangladesh and Nepal (Jones, 1982; Reeves and Brownell; 1989), with primary and secondary habitats in the Karnali River (Smith, 1993). Hoyt (2001) reports that 150+ whale watchers visited the area in 1998 and there is considerable potential for dolphin-watching ecotourism. For a recent review of the status and conservation of the Ganges River dolphin in the Karnali River Basin, see World Wildlife Fund (2006).

[2] For the most recent global statistics on whale watching worldwide, see Hoyt (2001).

Acknowledgements

The authors would like to acknowledge and thank three whale-watching companies on Vancouver Island for their volunteer participation in this study: at Stubbs Island Charters in Telegraph Cove, owner Jim Borrowman, head naturalist Jackie Hildering, and former naturalist Nik Dedaluk; at Jamie's Whaling Station in Tofino, owner Jamie Bray and former naturalist Jen Jackson; and at Springtide Charters in Victoria, owner Dan Kukat and former head naturalist Chelsea Garside. Thanks also to the whale watchers who participated in the study by taking the time to fill out questionnaires. University of Victoria student volunteers Leah Gabriel, Chelsea Garside and Kim Pearce entered all the survey data.

References

Amante-Helwig, V. (1996) Ecotourist's beliefs and knowledge about dolphins and the development of cetacean tourism. *Aquatic Mammals* 22, 131–140.

Baird, R.W. (2001) Status of killer whales, *Orcinus orca*, in Canada. *Canadian Field-Naturalist* 115, 676–701.

Baird, R.W. (2006) *Killer Whales of the World: Natural History and Conservation.* Voyageur Press, St Paul, Minnesota.

Bryan, H. (1977) Leisure value systems and recreational specialization: the case of trout fishermen. *Journal of Leisure Research* 9, 174–187.

Butler, R.W. (1980) The concept of a tourist area cycle of evolution: implications for management of resources. *Canadian Geographer* 24, 5–12.

Butler, J.R. and Fenton, G.D. (1987) Bird watchers of Point Pelee National Park, Canada: their characteristics and activities with special consideration to their social and resource impacts. *Alberta Naturalist* 17, 135–146.

Carlson, C. (1996) Whalewatching and its effects on the whales. *Whalewatcher* 30, 8–10.

Cole, D.N. and Stankey, G.H. (1998) Historical development of limits of acceptable change: conceptual clarifications and possible extensions. In: McCool, S.F. and Cole, D.N. (eds) *Proceedings – Limits of Acceptable Change and Related Planning Processes: Progress and Future Directions, May 20–22,1997, Missoula, MT.* US Department of Agriculture, Forest Service, Rocky Mountain Research Station, General Technical Report INT-GTR-371, Ogden, Utah.

Cole, J.S. and Scott, D. (1999) Segmenting participants in wildlife watching: a comparison of casual wildlife watchers and serious birders. *Human Dimensions of Wildlife* 4, 44–61.

Committee on the Status of Endangered Wildlife in Canada (COSEWIC) (1999) *Status assessment report, April 1999.* COSEWIC, Ottawa, Ontario, Canada.

Committee on the Status of Endangered Wildlife in Canada (COSEWIC) (2001) *Status assessment report, November 2001.* COSEWIC, Ottawa, Ontario, Canada.

Constantine, R. (1999) *Effects of Tourism on Marine Mammals in New Zealand.* Science for Conservation 106, Department of Conservation, Wellington, New Zealand.

Davis, D., Banks, S., Birtles, A., Valentine, P. and Cuthill, M. (1997) Whale sharks in ningaloo marine park: managing tourism in an Australian Marine Protected Area. *Tourism Management* 18, 258–271.

Decker, D.J. and Goff, G.R. (1987) *Valuing Wildlife: Economic and Social Perspectives.* Westview Press, Boulder, Colorado.

Dillman, D.A. (1978) *Mail and Telephone Surveys: The Total Design Method.* Wiley, New York.

Ditton, R.B., Loomis, D.K. and Choi, S. (1992) Recreation specialization: re-conceptualization from a social worlds perspective. *Journal of Leisure Research* 24, 33–51.

Donnelly, M.P., Vaske, J.J. and Graefe, A.R. (1986) Degree and range of specialization: toward a typology of boating related activities. *Journal of Leisure Research* 18, 81–95.

Duffus, D.A. (1996) The recreational use of grey whales in southern Clayoquot Sound, Canada. *Applied Geography* 16(3), 179–190.

Duffus, D.A. and Baird, R.W. (1995) Killer whales, whalewatching and management: a status report. *Whalewatcher* (Fall/Winter), 14–16.

Duffus, D.A. and Dearden, P. (1990) Non-consumptive wildlife-oriented recreation: a conceptual framework. *Biological Conservation* 53, 213–231.

Duffus, D.A. and Dearden, P. (1992) Whales, science and protected area management in British Columbia, Canada. *The George Wright Forum* 9, 79–87.

Duffus, D.A. and Dearden, P. (1993) Recreational use, valuation and management of killer whales (Orcinus orca) on Canada's Pacific Coast. *Environmental Conservation* 20(2), 149–156.

Dunlap, R.E. and Van Lierre, K. (1978) The new environmental paradigm: A proposed measuring instrument and preliminary results. *Journal of Environmental Education* 9, 10–19.

Finkler, W. (2001) The experiential impact of whale watching: implications for management in the case of the San Juan Islands, USA. MSc thesis, Department of Marine Science, University of Otago, Dunedin, New Zealand.

Finkler, W. and Higham, J. (2004) The human dimensions of whale-watching: an analysis based on viewing platforms. *Human Dimensions of Wildlife* 9, 103–117.

Ford, J.K.B., Ellis, G.M. and Balcomb, K.C. (2000) *Killer Whales*, 2nd edn. UBC Press, Vancouver, British Columbia.

Forestell, P.H. (1993) If Leviathan has a face, does Gaia have a soul? Incorporating environmental education in marine eco-tourism programs. *Ocean and Coastal Management* 20, 267–282.

Giroul, C., Ouellet, G. and Soubrier, R. (2000) *Etude des attentes et de la satisfactions de la clientele des croisieres aux baleines dans le secteur du parc marine du Saguenay-Saint-Laurent*. Departement des Sciences du Loisir et de la Communication Sociale, University du Quebec a Trois-Rivieres.

Hägerstrand, T. (1957) Migration and area. In: Hanenberg, D., Hägerstrand, T. and Odeving, B. (eds) *Migration in Sweden: A Symposium*. Lund Studies in Geography, Series B, Human Geography 13, C.W.K. Gleerup, Lund.

Hoyt, E. (2001) *Whale Watching 2001: Worldwide Tourism Numbers, Expenditures, and Expanding Socio-economic Benefits*. International Fund for Animal Welfare, Yarmouth Port, Massachusetts.

Hvenegaard, G.T. (1994) Ecotourism: a status report and conceptual framework. *The Journal of Tourism Studies* 5, 24–35.

Hvenegaard, G.T. (2002) Birder specialization differences in conservation involvement, demographics, and motivations. *Human Dimensions of Wildlife* 7, 21–36.

International Fund for Animal Welfare (IFAW) (1997) *Report of the International Workshop on the Educational Values of Whale Watching*, 8–11 May, Provincetown, Massachusettes.

Jones, S. (1982) *The Present Status of Gangetic susu*, Platanista gangetica *(Roxburgh) with Comments on the Indus susu*, P. minor *(Owen)*. FAO Advisory Committee on Marine Resources Research. Working party on Marine Mammals. FAO Fish. Series 4, pp. 97–115.

Koski, K.L. and Osborne, R.W. (2005) *The Evolution of Adaptive Management Practices for Vessel-based Wildlife Viewing in the Boundary Waters of British Columbia and Washington State: From Voluntary Guidelines to Regulations*. Presented at the Puget Sound Georgia Basin Research Conference. Available at: www.whale-museum.org/programs/soundwatch/soundwatch.html

Lee, J.-H. and Scott, D. (2004) Measuring birding specialization: a confirmatory factor analysis. *Leisure Sciences* 26, 245–260.

Lück, M. (2003) Education on marine mammal tours as agent for conservation—but do tourists want to be educated? *Ocean and Coastal Management* 46, 943–956.

Lusseau, D. (2003a) The effects of tour boats on the behavior of bottlenose dolphins: using Markov chains to model anthropogenic impacts. *Conservation Biology* 17, 1785–1793.

Lusseau, D. (2003b) Male and female bottlenose dolphins *Tursiops* spp. have different strategies to avoid interactions with tour boats in Doubtful Sound, New Zealand. *Marine Ecology Progress Series* 257, 267–274.

Lusseau, D. and Higham, J.E.S. (2004) Managing the impacts of dolphin-based tourism through the definition of critical habitats: The case of bottlenose dolphins (*Tursiops* spp.) in Doubtful Sound, New Zealand. *Tourism Management* 25, 657–667.

Manning, R.E. (1999) *Studies in Outdoor Recreation: Search and Research for Satisfaction*. Oregon State University Press, Corvallis, Oregon.

McFarlane, B. (1994) Specialization and motivations of birdwatchers. *Wildlife Society Bulletin* 22, 361–370.

McFarlane, B. (1996) Socialization influences of specialization among birdwatchers. *Human Dimensions of Wildlife* 1, 35–50.

Muloin, S. (1998) Wildlife tourism; the psychological benefits of whale-watching. *Pacific Tourism Review* 2, 199–213.

Neil, D., Orams, M. and Baglioni, A. (1996) Effect of previous whale-watching experience on participants knowledge of, and response to, whales and whale-watching. In: Colgan, K., Prasser, S. and Jeffery, A. (eds) *Encounters with Whales: 1995 Proceedings*, Australian Nature Conservation Agency, Canberra, Australia.

Orams, M. (1996) A conceptual model of tourist–wildlife interaction; the case for education as a management strategy. *Australian Geography* 27, 38–51.

Orams, M. (1997) The effectiveness of environmental education: can we turn tourists into 'greenies'? *Progress in Tourism and Hospitality Research* 3, 295–306.

Orams, M. (2000) Tourists getting close to whales, is it what whale watching is all about? *Tourism Management* 21, 561–569.

Orams, M.B. and Hill, G.E. (1998) Controlling the ecotourist in a wild dolphin feeding program: is education the answer? *The Journal of Environmental Education* 29, 33–38.

Otis, R.E. and Osborne, R.W. (2001) *Historical Trends in Vessel-based Killer Whale Watching in Haro Strait Along the Boundary of British Columbia and Washington State: 1976–2001*. Poster presented to the Society for Marine Mammalogy Conference, Vancouver, British Columbia.

Parsons, E.C.M., Warburton, C.A., Woods-Ballard, A., Hughs, A., Johnston, P., Bates, H. and Lück, M. (2003) Whale-watching tourists in West Scotland. *Journal of Ecotourism* 2, 93–113.

Reeves, R.R. and Brownell, R.L. (1989) *Susu-Platanista gangetica* and *Platanista minor.* In: Ridgway, S.H., and Harrison, S.R. (eds) *Handbook of Marine Mammals*, Vol 4: *River Dolphins and the Larger Toothed Whales*. Academic Press, London, pp. 69–100.

Reynolds, P.C. and Braithwaite, D. (2001) Towards a conceptual framework for wildlife tourism. *Tourism Management* 22, 31–42.

Ross, P.S. (2006) Fireproof killer whales (*Orcinus orca*): flame retardant chemicals and the conservation imperative in the charismatic icon of British Columbia, Canada. *Canadian Journal of Aquatic Sciences* 63, 224–234.

Ross, P.S., Ellis, G.M., Ikonomou, M.G., Barrett-Lennard, L.G. and Addison, R.F. (2000) High PCB concentrations in free-ranging Pacific killer whales, *Orcinus orca*: effects of age, sex and dietary preference. *Marine Pollution Bulletin* 40, 504–515.

Russell, C.L. and Hodson, D. (2002) Whale-watching as critical science education? *Canadian Journal of Science, Mathematics and Technology Education* 2, 485–504.

Schreyer, R., Lime, D.W. and Williams, D.R. (1984) Characterizing the influence of past experience on recreation behavior. *Journal of Leisure Research* 16, 34–50.

Scott, J. and Thigpen, J. (2003) Understanding the birder as tourist: segmenting visitors to the Texas Hummer/Bird Celebration. *Human Dimensions of Wildlife* 8, 199–218.

Scott, D., Ditton, R.B., Stoll, J.R. and Eubanks, T.L. Jr (2005) Measuring specialization among birders: utility of a self-classification measure. *Human Dimensions of Wildlife* 10, 53–74.

Smith, B. (1993) 1990 Status and conservation of the Ganges River dolphin (*Platanista gangetica*) in Karnali River, Nepal. *Biological Conservation* 66, 159–170.

Stankey, G.H., Cole, D.N., Lucas, R.C., Petersen, M.E. and Frissell, S.S. (1985) The Limits of Acceptable Change (LAC) System for Wilderness Planning. US Department of Agriculture, Forest Service, Intermountain Forest and Range Experiment Station, General Technical Report INT-176, Ogden, Utah.

Tourism Queensland (1999) *1998 Whale-watching Survey: Research Findings*. Available at: www.tq.com.au/qep/research/whalrsch/intro.htm

Trites, A.W. and Bain, D.E. (2000) *Short- and Long-term Effects of Whale Watching on Killer Whales* (Orcinus orca) *in British Columbia*. Paper presented to the IWC Workshop on the Long-Term Effects of Whale Watching. Adelaide, Australia. Available at: http://faculty.washington.edu/dbain/whalewatch.PDF

Watson, A., Roggenbuck, J. and Williams, D. (1991) The influence of past experience on wilderness choice. *Journal of Leisure Research* 23, 21–36.

Wellman, J., Roggenbuck, J. and Smith, A. (1982) Recreation specialization and the norms of depreciative behavior among canoeists. *Journal of Leisure Research* 14, 323–240.

Williams, R., Bain, D.E., Ford, J.K.B. and Trites, A.W. (2002) Behavioural responses of male killer whales to a 'leapfrogging' vessel. *Journal of Cetacean Research and Management* 4, 305–310.

Williams, R., Lusseau, D. and Hammond, P.S. (2006) Estimating relative energetic costs of human disturbance to killer whales (*Orcinus orca*). *Biological Conservation* 133, 301–311.

World Wildlife Fund (2006) *Status, Distribution and Conservation Threats of Ganges River Dolphin in Karnali River, Nepal*. WWF Nepal Program, Kathmandu. Available at: http://assets.panda.org/downloads/dolphin20report24may06.pdf

7 Captive Marine Wildlife: Benefits and Costs of Aquaria and Marine Parks

M. Lück

Introduction

The keeping of fish is an ancient pastime. Initially kept for food, fish were later kept in ponds for aesthetic reasons in places such as Rome, Sumeria, Greece and China (Bailey and Dakin, 1998; Markwell, in press). About 2000 years ago, goldfish and carp were bred in China and Japan, but it was not until the mid-1800s that aquaria, as we know them today, became common in Europe's households (Markwell, in press). In particular, R. Harrington was credited for a boost in the interest among the wider public, when he presented a paper to the Chemical Society in London in 1850, explaining how he was able to maintain a stable aquarium (Sandford, 2004). Shortly after Harrington's presentation, zoological organizations and entrepreneurs started building public aquaria in Europe and North America. For example, the London Zoological Society opened the first public aquarium in 1853, with a second aquarium opening in the Surrey Zoological Gardens shortly after (Sandford, 2004). Due to the easy access, most private aquarists initially kept native freshwater species. People in coastal areas tried to keep marine species as well, but the complexity of the marine ecosystems made this much more difficult (Markwell, in press; Sandford, 2004). Despite initial problems with keeping fish, plants and other creatures sustainably, aquaria became a fashionable feature in Victorian households. Ready-made tanks, as we know them today, were not on the market yet, but custom-made tanks were very popular (Sandford, 2004). The influence of the rapid advances of technology on marine tourism and recreation has been significant over the past decades (Orams, 1999). With better techniques for manufacturing plate glass, as well as the introduction of filter systems and heaters, the aquarium hobby has grown into the multi-million dollar trade of contemporary times. Equally, public aquaria benefited from technological improvements, and could keep fish in better conditions, for longer periods of time.

Closely related to the development of aquaria is the history of marine parks. Also called oceanaria, they are 'essentially theme parks with a marine focus, or aquaria with amusement park add-ons' (Rennie, in press). The history of marine parks goes back to the mid-1850s, when P.T. Barnum opened what was advertised as 'the first public aquaria in America' in the American Museum in New York City. For this, he bought a set of glass aquaria from the Royal Zoological Society in London, and brought those to New York (Betts, 1959). He also acquired the Aquarial Gardens in Boston, where he advertised 'The Whale Harnessed and Driven Around the Great Tank by a Young Lady'. For his aquaria, Barnum captured two beluga whales (*Delphinapterus leucas*) and brought them to New York City for display. They died only days later, because they were kept in a freshwater tank. Two other belugas survived somewhat longer in a second attempt with salt water (Whale and Dolphin Conservation Society, 2002). In 1913, the New York Museum captured and displayed five bottlenose dolphins (*Tursiops truncatus*). However, those dolphins perished soon as well, with the last one dying after only 21 months in captivity. In order to improve his marine exhibitions, Barnum acquired the rights to pump seawater directly from the sea, and was thus able to display living sharks, porpoises, sea horses and other fish (Betts, 1959). These '"Aquaria", including tanks and forty large cases made of marble, iron and glass, [were] said to excel those in London, Paris and Dublin' (Betts, 1959, p. 357).

Barnum was well known for his entrepreneurship, especially in the show world. He had previously acquired various museums and displayed a number of exotic exhibits, including elephants he had shipped from Ceylon (now Sri Lanka) and a 2000-year-old mummy, in a 'Great Asiatic Caravan, Museum and Menagerie' (Betts, 1959). This mix of curiosities, entertainment and display of exotic animals was the stepping stone for the development of marine parks. Today, these parks vary and might include a number of different features, such as amusement rides, live performances (both by humans and by animals), scientific research programmes, rescue and rehabilitation facilities for marine wildlife, as well as education programmes (Rennie, in press).

Today, marine parks are usually holding mammals, such as dolphins, whales and pinnipeds. They also have an important entertainment function, including shows with animals, and other rides and attractions. Aquaria are commonly smaller, and put significantly less focus on entertainment. They are often affiliated with universities, or not-for-profit research foundations, hold their own scientific libraries and have active research staff. Since they keep mostly smaller species, they are more likely to be able to keep them in an environment that is as close as possible to their natural environment. Of course, this is a black and white illustration, and the lines are not as clear-cut as it might appear.

It is very difficult to estimate the total number of aquaria and marine parks worldwide. According to McCormick (1993) there are more than 70 aquaria and marine parks in the USA and Canada alone, many of them newly built over the last two decades (e.g. the US$40 million Aquarium of the Americas in New Orleans). Aquae.com lists more than 200 aquaria around the globe (Aquae.com, 2006). Jiang *et al.* (in press) suggest that there are currently 14 parks worldwide that keep orcas. In addition, there are numerous aquaria and marine parks in many other countries, varying from very small facilities, such as the aquarium at Portobello, New Zealand, to large ocean parks, e.g. the Ocean Park in Hong Kong. There are also many aquaria in doctor's offices, restaurants, company receptions, etc., where people do not go specifically to watch fish, but they are in a commercial setting nevertheless. With this plethora of facilities offering a range of marine wildlife-viewing opportunities, concern has been growing about a number of issues, such as the ethics of keeping wild-life in captivity, the welfare of the animals and the potential educational benefits from the facilities (Hoyt, 1992; Williams, 2001; Rose *et al.*, 2006). Equally, there have been very positive views about the mandate of such facilities, and related benefits for conservation, rehabilitation, research and education (Hoyt, 1992; Alliance of Marine Mammal Parks and Aquariums, 2000). This chapter will address the most prominent of these contrasting viewpoints.

The Benefits and Costs of Aquaria and Marine Parks

It is evident that there is great public demand for aquaria and marine parks (Markwell, in press). Aquaria and marine parks have the following main func-tions: Entertainment and recreation, education, conservation and research (McCormick, 1993; Benbow, 2004). However, the lasting popularity of aquaria and marine parks (Markwell, in press; McCormick, 1993) brought with it a number of concerns. Considerable attention has been paid to investigating issues, such as the ethics of keeping marine mammals in captivity, welfare of captive marine mammals and the educational and conservational abilities of aquaria and marine parks (Hoyt, 1992; Williams, 2001; Rose *et al.*, 2006). Marine parks holding larger marine mammals are of particular concern to ani-mal welfare organizations and researchers (Jiang *et al.*, in press). The following section discusses the main benefits and costs attributed to aquaria and marine parks.

Jobs and economic benefits

Many of today's aquaria and marine parks are multi-million dollar businesses, particularly marine parks, with sophisticated entertainment regimes. The Vancouver Aquarium, for example, indicated total revenues of almost CAN$15 million in 2002 (US$11.2 million) (Vancouver Aquarium and Marine Science Centre, 2002). The theme parks of Anheuser Busch (including Sea World in San Diego, San Antonio and Orlando, Discovery Cove in Orlando, Busch Gardens in Tampa Bay and Williamsburg, as well as Water Country USA, Sesame Place and Adventure Island) generated a growth of 8% in 2002, despite the general decline in the tourism industry after the events of 11 September 2001 (Anheuser Busch, 2003). Aquaria and marine parks also create a large number of jobs, from occasional student jobs to highly paid scientific person-nel. While these are significant financial benefits, Whitehead (1990, p. 59) believes that 'we should not engage in activities which cause animals suffering and may harm natural populations simply to provide jobs and profits'. The issues around suffering of animals will be discussed below.

Education

Probably the most commonly cited justification for the existence of aquaria and marine parks today is education. In fact, most zoos, aquaria and marine parks define themselves as educational institutions (Falk and Adelman, 2003). In their annual report to the Congress, the Alliance of Marine Mammal Parks and Aquariums (2000, p. 2) states that the 'mission of educational exhibits and programming at Alliance member facilities is to enhance appreciation for and understanding of marine mammals and their ecosystems'. They list the following as goals of their education programmes (p. 2):

- Provide opportunities for visitors to expand their knowledge about marine mammal biology and natural history;
- Promote awareness of and sensitivity towards the marine environment and the relationships of marine biology and natural environments;
- Present information on marine conservation issues;
- Be available as marine science and environmental information resources to interested citizens, local schools, community groups and educators;
- Inspire visitors to embrace conservation behaviour.

In a study at various aquaria in the UK, however, Evans (1997, p. 239) contends that 'the aquaria managers did not believe that most visitors were interested in receiving educational information, in particular on conservation topics'. In contrast to this view, Evans found that the majority of visitors did indeed demand aquaria to improve their interpretation, with particular focus on conservation of the marine environment. In the debate about the educational value of aquaria and marine parks, there appears to be a distinct difference between those facilities which hold large marine mammals, and those that do not. While many researchers believe that aquaria can indeed fulfill an important educational role (Evans, 1997; Falk and Adelman, 2003; Markwell, in press), there is strong opposition to those facilities holding dolphins or whales. Along with criticism around the suitability of the facilities and programmes, opponents claim that there is little educational value in seeing large marine mammals in captivity. According to critics, the education at these facilities perpetuates an unrealistic view of animals 'enjoying' jumping through hoops, pulling their trainers through the water and performing all sorts of tricks, and is mainly based on an anthropogenic worldview (Whitehead, 1990; Williams, 2001; Rose *et al.*, 2006). Markwell (in press) contends that 'instead of providing visitors with a high quality educational experience, these critics argue that they instead entertain the crowds using dolphins and killer whales jumping through flaming hoops and balancing balls on their snouts'.

In addition, opponents claim that some of the information conveyed by the facilities is deliberately misleading, and occasionally even false (Hoyt,

1992). For example, scientific research suggests that the lifespan of male orcas in the wild averages 29.2 years, with a maximum of 50–60 years. Female orcas have an average lifespan of 50.2 years, with a maximum of 80–90 years (Williams, 2001). However, in an 'Educational Manual' of 1991, Marineland Niagara Falls states that 'killer whales may live for up to 50 years'. In a similar manual printed four years later, Marineland stated 'it is believed that the killer whale may live up to 35 years' (quoted in Williams, 2001, p. 51). The Alliance of Marine Mammal Parks and Aquariums (1999, p. 3) refers to a study showing that 'the average age of dolphins in marine life parks, aquariums, and zoos is similar to that of dolphins in their natural environment'. These statements suggest that, following the increasing pressure from environmental organizations and researchers, Marineland and other parks attempted to show themselves in a more positive light by reducing the lifespan of orcas in the wild.

Conservation

The mandate of conservation is often closely linked to education. In fact, many aquaria and marine parks indicate that they actively contribute to conservation through interpretation and education (see, e.g. the goals of the Alliance of Marine Mammal Parks and Aquariums in the previous section). The National Aquarium in Baltimore, for example, stated that conservation is a major reason for the industry's existence (Williams, 2001). Many aquaria, and few marine parks, are indeed emphasizing the importance of conservation and attempt to actively contribute to it. Especially, when it comes to endangered species, aquaria can be a safe haven for these animals. Captive breeding programmes, along with the reintroduction into the wild, can greatly benefit the well-being of both the species in question and the ecological balance in their natural habitat. Most aquaria have marine scientists on staff, who often lend their expertise when there is need, for example, in cases of whale strandings, or for ill or distressed wild marine animals (Vancouver Aquarium and Marine Science Centre, 2002; Sandford, 2004). Aquaria often make major efforts to rehabilitate stranded or injured marine animals.

Their facilities and expertise may help rescue these animals, and – in some cases – reintroduce them into the wild (Whitehead, 1990). Critics contend that due to a lack of systematic sustainable captive-breeding successes, very few animals in aquaria and marine parks are born and bred in captivity. The majority of displayed animals are caught in the wild, which is counteracting the goal of conservation. Being captured is not only highly stressful for these animals, but also potentially harmful for the other animals in the natural habitat. For example, the capture of one or more large marine mammals (dolphins or whales) can potentially leave a significant gap in the family of these highly social animals. Breeding success can be diminished, if important males or females are removed from the group (Whitehead, 1990; Hoyt, 1992; Williams, 2001; Markwell, in press). Capture procedures for cetaceans are often stressful and cruel (Hoyt, 1992; Jones, 2003). In particular, the practice of fisheries is used for the live capture of dolphins in Japan. With this method, large pods of dolphins are driven into shallow waters, where they are either selected for sale to aquaria and marine parks, or butchered and sold as meat (Jones, 2003).

Another goal of aquaria and marine parks is the conservation of endangered species (Benbow, 2004). Although this might be the case for many smaller species at the aquaria, there is concern that, in the case of cetaceans, this is just another marketing tool for marine parks to put themselves in a better light. In fact, most of the cetaceans, such as bottlenose dolphins (*T. truncatus*) and orcas (*Orcinus orca*), held in marine parks are not endangered species at all. Critics also contend that if a marine park has the conservation of the animals in mind, there is no need for ambitious training programmes and elaborate shows for pure entertainment purposes (Williams, 2001).

The unsustainable collecting practices used for capturing smaller fish are another concern. Especially in some developing countries, the use of dynamite to stun fish is as common as the use of cyanide to temporarily paralyse animals for easy collection for both public and home aquaria (Markwell, in press). The use of dynamite does not only stress the fish to be caught, but also has the potential to destroy large areas of the very sensitive coral reefs. About 85% of the international trade for aquarium fish originates in the Indo-Pacific region (Pratt, 1998). According to Pratt (1998), more than 1 million kg of cyanide have been used in the coral reefs in the Philippines to stun and capture ornamental aquarium fish destined for the pet shops and aquaria of Europe and North America since the 1960s. The practice of fishing with cyanide is also particularly destructive to the coral reef ecosystems in which it is practised. A large proportion (around 75%) of the fish caught dies within 48 h, which means that in order to compensate for this loss a much larger number of fish are being taken than actually needed. In addition, cyanide kills many fish, as well as coral and reef invertebrates, that are not targeted by the industry (Pratt, 1998). These practices are not only in contrast to the conservation efforts of the species collected for aquaria and marine parks, but also to the conservation of their habitat, including non-target species, and the ecological balance. Kolm and Berglund (2003) report a move away from cyanide and dynamite fishing at least in some parts of the world, and the use of the less destructive band fishing. However, their research suggests that these so-called non-destructive practices have significant effects on the populations of reef fish and sea urchins (Kolm and Berglund, 2003).

There are efforts to control and restrict the international trade of endangered species. Most developed countries have signed the Convention on International Trade in Endangered Species (CITES). However, in disaster-struck developing countries, the convention has little effect. The Solomon Islands in the South Pacific, for example, were hit hard by an ongoing low-level civil war and by the cyclone Zoe in 2002. A Western company in the Solomons had offered fishermen US$260 per healthy dolphin, which was shortly followed by offers as high as US$30,000 per dolphin by business people in Thailand and Taiwan (Ecott, 2003). Subsequently, more than 100 dolphins were caught and transferred to holding pens in the first half of the year 2002.

Research

Keeping fish in public aquaria and marine parks contributes to scientific research into the marine environment. Research is related to the study of animal and plant biology, ecology, reproduction, feeding and behaviour (Anon., 1995). Producing external publications and member newsletters, developing curricula and promoting advocacy are seen as additional research outputs of aquaria (McCormick, 1993). Indeed, many aquaria and marine parks allocate significant amounts of money to research, and support scientists with access and facilities. The Vancouver Aquarium and Marine Science Centre, for example, allocated almost CAN$100,000 to research in 2002, which equals 6% of their overall expenditures (Vancouver Aquarium and Marine Science Centre, 2002). Sea World does not have a large budget for scientific research itself, but they provide the funding for a research foundation. The Hubbs-Sea World Research Institute was established by the founders of Sea World, and is based at the Sea World park in San Diego, California (Hoyt, 1992). Hoyt (1992) acknowledges that orca studies in captivity have contributed important research work, including studies on reproductive research, hearing studies, underwater sound and vocalization studies, studies on tooth growth and genetic studies. The Alliance of Marine Mammal Parks and Aquariums (2000) underlines this, stating that research at member facilities helped understand the anatomy and the physiology of marine mammals, and treating sick and injured mammals from the wild. The development of vaccines and treatment methods, techniques for anesthesia and surgery, as well as tests on toxic substances all resulted from studies at marine parks and aquaria (Alliance of Marine Mammal Parks and Aquariums, 2000).

However, opponents are concerned that the artificial environment in marine parks puts marine mammals under stress, and thus their behaviour is very different from that in the wild. Williams (2001, p. 62) quotes the late Jacques Cousteau, saying that:

> [N]o aquarium, no tank or marineland, however spacious it may be, can begin to duplicate the conditions of sea. And no dolphin who inhabits one of those aquariums . . . can be described as a 'normal' dolphin. Therefore the conclusion drawn by observing the behaviour of such dolphins are often misleading when applied to dolphins as a whole.

Although acknowledging that at least some of the aquaria and marine parks are serious about research, they also claim that at least some of the research is driven by the agenda of the parks. For example, distribution studies of orca populations in the wild have been funded by Sea World at a time when they were planning to acquire new animals for their parks (Williams, 2001). Whitehead (1990) differentiates between whales and dolphins in captivity. While he sees little to no scientific significance of whales in captivity, he states that 'Dr Herman's work on captive bottlenose dolphins at the University of Hawai'i is amongst the most important marine mammal research being carried out' (p. 63). Since it is generally easier to replicate the natural environment for smaller species, aquaria are more successful in doing so. Research results tend to be of more validity in such settings.

Therapy programmes

Marine mammals, in particular dolphins, are increasingly used in therapy programmes (Whale and Dolphin Conservation Society, 1999). These programmes are far beyond the common educational programmes or education through interpretation during regular visits to aquaria or marine parks, and are designed for therapy for children with disabilities. Since the 1970s, several programmes have been developed in order to help autistic and mentally disabled children (Cochrane and Callen, 1998; Scope, 2002). Dolphin Assisted Therapy (DAT) is used to 'reward the individual for participating in the more traditional helping approaches and therapies, such as physio- and speech therapy' (Scope, 2002, p. 2). A variety of therapy centres have been established in different parts of the world, such as Florida, Hawaii, Switzerland, Israel and many more. The therapy is mostly used to work with children who suffer from cerebral palsy, or with autistic children (Dolphins Plus, 2002; Scope, 2002; The Alexander Trust, 2002). Most of the therapy programmes involve working with captive dolphins, which is a major concern and point of criticism (Cochrane and Callen, 1998). Environmentalists claim that, for example, a therapy project at an aquarium in Nürnberg, Germany, is just a PR project in order to justify the keeping of captured dolphins in Nürnberg's zoo (Lakotta, 2000). The Whale and Dolphin Conservation Society (WDCS) is reviewing details of the DAT programmes due to a growing concern for the well-being of both humans and dolphins. An increasing number of aggressive behaviours of dolphins in captivity, including biting and butting, has been recorded in North America (Whale and Dolphin Conservation Society, 1999).

Ethical concerns

In this volume, Garrod notes that despite the rapid growth of wildlife tourism worldwide, the ethical issues around human–wildlife interactions implicit in such activities have rarely troubled academics (Garrod, Chapter 14, this volume). A notable exception is a section on the rights of animals in Fennell's (2006) 'Tourism Ethics'. Not surprisingly, however, the ethical concerns about wildlife viewing in general have been one of the main concerns of animal rights and protection groups. Fennell (2006) reviews literature on animal rights and ethics in general, and concludes that there is a wide range of schools of thought. For example, Feinberg (1985, in Fennell, 2006) suggests that because animals have no duties like humans do, they in turn are not moral agents, and thus cannot have any rights either. Fennell goes on to explain that it is important to understand that animals do not have the ability to communicate pain and suffering. However, he points out that there is increasing evidence that animals do indeed feel pain, and communicate it in similar ways as humans do, through crying out, attempts to escape the situation and examining the affected part. Thus, it is today accepted that animals do indeed have rights, which is underlined in various animal protection laws and regulations in many countries. The problem is that animals cannot claim these rights, and thus are dependent on humans to do this on their behalf (Fennell, 2006). Fennell (2006, p. 184) concludes that 'we have the duty not only to these animal populations, but also to human future generations to conserve the biodiversity on the basis of rights'.

It does not take much to imagine that holding a 9 m orca in a small, barren concrete tank inevitably causes some sort of suffering for the animal. No tank can be large enough to create living conditions that are similar to those in the ocean. Orcas travel an average of 150 km a day (Berghan, 1998), live in highly social groups (pods) and hunt cooperatively for food. No captive orca has access to any of these features. Thus, it is easy to understand why opponents of keeping orcas, or other cetaceans for that matter, claim that this is unethical. Williams (2001), for example, questions whether humans morally have the right to remove animals from their natural habitat and confine them, even if there are educational, entertainment and financial benefits.

The American Association of Zoological Parks and Aquariums' (AAZPA) Ethics and Law Working Group stated that the display industry acknowledges a 'moral obligation' to show 'compassion and humane treatment to animals in captivity' (in Hoyt, 1992, p. 77). The interesting part of this statement is that keeping animals in captivity in general is not morally wrong. This illustrates the problem at hand: as mentioned above, it is easy to understand why it could be morally wrong to keep large cetaceans in aquaria and marine parks, but what about smaller species, such as small, fish, sea horses, coral and many more. If an aquarium is able to replicate the natural environment to a large extent, i.e. to create a complete biosphere, are the animals in this biosphere suffering, and thus, would this be morally wrong? It is in the very nature of human existence that we all have different moral standards and thresholds. What is acceptable for some might be unacceptable for others. The many factors influencing these variations in standard (culture, upbringing, education, etc.) will not be discussed in this chapter. It shall just be highlighted that, depending on the moral standards of individuals or societies, aquaria and marine parks do have a moral right to exist – or not.

Conclusion

Overall, keeping any animal in captivity is a contentious issue itself, no matter how big or small the animals or the facilities are, and how well they are treated. Our morals and ethics are deeply grounded in our upbringing, culture, education and other determinants (Fennell, 2006), and thus each individual has his or her own view about this issue.

Most marine parks and aquaria define themselves as educational institutions, and claim that they help elevate the profile of marine conservation through public education (Evans, 1997; Falk and Adelman, 2003). However, do visitors indeed learn during their visits, and does this learning contribute to their environmental dispositions? Falk and Adelman (2003) suggest that this depends widely on the type of visitor, how much prior knowledge they have and how interested they are. For example, in a study at the National Aquarium in Baltimore, they found that regardless of entering knowledge, only individuals with moderate or high interest showed significant gains in knowledge. They conclude that the majority of visitors indeed possess low to moderate knowledge, and high to extensive interest, and thus are the main beneficiaries of museum experiences (Falk and Adelman, 2003). In line with tourists on swim-with-wild-dolphins tours (see Chapter 18, this volume), Evans (1997) found that visitors to aquaria also desired an increase in the levels of interpretation. In particular, again in line with the study on swim-with-wild-dolphins tours, they asked for more information about wider marine-related issues, such as information about conservation and how they could contribute to the conservation of the marine environment (Evans, 1997).

Many children observed at Marineland, Canada, were very enthusiastic about seeing orcas and beluga whales through the large underground viewing panes (see Fig. 7.1). Most likely, most of these children (and adults!) would never see a living orca or beluga whale in their lives if it was not for Marineland.

Many reasons lead to this conclusion, such as a lack of interest, the families probably not holidaying at a location where whale watching is possible, the often relatively high price for whale-watch tours, to name but a few. However, does this justify us keeping these large creatures in tanks that are barren, far too small and as close to their natural environment as a prison cell would be to our regular living spaces? Do we really have to see every animal in real life, or is it alright to 'just' watch them on TV, in cinemas, museums, etc.? Or maybe we will have to look at entirely different concepts, such as the recent attraction opened at the Detroit Zoo. Visitors have the opportunity to dive into the ocean depth in a Wild Adventure Simulator Deep Sea ride. Based on NASA's flight simulator technology, the attraction uses 70 mm film, elaborate sound systems and almost 2 g of motion (Travelwire News, 2004). Visitors go on a multi-sensory journey that allows them to experience an environment that would otherwise be inaccessible to them.

Fig. 7.1. Orca at Marineland, Niagara Falls, Canada. (Photograph M. Lück.)

I conclude this chapter with an astonishing statement by William Donaldson, president of the Zoological Society in Philadelphia: 'the studies we have conducted . . . show that the overwhelming majority of our visitors leave us without increasing either their knowledge of the natural world or their empathy for it. There are even times when I wonder if we don't make things worse by reinforcing the idea that man is only an observer in nature and not a part of it' (quoted in Williams, 2001, p. 53).

References

Alliance of Marine Mammal Parks and Aquariums (1999) *Fact Book on Marine Mammals*. Alliance of Marine Mammal Parks and Aquariums, Alexandria, Virginia.

Alliance of Marine Mammal Parks and Aquariums (2000) *Annual Report to Congress 2000*. Alliance of Marine Mammal Parks and Aquariums, Alexandria, Virginia.

Anheuser Busch (2003) *Anheuser Busch Companies Annual Report 2002*. Anheuser Busch, St Louis, Missouri.

Anon. (1995) *The Complete Aquarium Guide: Fish, Plants and Accessories for Your Aquarium*. Könemann Verlagsgesellschaft, Cologne, Germany.

Aquae.com (2006) *Public Aquariums* [webpage]. Available at: http://www.aquae.com/Aquariums_Ultra_Quick.html

Bailey, M. and Dakin, N. (1998) *The Aquarium Fish Handbook*. Caxton, London.

Benbow, S.M.P. (2004) Death and dying at the zoo. *The Journal of Popular Culture* 37(3), 379–398.

Berghan, J. (1998) *Marine Mammals of Northland*. Department of Conservation, Wellington, New Zealand.

Betts, J.R. (1959) P.T. Barnum and the popularization of natural history. *Journal of the History of Ideas* 20(3), 353–368.

Cochrane, A. and Callen, K. (1998) *Beyond The Blue: Dolphins and Their Healing Powers*. Bloomsbury Publishing, London.

Dolphins Plus (2002) *Dolphins Plus Key Largo, FL: Therapy Programs* [webpage]. Island Dolphin Care Inc. Available at: http://www.pennekamp.com/dolphins-plus/therapy.htm

Ecott, T. (2003) Dolphins are victims in an island war. *Conde Nast Traveller* (November), 33–34.

Evans, K.L. (1997) Aquaria and marine environmental education. *Aquarium Sciences and Conservation* 1(4), 239–250.

Falk, J.F. and Adelman, L.M. (2003) Investigating the impact of prior knowledge and interest on aquarium visitor learning. *Journal of Research in Science Teaching* 40(2), 163–176.

Fennell, D.A. (2006) *Tourism Ethics*. Channel View Publications, Clevedon, UK.

Hoyt, E. (1992) *The Performing Orca – Why the Show Must Stop*. The Whale and Dolphin Conservation Society, Bath, UK.

Jiang, Y., Lück, M. and Parsons, E.C.M. (in press) Public awareness and marine mammals in captivity. *Tourism Review International, Special Issue Zoos, Aquaria, and Other Captive Wildlife Attractions*.

Jones, H. (2003) Dolphin massacre. *Faze*, 22–23.

Kolm, N. and Berglund, A. (2003) Wild populations of a Reef fish suffer from the nondestructive aquarium trade fishery. *Conservation Biology* 17(3), 910–914.

Lakotta, B. (2000) Doktor Flipper. *Der Spiegel*, 192–195.

Markwell, K. (in press) Aquaria. In: Lück, M. (ed.) *Encyclopedia of Tourism and Recreation in Marine Environments*. CAB International, Wallingford, UK.

McCormick, D. (1993) The age of aquariums. *Sea Frontiers* 39(2), 23–38.

Orams, M.B. (1999) *Marine Tourism: Development, Impacts and Management.* Routledge, London/New York.

Pratt, V.A. (1998) *Poison and Profits: Cyanide Fishing in the Indo-Pacific* [webpage]. Encyclopedia.com. Available at: http://www.encyclopedia.com/doc/1G1-21222051.html

Rennie, H. (in press) Marine parks (Oceanaria). In: Lück, M. (ed.) *Encyclopedia of Tourism and Recreation in Marine Environments.* CAB International, Wallingford, UK.

Rose, N.A., Farinato, R. and Sherwin, S. (2006) *The Case Against Marine Mammals in Captivity,* 3rd edn. The Humane Society of the United States, and World Society for the Protection of Animals, Washington, DC and London.

Sandford, G. (2004) *Aquarium Owner's Manual.* Dorling Kindersley, London.

Scope (2002) *Dolphin Therapy* [webpage]. Scope Organization. Available at: http://www.scope.org.uk/cgi-bin/eatsoup.cgi?id = 6127

The Alexander Trust (2002) *The Alexander Trust: Supporting Special Needs Children and Their Families* [webpage]. The Alexander Trust. Available at: http://www.execulink.com/~road/trustmain.html

Travelwire News (2004) *The Adventures of the Deep Sea Come to the Detroit Zoo* [webpage]. Travelwire News. Available at: http://www.travelwirenews.com/cgi-script/csArticles/articles/000002/000285.htm

Vancouver Aquarium and Marine Science Centre (2002) *2002 Annual Report.* Vancouver Aquarium and Marine Science Centre, Vancouver, British Columbia.

Whale and Dolphin Conservation Society (1999) Dolphin interaction programmes. *Whale and Dolphin Conservation Society Magazine* (October), 10.

Whale and Dolphin Conservation Society (2002) *History of Captivity* [webpage]. Whale and Dolphin Conservation Society. Available at: http://www.wdcs.org/dan/publishing nsf/allweb/281C1D97F10E9573802568DD00306F2C

Whitehead, H. (1990) *The Value of Oceanaria.* Paper presented at the Whales In Captivity. Right or Wrong? Symposium, Ottawa.

Williams, V. (2001) *Captive Orcas Dying to Entertain You: The Full Story.* WDCS, Bath, UK.

II The Impacts of Tourist Interactions with Marine Wildlife

8 The Economic Impacts of Marine Wildlife Tourism

C. CATER AND E. CATER

Economic Impacts and Values

It is important to establish from the outset that the economic impact of marine wildlife tourism is only part of a much bigger picture in terms of its overall economic value. Wells (1997) distinguishes between the economic impact of nature tourism, which he defines as the amount of money spent by nature tourists in the economy on travel, accommodation, food, souvenirs, etc., and the total economic value, which covers 'the broader economic benefits of conservation which can be associated with a nature tourism destination. Direct use by tourists is only one of the economic values which flow from nature tourism destinations' (Wells, 1997, p. 9). Furthermore, the direct effects are only one of the three classes of multiplier effects in the economy: the other two being indirect effects arising from establishments which receive the tourist expenditure purchasing goods and services from other sectors within the local economy; and induced effects which occur from local residents spending their wages, salaries, distributed profit, rent and interest on goods and services in the local economy (Cooper et al., 1998). These positive multiplier effects are, however, limited by leakages which reduce the positive economic impacts of tourism.

However, these positive impacts from economic multipliers are only a partial reflection of the total economic value of nature tourism because there are also significant non-use values to add into the equation. These values include the *existence value* which is the amount individuals would be prepared to pay to know that the area or species continues to exist (Tisdell, 2003), described by Emerton and Tessema (2001, p. 5) as the intrinsic value to people 'regardless of the direct and indirect benefits they gain from it including cultural, scientific, aesthetic, heritage and bequest significance', and the *option value* which is 'the amount that individuals would be prepared to pay to safeguard an asset for the option of using it at a future date' (Wells, 1997, p. 21). Future possible tourism uses may include some which are not known at present, but is evident when we consider that; for example, the proliferation of scuba-diving across the globe could not have been envisaged early last century. Despite the fact that the difference between the amount that individuals would be willing to pay and the amount that they actually pay constitutes foregone income to the destination (therefore known as 'consumer surplus'), decision makers are less interested in these non-use values because of the difficulty in capturing or using them in practical terms (Wells, 1997). Pendleton and Rooke (2006) suggest that non-market use values for a scuba-diving or snorkelling day in warmer waters ranges from US$3 to US$199 per day for snorkelling and US$31 to US$319 per day for scuba-diving, with the consumer surplus for non-residents generally exceeding that for residents. They cite the work of Leeworthy *et al.* in Florida who estimated diving consumer surplus to be in the range of US$3 for residents and from US$8 to US$16 for non-residents.

Attempts to derive estimates of this missing potential revenue have used willingness to pay (WTP) or contingent valuation methodology to justify the introduction of, or increase in, user fees at varying marine tourism locations across the globe. One of the earliest such exercises was conducted by Dixon *et al.* (1993) who surveyed diver attitudes to the introduction of user fees in the Caribbean island of Bonaire. Mathieu *et al.* (2003) used contingent valuation methodology in a survey to estimate tourists' WTP on three different islands in the Seychelles in 1998. While they found that WTP values were higher than the current Rs 50 per visit to the Marine National Parks, and that, therefore, entrance fees could be increased without a concomitant reduction in visitation, they caution that not only might their respondents have had many different motivations, 'acting as both consumers and citizens in stating their preferences' (p. 386) (the latter, in particular, referring to their attitudes to existence value), but also that they were only sampling people who had sufficient income to travel to the Seychelles in the first place.

In this chapter we examine the complexity of economic impacts of marine wildlife tourism, particularly on the communities that host this activity. While we have recognized that the economic impact of marine wildlife tourism is only a partial reflection of its true economic value we inevitably have to focus on the former because of the above considerations. We show how these impacts are frequently uneven in distribution, often a reflection of power differentials. Some examples of more equitable marine tourism development are examined, in addition to the funding regimes for the marine protected areas (MPAs) that are discussed elsewhere in this book. An assessment of net economic impacts is arrived at through a consideration of the economic costs as well as benefits of marine wildlife tourism.

Economic Costs

In the same way that direct and indirect effects can be discerned in the economic benefits of marine wildlife tourism so, too, are they manifest in the costs of establishing and maintaining nature tourism destinations and attractions. The direct costs are those involved in 'the purchase of land, preparation of management plans, capital expenditures, development and maintenance of roads and facilities, and all recurrent management and administration costs' (Wells, 1997, p. 21). They thus involve direct physical expenditures on equipment, infrastructure and human resources (Emerton and Tessema, 2001). Because marine wildlife tourism takes place in an environment in which humans do not live, and consequently in which they are dependent on equipment to survive (Orams, 1999), it is likely that expenditure on specialist equipment and watercraft, in some cases hugely capital-intensive glass-bottomed boats, semi-submersibles and tourist submarines, will considerably augment leakages. The very considerable capital costs of entry (e.g. a minimum of US$4.5 million for a tourist submarine), coupled with stringent maintenance and safety requirements, put this form of entrepreneurship way beyond the realms of truly local involvement. Even the cost of conventional craft may be prohibitive; Warburton (1999) describes how in the cetacean-viewing industry of the island of Mull, in the West of Scotland, the cost of a boat precludes many from entering the sector.

The indirect costs concern the negative impacts which arise, such as property damage or personal injuries caused by wildlife. While these are perhaps less evident than in terrestrial locations where crop destruction and predation of livestock on the margins of National Parks has been widely documented (see, e.g. Newmark *et al.*, 1994), recent concern about shark-cage diving leading to aggressive behaviour by sharks is the evidence of such indirect costs. Frequently, a procedure known as chumming, usually using a 'soup' made of blood and fish scraps, is used to entice sharks, in order to facilitate close-quarter encounters with tourists lowered in heavy-duty shark cages. Such shark-cage diving experiences are on offer at a number of locations across the globe, and critics have attributed a number of recent shark attacks, for example, in the Western Cape of South Africa, to a Pavlovian response whereby sharks associate humans with food, although studies examining this potential correlation have proved inconclusive. While operators defend their activities by claiming that they fulfil an educational purpose, the ethics of disturbing the natural balance and of conditioning behaviour must be under scrutiny, and several locations across the globe such as Florida, Hawaii, the Cayman Islands and the Maldives have placed bans on the feeding of sharks in the wild. Another example of indirect costs arising from tourist use of marine resources is that divers may cause damage to fishing equipment. At Apo Island in the Philippines, local fishers reported damage to fishing traps, also claiming that fish had been driven away from fishing grounds (Raymundo, 2002). Similarly, in Bonaire, the Netherlands Antilles, Dixon *et al.* (1993) reported anecdotal information that fish traps set by local fishermen became detached from their moorings due to diving activity. In terms of costs we also need to consider how marine wildlife tourism infringes on maintaining the habitat necessary to maintain populations as well as any negative effects on the species due to viewing. Duffus and Dearden (1993, p. 151), for example, examine the 'harassment issue' with regard to whale watching (see Constantine and Bejder, Chapter 17, this volume).

A third category of economic costs is that of *opportunity costs* which cover the value of the benefits foregone as a result of the utilization of the location and its resources for marine wildlife tourism. This is a particularly contentious issue in relation to the designation of MPAs where, although tourists benefit from enhanced viewing prospects, 'there is a high opportunity cost to conservation for many MPAs . . . in terms of resource utilisation activities foregone or precluded' (Emerton and Tessema, 2001, p. 2) for local people. This, of course, applies particularly to fishing, and all around the world there are examples of resistance by local fishers to the designation of MPAs, in particular no-take areas, on the grounds of impacts on local livelihood. This move has been resented and resisted by fisherfolk, dependent on fishing for sustenance, livelihood and recreation in many locations around the world. Emerton and Tessema (2001) describe how the opportunity costs of fishing activities foregone through the designation of the Kisite Marine National Park and Mpunguti National Reserve in Kenya (some US$172,000) overshadowed the estimated US$39,000 in local benefits accrued in 1998. The banning of all commercial and recreational fishing boats from one-third of the Great Barrier Reef in Australia (up from only 4.5%) in 2004 was heavily criticized by the fishing industry, which declared that hundreds of jobs would be lost (CNN, 2003). Similarly, the call by New Zealand's Conservation Minister for 10% of coastal waters to be designated as marine reserves, in which fishing is banned, was countered by the fishing industry which declared that New Zealand's exclusive economic zone (EEZ) is already protected by the Fisheries Act, with its focus on sustainability and restriction on catches (Thomas, 2002).

The opportunity costs may not be measured solely in financial terms. In largely subsistence economies the question of local sustenance arises. A particularly controversial example of marine resources foregone is the advocacy of whale-watching tourism in Tonga. Orams (2002a, p. 376) reports the findings of a survey of visitors conducted in 1999 where 65% of yacht-based respondents and 73% of aircraft-borne respondents agreed that 'they would be less likely to visit Vava'u if whales were hunted there'. However, Evans (2005) examines how the loss of whale meat produced for domestic consumption by indigenous Tongan whalers, who were not in themselves significant contributors to the drastic decline in humpback whale stocks (caused by international commercial operators), has had significant consequences for both the national economy and for the health of individual Tongans. He argues that 'whale-watching tourism is frequently presented as the economic and moral antithesis of whaling, and thus whale watching advocates systematically preclude development options that include the consumptive use of whales . . . whaling is a moral, not economic or ecological issue . . . the suppression of any serious debate of this issue is a product of western ethnocentrism and a contemporary form of cultural imperialism' (Evans, 2005, p. 49). Furthermore, he questions the viability of whale-watching tourism in terms of an economic development strategy for Tonga given the competition from more accessible and more firmly established sites in New Zealand. However, Orams (2002a) suggests that the total economic impact (direct, indirect and induced) of whale watching in Vava'u (the northern island group of the Kingdom of Tonga) could exceed US$700,000. In Australia, the annual AUS$15 million injection from 75,000 whale-watchers to Hervey Bay leads to estimates that each whale is 'worth' US$100,000 to the local economy (C. Bulbeck, 2005, Hervey Bay City Council, personal communication). Other estimates put the national significance of whale watching in 2003 at over AUS$300 million, up from AUS$46 million in 1991 (Club Marine, 2005).

The Galápagos Islands are probably the most graphic example of the conflict between conservationists, marine ecotourists and fishermen (Buckley, 2003). In the year 2000, islanders took giant tortoises hostage in protest against the designation of a marine reserve 40 miles offshore, restricting their lucrative catch of sharks and sea cucumbers. Shark fins fetch as much as £66/ kg in Asia, while Galápagan fishermen, who could sell as many as 2000 sea cucumbers at 60 pence apiece were 'doing as well as a dope dealer selling cocaine on the mainland' (McCosker cited in Bellos, 2000). As the Galápagos Islands received 90,500 tourists in 2003 (Galápagos Conservation Trust, 2005) to appreciate the marine and island ecology of this world-renowned destination, the environmental damage as well as the adverse publicity being generated by the fishermen was of considerable concern. An example of the resolution of these types of conflict is that of the island of St Lucia, where fishermen complained of severe declines in their catches as a result of the designation of no-take zones within the Soufriere Marine Management Area. As a result, they were compensated the equivalent of US$150 a month for a year, and part of one reserve was reopened for pot fishing. The year's compensation allowed for a period of adjustment while fishermen became more knowledgeable about the benefits of the reserves (MPA News, 2002).

The consideration of economic benefits and costs is further complicated by discontinuities in the system: those who benefit from marine wildlife tourism are frequently not those who shoulder the costs. It is, therefore, essential that we consider distributional aspects, which are bound up with the political economy of marine wildlife tourism.

The Political Economy of Marine Wildlife Tourism

Prospects of, and for, marine tourism at the local level are linked with multi-scale political-economic and ecological processes. Therefore, it is vital that we recognize the need to go 'beyond single geographical scale factors influencing land and resource use (e.g. the village) to consider the many regional, national and international dimensions' (Zimmerer and Bassett, 2003, p. 288). Young makes the case for a political ecology approach for comparatively assessing local patterns of resource use with reference to marine ecotourism, declaring that 'As a multiscalar, contextual approach to understanding how markets, policies and political processes shape nature–society relations, political ecology provides a useful framework' (Young, 2003, p. 45). She highlights how 'A growing number of studies use a political ecology approach to examine the relationship between access conflicts in the commons and ecological change in aquatic habitats and wildlife, particularly in marine environments' (Young, 2003, p. 31). Young's own study of marine ecotourism in Baja California, examines how 'the multi-million dollar whale-watching industry there has become dominated by operators based in the United States . . . In 1994, the

Mexican Ministry of Tourism estimated that, in one weekend during the gray whale season, 30 planes of United States origin landed on the airstrip' (1999, pp. 601–602). She also examines the national scale, revealing how the two main Mexican federal agencies, which are legally empowered to both monitor tourism activities around gray whales and enforce laws that restrict such activities, are overcentralized, and how government decision makers (based in Mexico City) are unfamiliar with local ecological and social conditions. Young also points to the fact that 'insufficient funding for field personnel, facilities and equipment impede effective regulation of local activities in both areas' (Young, 1999, p. 609).

The multiscalar approach of political ecology is, therefore, of value in reminding ourselves that 'the narrative of globalization downplays the importance of national dynamics, failing to adequately address the symbiotic relationship between national and international institutions and elites' (Walley, 2004, pp. 262–264). Rudkin and Hall (1996, pp. 203–204), for example, describe how ecotourism development in the Solomon Islands has 'primarily been driven by Western consultants . . . operating in conjunction with the local business and political elite'.

Economic Impacts and Scale

It is evident from our discussion above that, while not necessarily mutually exclusive, the economic benefits and costs from marine wildlife tourism accrue differentially between various scalar levels and interests. At the international level, it is evident that, as Wells (1997, p. 33) declares 'Many of the benefits of conserving wildlife go to the world as whole, while the costs are usually borne at national and local levels . . . Costs usually result from loss of access to protected lands and damage caused by wildlife. The heaviest burden tends to be borne by poorer countries and especially by impoverished people living in rural areas of these countries in the proximity of protected areas.' Concerning the national scale, the emphasis placed on tourism development by many countries serves to illustrate its significance to national economies, particularly, in terms of foreign exchange earnings. Dixon *et al.* (1993) describe how Bonaire's economic mainstay is tourism, particularly related to scuba-diving, with total government revenue in 1991 from direct and indirect taxes related to tourism of US$8.4 million, with taxes levied directly on visiting divers estimated at US$340,000. The significance to regional and provincial economies can be even more pronounced. From surveys conducted as long as 20 years ago, Duffus and Dearden (1993) estimated that CAN$4,000,000 was injected into the Vancouver Island economy from whale-watchers at Johnstone Strait. Pendleton and Rooke (2006) estimate that diving in California, statewide, probably generates between US$138 and US$276 million, with a similar magnitude of expenditure associated with snorkelling.

At the local scale, however, the picture becomes more complex. In general it can be said that local populations receive nothing like their due share of economic benefits from marine wildlife tourism. For example, many residents of the Galápagos Islands, Ecuador, find it difficult to benefit. Trends suggest that not only are the Galápagos becoming a premium nature tourism destination, visited mainly by well-off people but also that less than 15% of foreigners' expenditures are estimated to reach the islands. In practice, local benefits are confined to employment on a very modest scale, with most tourism benefits leaking out to the national or international level (Wells, 1997). However, it is undeniable that there are those in the community who benefit while others lose out and these divisions are sharply drawn along the lines of ethnicity, class, gender and age as discussed below.

The Economic Impacts of Marine Wildlife Tourism on Local Communities

Possibly the most useful way of assessing economic impacts of marine wildlife tourism on local communities is through the Sustainable Livelihoods Approach (SLA), discussed in detail in a marine context in Cater and Cater (2007). At the heart of the SLA lies an analysis of five types of assets upon which people draw to build their livelihood (Sustaining Livelihoods in Southern Africa, 2002). These are: natural capital (the natural resources stocks upon which people draw for livelihood); human capital (the skills, knowledge, ability to labour and good health important to be able to pursue different livelihood strategies); physical capital (the basic enabling infrastructure such as transport, shelter, water, energy and communications); financial capital (the financial resources which are available to people such as savings, credit, remittances or pensions, which provide them with different livelihood options); and social capital (the social resources such as networks, membership of groups, relationships of trust upon which people draw in pursuit of their livelihood). However, it has been suggested that to this classic pentagon should be added cultural capital, which can be defined as the cultural resources (heritage, customs, traditions) which are very much a feature of local livelihood (although this should be secondary) are major tourist attractions in themselves (Glavovic *et al.*, 2002; Sustaining Livelihoods in Southern Africa, 2002). Of these six livelihood assets (Fig. 8.1) it is probably physical capital and financial capital with which we are most concerned in assessing economic impacts, but the inextricability of all these forms of capital should be borne in mind, for example, the interplay between health as an element of human capital and worker productivity.

Physical capital

Marine wildlife tourism may act as a catalyst, providing the incentive for the improvement of infrastructure which will not only benefit the tourists, but also the local population, in the case of electricity, safe water supply and improved roads. On the island of Manono, Western Samoa, home-stay visitation by

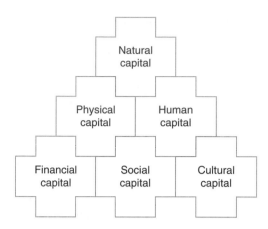

Fig. 8.1. The building blocks of the Sustainable Livelihoods Approach.

American elder hostellers prompted the construction of wharves on the shore as it was too difficult for older people to wade and climb the rocks. Flush toilets and showers, a new concept to the families concerned, were also necessary in the households visited. There is the clear danger, however, that enhancing one livelihood asset, in this case, increased access to physical capital, may mean a concomitant erosion of another. In the case of Manono, there would have been the problem of the reduction of financial assets should the islanders have had to provide and finance improved sanitation. This was circumvented by public works providing a design complete with a septic tank built by the families themselves. The necessary appliances were financed by a revolving fund from Australia (Ala'ilima and Ala'ilima, 2002). A further consideration is that, however low-key and small-scale the marine tourism development, the question of access means that frequently physical capital is enhanced at the cost of natural capital. De Haas (2002) describes the situation of small-scale ecotourism on the island state of Niue in the South Pacific where concrete tracks, which clearly detracted from Niue's natural resources, were built across the island to allow for easy access to coastal areas.

Financial capital

It is important to recognize that coastal communities in the developing countries undertake a variety of income-generating activities, in particular fishing, and that marine wildlife tourism must take its place alongside them, viewed as a complement or a supplement, not an alternative. Warburton (1999) describes how, although whale-watching businesses in Mull are relatively profitable, most of the operators have alternative sources of income, either during or out of season such as tourist accommodation, farming or fishing.

There are many examples across the world where marine tourism has proved a valuable supplement to the financial assets of coastal livelihood, in particular, where it has occurred within a community-based coastal resource management (CBCRM) programme. One of the most successful CBCRM projects is that of the award-winning Olango Birds and Seascape Tour (OBST) Project in the Philippines. Faced with severely depleted fish stocks (the average daily fish catch having dropped from around 20 kg per fisher in 1960 to less than 2 kg by 2000), Olango fishermen turned to cyanide fishing to supply the aquarium trade as a source of income (oneocean, 1999). The women stayed at home to make shell-craft, but their income was minimal and their livelihood threatened by an over-saturated market and a dwindling supply of shells. OBST, owned and operated by the Suba, Olango Ecotourism Cooperative, was initiated in 1998 with the help of the Philippine Coastal Resource Management Programme. The villagers provide tours, such as canoeing through an island seascape, snorkelling and diving in a protected marine sanctuary, visiting seaweed farms, interacting with the community and guided birdwatching in the Olango Island Wildlife Sanctuary (a RAMSAR site of international significance because of its high biodiversity and critical feeding and roosting site for tens of thousands of shorebirds). The men, who are mainly involved in paddling the day visitors and guiding, formed a Paddlers' Group. They set up and implemented guidelines for accrediting, orienting, assigning and monitoring paddlers for each tour. The roles of the Women's Group include cooking, purchasing, physical arrangements, cookery and shell-craft demonstration, and book-keeping. The villagers, as owner–manager–operators of this venture, and therefore economic beneficiaries of the project (community service fees, product sales and profit margins account for 20–50% of the tour price), consequently appreciate the value of the Coastal Resource Management Project (Flores, no date).

Economic Equity

Although a wide section of the community benefits in Suba, there are examples where attempts to develop community tourism have either exacerbated or caused conflicts. It is naïve to think that all in the community will benefit equally. Coastal communities are highly heterogeneous, their members sharply differentiated by demographic and socio-economic characteristics. Borrini-Feyerabend's (1996) criterion of equity in access to the resources and the distribution of their benefits may remain an unattainable ideal, with elite capture of the benefits from marine ecotourism being a frequent phenomenon. Indeed, it has been argued that ecotourism may even exacerbate, or even create, divisions. Entus (2002) describes how:

> [M]any projects which have set out to be community-based . . . have, at some point or another in their evolutionary cycles, engendered or exacerbated pre-existing internal divisions of power, and led to the formation of new business elites who represent but a small fraction of the 'local community', so that they end up catering primarily to those interests rather than those of the community at large, leaving the latter to pay the costs of development without also sharing worthwhile benefits.

This concern illustrates a manifest power differential not only between the different types of stakeholder but also within the local community itself, it is far from a homogeneous construct and, as Burkey (1993) argues, there is a need to demystify the harmony model of community life. Community members are differentiated by ethnicity, class, gender and age.

Ethnicity

In terms of ethnicity, we need to recognize that, while members of the coastal community may be local residents they may be 'outsiders' either in the sense that they are economic migrants, or that they are outside investors. In both instances qualities are imported which 'do not and cannot stem from the group itself' (Taylor, 1995, p. 488). Stonich et al. (1995) describe how, in the Bay Islands, Honduras desperately poor ladinos (Spanish speakers) from the mainland seeking a better life migrated to the islands where the rapid growth of tourism had brought increased prosperity. These migrants helped escalate the local population to the level at which the islands' fresh water supply, food and land resources were jeopardized. At Sandy Bay, they lived in a 'shabby ghetto of small wooden structures built on stilts, above a lagoon filled with human waste and other garbage' (p. 22). A similar situation occurred at Ambergris Caye, off the coast of Belize, where the rapid construction of hotels and condominiums in the late 1980s and early 1990s caused low paid and unemployed migrants (again predominantly Spanish-speaking) to move to San Pedro from the mainland of Belize and the rest of Central America in search of employment. Adequate accommodation and infrastructure were not available and so generally substandard housing was built on infilled mangrove swampland (McMinn and Cater, 1998). In both cases, the indigenous islanders were better placed to take advantage of new economic opportunities provided by the growth of the tourist industry, even if the poorest received only marginally better benefits. Shah and Gupta (2000) distinguish between poor, unskilled migrants seeking employment in tourism and outside entrepreneurs with better access to skills and capital than the locals. With respect to these outside entrepreneurs, Place (1991) describes how the rapid increase of visitors to view the nesting sites of the green turtle at Tortuguero, Costa Rica, actually had the net effect of reducing the opportunity for villagers to be involved in the business other than as menial employees. The pace of outside investment, in particular from the capital, San Jose, was too fast to permit villagers to accumulate sufficient capital to invest in the construction of tourist facilities.

A different slant on ethnicity is provided by the example of Kaikoura, New Zealand. Through a range of tourist developments in Kaikoura, including the award-winning Whale Watch Kaikoura, local Maori moved from a position of relative powerlessness and low economic status to become a major employer and economic force in the community. Whale Watch, a not-for-profit Maori enterprise is the sole sea-based whale-watch operator in the area – a monopolistic position supported by the New Zealand Court of Appeal in 1995 – which has caused resentment from other locals (Orams, 2002b). Maori use their position to defend their monopoly, which, unfortunately, adds a political and racial dimension to this strategy, whereby any criticism of this position is construed as racist (Horn et al., 1998).

Class

Often closely allied to the question of ethnicity is that of social class. There is, unfortunately, no substantiation with hard facts to guarantee the claim that marine wildlife tourism generally contributes to a more equitable distribution of tourism income and a reduction in poverty. At both Tortuguero, Costa Rica (Place, 1991) and in the Bay Islands, Honduras (Stonich *et al.*, 1995) those members of the community who did benefit from tourism were those who started out wealthier than most and who could, therefore, take advantage of emerging opportunities because they had sufficient income to invest in tourist-related enterprises. In both these examples, the divisions have, as Entus (2002) suggests, consequently been exacerbated.

Young (1999) identifies two major problems regarding the distribution of benefits from whale watching in Baja California, Mexico. The first is that outside tourism companies, who organize the activity and also often use outside whale-skiff drivers, are the main beneficiaries, with only as little as 1.2% of revenues accruing locally. The second is that she comes to the conclusion that many of the same problems of managing common-pool resources encountered in fishing are emerging in marine ecotourism. Even with the organization of a local tourism cooperative in one of the villages there is the problem of inequitable distribution of benefits. One of her respondents declared 'The president of the cooperative is managing it as if he were the owner. [People who rent out their privileges as whale-skiff drivers to others] should give those privileges away to other families who really need the money . . .' (p. 604). Young also describes how tensions flared during the 1994 season, when a new group of 31 aspiring skiff guides challenged the capacity of the cooperative to manage local whale-watching activities.

Not only are there marked divisions between those in the community with privileged status and the poor, but even amongst the poor, lines of division are sharply drawn according to access to resources, markets and employment, whether formal or informal. In the case of coastal fisheries in the developing countries, for example, the situation may be similar to that described by Ellis and Allison (2004) for the African lakes and wetlands where wealthier households own assets related to fishing (boats, nets, traps), as well as coastal land and businesses, and may have control over the best fishing areas. Middle-income households often own land, but have not generated sufficient capital to own substantial fishing-related assets, although they may share these. Lower-income households may have access to land for subsistence cropping but have access to fishing opportunities only as crew labourers on boats owned by others. It is obvious, therefore, that similar groups in coastal areas will be differentially placed with regard to the impact of marine wildlife tourism on their livelihood.

Gender

There are also clear divisions within communities attributable to gender. Flintan (2003) describes how the collection of natural resources is gender differentiated. While fisheries tend to be male dominated, women are becoming increasingly involved in the processing of natural resources as opportunities are opened for diversification of livelihood. Off the east coast of Unguja, Zanzibar, for example, the overwhelming majority of seaweed farmers are women (Pettersson-Löfquist, 1995). While the men may benefit from both supplying fish and by acting as guides and boatmen for tourists, the women face a scenario of conflicting use: seaweed cultivation is not the most visually aesthetic resource use. Flintan (2003) suggests that in Integrated Conservation and Development Projects, already existing gender inequalities may be increased as a number of opportunities have been opened up for men but not women.

One of the ways in which marginalized sections of the community, including the elderly and disabled, can share in the capture of ecotourism revenue is through the sale of tourist merchandise. Healy (1994) summarizes the advantages of home and village-based handicraft production under five headings: compatibility with rural activities; economic benefits (particularly a more equitable distribution); product development; sustainability; and tourist education. However, careful thought needs to go into the choice of product. For example, the soap production by women at Olango, which has a limited domestic market, might be enlarged to the tourist market if packaging included information on the bird sanctuary.

The Economics of Marine Protected Areas

Financing MPAs

Although there are a number of alternative methods of financing protected areas, the principal ways are through government support or through revenues relating to activities within the protected area. Font *et al.* (2004) describe the main mechanisms used by protected areas to raise funds from tourism under six headings: entrance fees; user fees (such as dive fees); concessions and leases which involve payment for permission to operate within the protected area (such as licences for dive boat or kayaking operations); direct operations by the protected area management themselves; taxes, such as a dedicated conservation tax or a room tax, part of which is earmarked for conservation; and volunteers offering their services for free or for basic living expenses as well as donations given to support the protected area. As discussed in the introduction to this chapter, entrance fees or user fees set at an appropriate level are the most commonly utilized mechanisms of capturing a larger share of the economic value of tourism in protected areas. Although, in theory, they are one of the best ways of generating income which can constitute a substantial proportion of operational costs, in practice only a minority of MPAs levy such charges, and even if they do, the fee level is set below that which users would be willing to pay. This is particularly so in the case of the less economically developed countries. Green and Donnelly (2003) describe how only 25% of MPAs in the Caribbean and Central America containing coral reefs charge divers an entrance or user fee which is usually US$2–3 per dive or diver. As surveys conducted in

Curacao, Jamaica and Bonaire indicate a WTP of around US$25 per person it is clear that the potential revenue is not being realized. Green and Donnelly point out that, if the 3.75 million divers visiting MPAs in the Caribbean region (excluding Florida) annually were to pay this higher amount, 78% of the financial shortfall they currently face could theoretically be raised. While they recognize the practical constraints of introducing and maintaining a fee collection system, as well as the political and socio-economic factors that may militate against it, they point to successful implementation elsewhere in the Caribbean. At Bonaire Marine Park, in the Netherlands Antilles, revenue generated by the US$10 fee per diver per year now finances a large share of management costs (Green and Donnelly, 2003).

The US$17 a year, or US$5.50 per day, entrance fee system at the award-winning Bunaken National Marine Park in Indonesia, modelled on Bonaire's diver fee system, succeeded in doubling revenues in 1 year and collected US$11,000 in 2002 (Spergel and Moye, 2004). Of course, the earmarking of such fees, as discussed below, is a crucial factor. As Tisdell (2003) points out, the funds available to a protected area will depend on institutional arrangements. In Kenya, for example, KWS Shimoni, responsible for Kisite Marine National Park and the adjacent Mpungati Marine National Reserve, remitted US$130,000 generated from entrance fees to central coffers in 1998. Only 15% of these earnings were returned to KWS Shimoni who, of course, bore the lion's share of management expenditures on the National Park (Emerton and Tessema, 2001).

The largest MPA in the world, the Great Barrier Reef Marine Park (GBRMP), levies an Environmental Management Charge of AUS$5 per tourist per day. In 2002/03 the total income of AUS$6.7 million from the charge covered approximately 20% of the budget of the GBRMP Authority (Spergel and Moye, 2004), with the bulk of management costs met by the Australian taxpayer (Buckley, 2003). However, it is estimated that marine park tourism generates AUS$2 billion per annum for the Queensland State regional economy. When viewed from this perspective, it becomes apparent that State and Federal governments treat GBR tourism as a 'cash cow', which is actually underfunded by the various authorities (Mules, 2004). The shortfall then lies with operators, many of who engage in programmes that are outside of any formal accounting, such as population control of Crown of Thorns starfish. Initiatives such as these are essential to the long-term viability of the tourist resource, and illustrate industry interest in its maintenance.

A few, high profile, charismatic sites around the world are able to command much higher fees. Visitors to the Galápagos Islands National Park are willing to pay the US$100 entry fee because of its uniqueness. Another world-class location is that of the Tubbataha Reefs National Marine Park in the Philippines, a World Heritage site, where foreign scuba-divers pay a US$50 reef conservation fee (Spergel and Moye, 2004). Lindberg and Halpenny (2001) present a very useful country review of protected area visitor fees which includes those of a number of MPAs across the globe.

Although it would appear that raising fees is a viable prospect for MPAs the implications need to be thought through carefully. The advantages are that net revenues may increase while simultaneously visitor numbers may be reduced, thereby reducing visitors' total environmental impact. Wells (1997) cites the examples of Hurghada and Sharm el Sheik in Egypt, where the former displays 'mass' tourism density with unlimited reef use, three times that at Sharm. Hotels at Sharm el Sheik, where fees are charged and tourism thus regulated, are able to charge double those of Hurghada. The disadvantages are that such moves may be seen as exclusionary and elitist and, also, if fees are increased the financial benefit may be outweighed by the cost of reduced visitor spending in the broader economy (Wells, 1997).

The need for earmarking

Just how the revenue generated by fees is allocated needs very careful consideration. It is important to note that there is increased public support for park fees if income is returned directly to the parks rather than accruing to central treasury. There is a circular and cumulative effect when this happens, as consequent enhancement of services and facilities both maintain and increase visitor satisfaction and, in turn, increase revenues (Lindberg and Halpenny, 2001). Conversely, if local communities living adjacent to MPAs see little or no economic benefit then support will wane. As Emerton and Tessema (2001, p. 17) declare: 'Among the communities who live around MPAs there is rapidly growing pressure on land and resources, and on available sources of income and employment. These communities are becoming less and less willing, and less able to afford, to support MPAs in which they have no economic stake and yield them no tangible benefits.'

In Bunaken, when the nature reserve was upgraded to the status of a marine national park in the late 1980s, control over the park, including the authority to collect fees, passed to the central government. The instigation of the multistakeholder Bunaken National Park Management Advisory Board (BNPMAB) to manage the protected area has, however, resulted in a remarkable turnaround that serves as a model not only for Indonesia but also globally. BNPMAB has adopted a participatory and consultative approach to managing the entrance fee system which was inaugurated in 2001. Instead of all user fees passing directly to central government, 80% of revenues are retained by the park management board, with 20% divided between local, provincial and national government (MPA News, 2004). A small grants programme implemented by the board ensures that, of the funds retained by the board, 30% are returned to the community in the form of small-scale conservation and community development projects which they propose and implement themselves. The significance of this achievement is marked by the fact that the International Coral Reef Action Network (ICRAN) chose Bunaken as its Asian demonstration site for sustainable reef tourism.

Differential charging

It is common for marine parks to charge different fees for foreigners than for nationals. Lindberg and Halpenny (2001) cite the cases of Belize where foreigners pay US$2.50 at Hol Chan and US$5 at Half Moon Caye, with Belizeans paying nothing; at Ras Mohammed in Egypt foreigners pay US$5 while Egyptians pay US$1.20; and Bunaken Marine Park where foreigners pay US$7 while locals pay about US$0.25. They point out that these differences can encourage greater interest in conservation and national parks, but more fundamental moral justification, particularly evident in Third World destinations, comes from the manifest income differences between the two groups of visitors. Differential charging is also vindicated by the fact that local people may already contribute to park management costs through government taxation. Lindberg and Halpenny also highlight the fact that divers are frequently charged more than snorkellers; the Soufriere Marine Management Authority in St Lucia charges divers US$4 a day (US$12 per year) as opposed to snorkellers US$1 a day. This is based on the premises that divers are wealthier than snorkellers, that diving is a more specialized activity and that divers may stay longer, so a flat fee is spread over a number of days.

What Lies Beneath?

It is clear from the discussion in this chapter that the economic impacts from marine wildlife tourism are far from simple, and manifest themselves in many ways and at multiple scales. However, recognition of this complexity means that there is much to be learned from the economic perspective, which should inform and ground approaches to marine wildlife management. It is essential to adopt a holistic viewpoint that takes into account the various levels and stakeholders that are involved and impacted by the development of marine tourism. We should be aware of complementary and conflicting economic activities that may assist or prejudice the very existence of this sector. Attention must also be paid to the non-monetary aspects of economic impacts, which may often outweigh the financial flows that are superficially more obvious. This is evident when it is considered that financial returns from marine wildlife tourism may well be less than the opportunity cost of activities foregone (e.g. the return from fisheries in the example of Kisite Marine National Park cited above); but total economic value may be greater than the opportunity cost. Indeed, the complexity of economic impacts hinders transparency, making complete accounting difficult in a situation where imperfect and incomplete information is available for policy making. The Great Barrier Reef Marine Park, being portrayed as a cost to the Australian taxpayer, whilst contributing billions in tourist revenue (and, one might add, of priceless destination marketing value), is one example of this. More complete accounting models need to be employed that are able to take such nuances into account. Despite the difficulty in their estimation, the non-use values discussed earlier should also be included. As a further example of this complexity, how might we account for the recognized, but rarely valued, physical and mental health benefits to tourists engaging in active wildlife tourism? Such acknowledgement will become an area ripe for enquiry with rapidly ageing populations and consequent implications for health budgets faced by many developed nations.

More complete accounting procedures would not only take into account the economic impacts which are frequently overlooked but also, in order to understand the relative costs and benefits of marine wildlife tourism, their distributional aspects. An appreciation of relationships of power will lead to better-informed decision making through a more detailed understanding of who really benefits and who loses, how and why. This attention to the distributional aspects of economic impacts is perhaps not limited to the human realm, as there should also be concern for the 'stars of the show'. Arguably, without ensuring the economic welfare of the wildlife that drives the entire industry, there will be no product to attract tourists. While, as Holden (2003, p. 105) suggests, the acceptance of non-anthropocentric ethics which recognize the value of nature in its own right would currently imply a significant conceptual shift, 'The fact that the natural environment can be given an economic value in a conserved state, through its use for tourism, means that environmentalists are not forced to fight conservation battles based upon the mere esoteric and altruistic concept of its intrinsic value'. The scientific approach to wildlife tourism management that is advocated in this book should contribute to improved welfare for marine species and their habitats, which will help to ensure sustainable opportunities for the future.

References

Ala'ilima, F. and Ala'ilima, L. (2002) Manono: an experiment with community based ecotourism. *Sixth International Permaculture Conference Proceedings Chapter 6.* Available at: http://www.rosneath.com.au/ipc6/ch06/alailima

Bellos, A. (2000) Galápagos turmoil: tortoise dragged into fishing war. *The Guardian* 30, December.

Borrini-Feyerabend, G. (1996) *Collaborative Management of Protected Areas: Taking the Approach to the Context in Issues in Social Policy.* IUCN, Gland, Switzerland.

Buckley, R. (2003) *Case Studies in Ecotourism.* CAB International, Wallingford, UK.

Bulbeck, C. (2005) *Facing the Wild: Ecotourism, Conservation and Animal Encounters.* Earthscan, London.

Burkey, S. (1993) *People First.* Zed Books, London.

Cater, C. and Cater, E. (2007) *Marine Ecotourism: Between the Devil and the Deep Blue Sea.* CAB International, Wallingford, UK.

Club Marine (2005) Whale Ahoy! *Club Marine* 20(4), 146–154.

Cooper, C., Fletcher, J., Gilbert, D., Shepherd, R. and Wanhill, S. (1998) *Tourism: Principles and Practice.* Prentice-Hall, Harlow, UK.

CNN (2003) Barrier Reef eco-plan unveiled. *CNN December 3, 2003.* Available at: http://www.cnn.com/2003/WORLD/asiapcf/auspac/12/03/australia.reef/index.html

De Haas, H.C. (2002) Sustainability of small-scale ecotourism: the case of Niue, South Pacific. *Current Issues in Tourism* 5(3 & 4), 319–337.

Dixon, J.A., Scura, L.F. and van't Hof, T. (1993) Meeting ecological and economic goals: marine parks in the Caribbean. *Ambio* 22(2–3), 117–125.

Duffus, D.A. and Dearden, P. (1993) Recreational use, valuation, and management, of killer whales on Canada's Pacific Coast. *Environmental Conservation* 20(2), 149–156.

Ellis, F. and Allison, E. (2004) Livelihood diversification and natural resource access. FAO Livelihood Support Programme (LSP) Working Paper 9. FAO, Rome.

Emerton, L. and Tessema, Y. (2001) *Economic Constraints to the Management of Marine Protected Areas: The Case of Kisite Marine National Park and Mpunguti Marine National Reserve, Kenya.* IUCN Eastern Africa Regional Office, Nairobi.

Entus, S. (2002) Re: participative (business) community development. Discussion list available at: trinet@hawaii.edu

Evans, M. (2005) Whale-Watching and the Compromise of Tongan Interests through Tourism. Available at: http://www.sicri.org/assets/downloads/SICRI05_PDF/SICRI2005_Evans.pdf

Flintan, F. (2003) Engendering eden volume 1: women, gender and ICDPs: lessons learnt and ways forward. *IIED Wildlife and Development Series No 16.* IIED, London.

Flores, M. (no date) *Six Steps in the Making of the Olango Birds and Seascape Tour.* Available at: http://www.oneocean.org/ambassadors/migratory_birds/obst/six_steps_in_the_making.html

Font, X., Cochrane, J. and Tapper, R. (2004) *Tourism for Protected Area Financing: Understanding Tourism Revenues for Effective Management Plans.* Leeds Metropolitan University, Leeds, UK.

Galápagos Conservation Trust (2005) Frequently Asked Questions. Available at: http://www.gct.org/faq.html

Glavovic, B., Scheyvens, R. and Overton, J. (2002) Waves of adversity, layers of resilience: exploring the sustainable livelihoods approach. *Paper Presented at the Development Studies of New Zealand Conference, 2002.* Available at: http://devnet.massey.ac.nz/papers/Glasovic,%20Overton%20&%20Scheyvens.pdf

Green, E. and Donnelly, R. (2003) Recreational scuba diving in Caribbean marine protected areas: do the users pay? *Ambio* 32(3), 140–144.

Healy, R. (1994) Tourist merchandise as a means of generating local benefits from ecotourism. *Journal of Sustainable Tourism* 2(3), 137–151.

Hervey Bay City Council (2005) Available at: http://www.herveybay.qld.gov.au/

Holden, A. (2003) In need of new environmental ethics for tourism? *Annals of Tourism Research* 30(1), 94–108.

Horn, C., Simmons, D.G. and Fairweather, J.R. (1998) Evolution and change in Kaikoura: responses to tourism development. *Tourism Research and Education Centre Report No. 6.* University of Lincoln, Lincoln, New Zealand.

Leeworthy, V.R., Johns, G.M., Bell, F.W. and Bonn, M.A. (2001) Socioeconomic study of reefs in Southeast Florida. *Hazen and Sawyer Final Report.* National Oceanic and Atmospheric Administration.

Lindberg, K. and Halpenny, E. (2001) Protected area visitor fees summary. Available at: www.ecotourism.org/pdf/protareasfeesoverview.pdf

Mathieu, L.F., Langford, I.H. and Kenyon, W. (2003) Valuing marine parks in a developing country: a case of the Seychelles. *Environment and Development Economics* 8, 373–390.

McMinn, S. and Cater, E. (1998) Tourist typology: observations from Belize. *Annals of Tourism Research* 25(3), 675–699.

MPA News (2002) Financial support for fishermen who are affected by Marine reserves: examining the merits. *MPANews* 3(11), 1–3.

MPA News (2004) Tools and strategies for financial sustainability: how managers are building secure futures for their MPAs. *MPA News* 5(5), 1–4.

Mules, T. (2004) *The Economic Contribution of Tourism to the Management of the Great Barrier Reef Marine Park: A Review.* Prepared for Queensland Tourism Industry Council, Queensland, Australia.

Newmark, W.D., Manyanza, D.N., Gamassa, D.-G.M. and Sariko, H.I. (1994) The conflict between wildlife and local people living adjacent to protected areas in Tanzania: human density as a predictor. *Conservation Biology* 8, 249–255.

oneocean (1999) At Olango challenge spells opportunity. *Overseas: The Online Magazine for Sustainable Seas* 2(2). Available at: http://www.oneocean.org.oversear/feb99/at_olango_challenge_spells_opportunity.html

Orams, M. (1999) *Marine Tourism: Developments, Impacts and Management*. Routledge, London.

Orams, M. (2002a) Humpback whales in Tonga: an economic resource for tourism. *Coastal Management* 30, 361–380.

Orams, M. (2002b) Marine ecotourism as a potential agent for sustainable development in Kaikura, New Zealand. *International Journal of Sustainable Development* 5(3), 338–352.

Pendleton, L.H. and Rooke, J. (2006) Understanding the potential economic impact of SCUBA diving and snorkeling: California. Available at: http://linwoodp.bol.ucla.edu/dive.pdf

Pettersson-Löfquist, P. (1995) The development of open water algae farming in Zanzibar: reflections on the socio-economic impact. *Ambio* 24(7/8), 487–491.

Place, S. (1991) Nature tourism and rural development in Tortuguero. *Annals of Tourism Research* 18, 186–201.

Raymundo, L.J. (2002) Community-based coastal resources management of Apo Island, Negros oriental, Philippines: history and lessons learned. Available at: http://www.icran.org/SITES/doc/ws_apo.pdf

Rudkin, B. and Hall, C.M. (1996) Unable to see the forest for the trees: ecotourism development in Solomon Islands. In: Butler, R. and Hinch, T. (eds) *Tourism and Indigenous Peoples*. Thomson, London, pp. 203–226.

Shah, K. and Gupta, V. (2000) Tourism, the poor and other stakeholders: experience in Asia. *Fair Trade in Tourism Project*. ODI, London.

Spergel, B. and Moye, M. (2004) *Financing Marine Conservation: A Menu of Options*. WWF Center for Conservation Finance, Washington, DC.

Stonich, S., Sorensen, J.H. and Hundt, A. (1995) Ethnicity, class, and gender in tourism development: the case of the Bay islands, Honduras. *Journal of Sustainable Tourism* 3(1), 1–28.

Sustaining Livelihoods in Southern Africa (2002) Social capital and sustainable livelihoods. Issue 5 March 2002. Available at: http://www.cbnrm.net/pdf/khanya_002_slsa_issue05_sc.pdf

Taylor, G. (1995) The community approach: does it really work? *Tourism Management* 16(7), 487–489.

Thomas, L. (2002) Fish-hooks in the case for reserves. *New Zealand Herald*, 31 January.

Tisdell, C. (2003) Economic aspects of ecotourism: wildlife-based tourism and its contribution to nature. *Sri Lankan Journal of Agricultural Economics* 5(1), 83–95.

Walley, C.J. (2004) *Rough Waters: Nature and Development in an East African Marine Park*. Princeton University Press, Princeton, New Jersey.

Warburton, C. (1999) Tourism and whale-watching on the island of Mull, West Scotland. Available at: http://www.whaledolphintrust.co.uk/research/documents/Marine_Ecotourism.pdf

Wells, M.P. (1997) Economic perspectives on nature tourism, conservation and development. Environment Department Papers No. 55. Environmental Economics Series. Environmentally Sustainable Development. The World Bank. Available at: http://www.icrtourism.org/Publications/Economicperspectivestourism.pdf

Young, E. (1999) Balancing conservation with development in small-scale fisheries: is ecotourism an empty promise? *Human Ecology* 27(4), 581–620.

Young, E.H. (2003) Balancing conservation with development in marine-dependent communities. In: Zimmerer, K.S. and Bassett, T.J. (eds.) *Political Ecology: An Integrative Approach to Geography and Environment-Development Studies*. Guilford Press, New York, pp. 29–49.

Zimmerer, K.S. and Bassett, T.J. (2003) Future directions in political ecology: nature–society fusions and scales of interaction. In: Zimmerer, K.S. and Bassett, T.J. (eds.) *Political Ecology: An Integrative Approach to Geography and Environment-Development Studies*. Guilford Press, New York, pp. 275–295.

9 Effects of Human Disturbance on Penguins: The Need for Site- and Species-specific Visitor Management Guidelines

P.J. SEDDON AND U. ELLENBERG

Introduction

The spectacular growth of nature-based tourism in recent decades is the inevitable product of increasing disposable income and leisure time, married to our fascination with wildlife in wild places. Few things delight and inspire as much as wild animals en masse, hence the tourist-pulling power of wildlife aggregations such as the flamingos of Lake Victoria, the wildebeest migrations of East Africa or the unique sights, sounds and smells of seabird breeding colonies. If anything can match our love of wildlife writ large it is the charisma of select species, such as the large mammals of land or sea. More than these, we seem drawn to species that show human characteristics – the monkeys and apes command our attention, for example. Few taxa, however, represent the combination of anthropomorphic characters with dense natural concentrations in accessible locations – few indeed, possibly only penguins.

Two factors make penguins a perfect tourist attraction: first, they must come ashore to breed, and anywhere that is accessible to the stumpy legs of a penguin will be accessible to humans; and second, their short stature, bipedal gait, tuxedo-like coloration and general demeanour beg for us to look upon them as little people. It is a common perception that penguins are little worried by the proximity of large groups of humans. In part this is driven by people's experiences with more tolerant species in well-regulated settings, such as Little penguins (see Appendix 9.1 where all common and scientific names of species mentioned in the text are provided) on Phillip Island, Australia, and in part by the apparent absence of overt behavioural responses to human approach, particularly by nesting penguins. A little over a decade ago these perceptions lead some researchers to suggest that the direct impact of tourism on penguins can be slight if visitor activities are well managed (Boersma and Stokes, 1995). In contrast, at around the same time, scientists working on different species in different environments concluded that: 'we believe that tourism does adversely

Appendix 9.1. Common and scientific names of species
mentioned in this chapter.

Common name	Scientific name
Little penguin	*Eudyptula minor*
Yellow-eyed penguin	*Megadyptes antipodes*
Fiordland penguin	*Eudyptes pachyrhynchus*
Royal penguin	*Eudyptes schlegeli*
Rockhopper penguin	*Eudyptes chrysocome*
Adélie penguin	*Pygoscelis adeliae*
Gentoo penguin	*Pygoscelis papua*
Chinstrap penguin	*Pygoscelis antarctica*
Emperor penguin	*Aptenodytes forsteri*
King penguin	*Aptenodytes patagonicus*
African penguin	*Spheniscus demersus*
Humboldt penguin	*Spheniscus humboldti*
Magellanic penguin	*Spheniscus magellanicus*
Weka	*Gallirallus australis*
Skua	*Catharacta* sp.
Giant Petrel	*Macronectes* sp.
Southern elephant seal	*Mirounga leonina*

affect breeding penguins, almost irrespective of how "well-behaved" the tourists are' (Culik and Wilson, 1991).

So which position is correct and how should responsible tourism managers respond? Well, unhelpfully, both may be correct, but fortunately the last 10 years have seen some powerful research techniques applied to the problem of investigating human disturbance effects on penguins and we are closer to being able to formulate sensible species-specific guidelines to manage visitors.

In this chapter we will provide a summary of the state of knowledge of how human disturbance affects penguins. We start by reviewing the full spectrum of human-related disturbance and consider where tourism fits in. We provide a general framework for understanding disturbance effects and consider the challenges inherent in studying such effects. We then summarize the current state of knowledge of penguins' behavioural and physiological responses to disturbance, and conclude by identifying future needs and the implications of research findings for tourism management. We believe that effective mitigation of the harmful effects of human disturbance that may accompany tourism will only arise from detailed species-specific guidelines derived from rigorous research.

Types of Human Disturbance

Penguins, as much as any other living thing, must cope with a number of external factors. These may be broadly divided into anthropogenic (related to human activity) or non-anthropogenic; the latter category including such challenges as extreme weather events, natural predators, habitat change and environmental variation over short or longer time scales. Anthropogenic impacts may be indirect or direct. Indirect human impacts encompass introduced predators, habitat modification, marine pollution, competition or other disruption of food sources and human induced climate change. Direct impacts relate to human disturbance, defined as any human activity that changes the contemporaneous behaviour or physiology of one or more individuals (Nisbet, 2000). Direct human activity may be beneficial, such as conservation management, but seems more often to carry at least the potential for harmful impacts, ranging from direct persecution through harvest, harassment or by-catch, to incidental disturbance, perhaps from vehicular activities, through to the possible deleterious effects of even well-meaning disturbance by tourists, researchers and managers. This last point is our focus here.

It could be argued that what is done to individual penguins in the name of science has the very real potential to be vastly more intrusive than anything any even half-way respectable tourism operation would allow. In the pursuit of knowledge, penguins are caught and handled, weighed and measured, blood-sampled, induced to regurgitate their last meal and fitted with metal flipper rings and backpack-mounted devices to monitor their movements. Many of these types of manipulations have been shown to carry energetic and survival costs for the penguins involved (Wilson *et al.*, 1989; Petersen *et al.*, 2005). Growing awareness of this problem, along with the development of less intrusive research techniques (such as the use of automatic weighbridges and implanted transponders) has greatly reduced the impact of scientific investigations. Today, in most countries, researchers have to go through a rigorous process of obtaining permits and ethics approvals for any proposed investigations. The value of possible research findings is carefully weighed against potential costs, and if permits are granted research activities are closely monitored to ensure that potential impacts are minimized.

In comparison, tourism ventures usually operate with less rigorous controls, if any, of their effects on target colonies. Furthermore, there is an important difference in the scale and duration of research interventions and those of larger-scale tourism. Research manipulations focus on a selected sample of individual birds, with controls in place specifically to assess impacts and adjust protocols accordingly; are often of short duration, one-off or short term; are regular events controlled as to observer, timing and methods. In contrast, penguin tourism involves much larger numbers of people, with exposure to a greater proportion of the penguin population; is of longer duration and longer term; and is less regular and standardized. Clearly there is also a big difference between well-regulated and essentially unregulated tourism, the latter characterized by a lack of adequate visitor information, supervision or enforcement of effective guidelines for visitor behaviour.

Challenges in Measuring Human Disturbance

Any stimulus may evoke a reaction of some kind and when assessing the effects of human disturbance a central challenge is to distinguish between responses

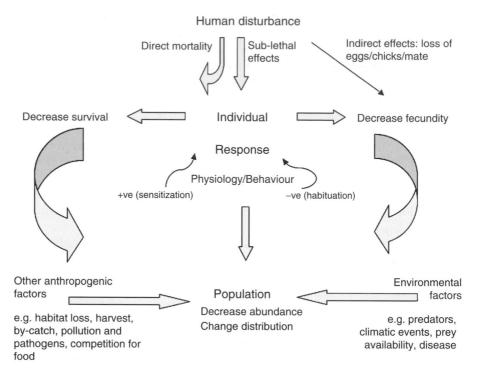

Fig. 9.1. A framework for considering human disturbance on penguins.

and impacts. A simple response may be short-term and, in isolation, effectively inconsequential, whereas an impact will have longer-term negative implications for the fitness of the individual concerned. If enough individuals experience impacts there may be population-level or even species-level consequences. Thus any study of human disturbance must attempt to do more than simply document responses; it must explore the degree to which any response has the potential to reduce, even to a tiny degree, an individual's probability of survival and reproduction (see Fig. 9.1).

A second, not insignificant challenge for studies of human disturbance is to measure responses in an unobtrusive and unbiased way. Much innovative science has gone into addressing this problem in human disturbance studies of penguins – how to remove the confounding effects of observer, manipulation and instrumentation impacts from the measured disturbance response?

Disturbance will act on individual penguins and evoke a response that may be behavioural or physiological or both. Responses may be neutral, having no fitness consequences, or deleterious. These harmful effects work in two ways: they may decrease survival and/or decrease fecundity. In addition, human disturbance may directly decrease survival if the penguin is harmed in any way, regardless of the bird's responses; similarly human disturbance may directly decrease reproductive output of an individual by disrupting a breeding attempt at any stage, e.g. by crushing or collapsing of nests and contents, or by causing the loss of a mate. Responses of individual penguins may have a cumulative effect at the population level, whereby avoidance behaviours and disruption of breeding may result in changes in onshore distributions, and reduced survival or fecundity of individual birds may result in population declines. Human disturbance effects are additive to other deleterious anthropogenic factors and natural perturbations, both of which may act to increase the magnitude of population declines and shifts. In addition, it is useful to consider two types of response feedback loops. The first may be considered a positive feedback whereby responses to future disturbance will be enhanced, resulting in sensitization. Alternatively, the feedback may be negative, thus a similar degree of disturbance in the future will elicit a reduced response via habituation. The degree to which disturbance may result in sensitization or habituation is not well understood, but seems likely to be species and stimuli specific, and will have major implications for whether disturbance will have individual-level or population-level impacts.

Techniques and Findings

Behaviour

Monitoring behaviour before, during and after human presence by a well-hidden observer or camera unit is one way to measure responses to human disturbance while keeping the observer effects to a minimum. Agitated birds may display a number of behaviours indicative of heightened alertness or alarm, the most unequivocal of which is to flee from or attack the source of the disturbance. The distance at which a bird will react can be quantified experimentally using standardized approach protocols, and findings are often used to define minimum approach distance guidelines, or setback distances, to manage human visitors. However, penguins show little behavioural reaction to human presence at their breeding sites (Culik and Wilson, 1991; Nimon *et al.*, 1995), which is often mistaken for habituation (see Box 9.1: Habituation Potential). Evolutionarily this lack of externally manifest stress makes sense: in the absence of land-based predators there is no selective advantage to display visible alarm reactions to human approach (Wilson *et al.*, 1991) and tending eggs or young chicks has absolute priority under often adverse breeding conditions. Even if flight reactions were possible at lesser costs, e.g. for older chicks, energy conservation seems to be more important. Walker *et al.* (2005) observed in Magellanic penguin chicks that were close to fledging, behavioural habituation to human presence despite an unaltered physiological stress response. As long as we bear in mind that the observed behaviour is only the tip of the iceberg and thus results have to be treated with caution, measuring behaviour can provide important information about disturbance impact, particularly in situations where physiological measures are difficult to obtain. The evidence suggests that the distance at which a penguin will tolerate the proximity of a human varies not only with the type of disturbance (number or behaviour of humans), but also according to penguin species, and intra-specifically according to individual, age, condition, current behaviour, stage of breeding and previous experience with humans. Hence, one local short-term behavioural study cannot provide a basis for

Box 9.1. Habituation potential.

Higher tolerance levels by penguins that have been frequently exposed to human activity have been reported in several studies. For example, African penguins showed reduced responses following regular disturbance on landing beaches (van Heezik and Seddon, 1990), behavioural responses of Gentoo penguins to visitation were significantly stronger in off-station colonies (Holmes *et al.*, 2006), and Magellanic penguins at Punta Tombo responded less in tourist-exposed parts of the colony (e.g. Yorio and Boersma, 1992). However, recorded differences in tolerance may be the consequence of shyer individuals leaving the area or failing to reproduce. In a breeding area regularly traversed by humans very few Humboldt penguins occupy nests despite abundant nesting opportunities (Ellenberg *et al.*, 2006); and lower nesting densities and lower variability of stress response in a tourist area led Fowler (1999) to suggest avoidance behaviour rather than habituation may have caused the reduced hormonal stress responses observed in the tourist area.

Habituation can be defined as a 'reduced response to repeated stimulation not attributable to fatigue or sensory adaptation' (Domjan, 2003). To date, only one study has been able to demonstrate actual habituation of individual penguins (Walker *et al.*, 2006). During experiments the behavioural responses of naïve penguins to human proximity declined within 10 days to levels equivalent to those measured for tourist-exposed birds. Interestingly, physiological stress responses did not exhibit a decline of similar magnitude. Although Walker and colleagues demonstrated habituation of Magellanic penguins to repeated short and consistent human presence, 3 years of exposure to 1 h of unregulated visitation per day was insufficient to result in habituation at the same colony (Fowler, 1999). Habituation appears to require a maximum of predictable low-level disturbance. However, what 'low level disturbance' actually is appears to be dependent on species and location. Although Magellanic penguins habituated to standardized daily visits at close proximity (<5 m), Humboldt penguins did not show signs of habituation to the same person passing at 20 m or 50 m from the nest (Ellenberg *et al.*, 2006). Thus, habituation by penguins to even apparently minor human disturbance cannot by assumed.

proper visitor management guidelines, nor can results for one species be applied to another species in a different setting.

For example, Adélie penguins tending large chicks late in the season will flee if approached to within ~6 m, whereas when chicks are young the attending parent will tolerate approach to within ~1 m (Wilson *et al.*, 1991). Birds not tied to a nest site show much more marked avoidance behaviour. Adélie penguins commuting between landing beach and colony will deviate up to 70 m off established paths to avoid a solitary human standing 20 m away, and this deviated route will be maintained for several hours after the person has left resulting in an estimated extra 840 penguin km covered by the 12,000 birds on the track over a 10 h observation period (Culik and Wilson, 1995). On the Falkland Islands Gentoo, Magellanic and King penguins show avoidance behaviours in response to tourists near penguin access paths (Otley, 2005). Yellow-eyed penguins will delay even coming ashore if people are present at or near landing sites (Wright, 1998) (see Fig. 9.2), and delayed landings during the chick rearing

Fig. 9.2. Yellow-eyed penguins run the tourist gauntlet at a beach in Southern New Zealand. These two penguins have waited for more than an hour in the surf before they found a gap in the flow of tourist visitors that they considered big enough to enable them to cross the beach and return to their nest sites (Photograph Hermann Ellenberg).

period may disrupt chick feeding and result in reduced fledging weights with possible implications for post-fledging survival (McClung *et al.*, 2004).

There is evidence that Chinstrap penguins make risk-based assessments of human disturbance, treating humans as potential predators, thus different types of human approach and varying proximity to subcolonies will induce different types of responses in non-breeding birds (Martin *et al.*, 2004). In African penguins, a gradual approach (with regular stops) caused less disturbance than a person approaching at a steady pace (van Heezik and Seddon, 1990), and in Humboldt penguins walking clearly past the bird (tangential approach) was less intrusive than a direct approach (Ellenberg *et al.*, 2006).

To make things even more complex, the condition and health status of a bird appears to affect its behavioural response to a potential threat. Birds whose condition had been experimentally enhanced showed greater responsiveness to standardized human disturbance (Beale and Monaghan, 2004), suggesting they could afford the energy to be vigilant or to flee at greater distances and for longer, whereas birds in control groups had to prioritize energy conservation. On the other hand, unguarded Chinstrap penguin chicks fled later and for shorter distances when they were in good health and condition, presumably because these chicks were able to defend themselves more efficiently against predator attack (Martin *et al.*, 2006).

Emperor penguins showed increased vigilance when exposed to helicopter overflights at 1000 m altitude, with 69% of chicks walking or running away from the disturbance (Giese and Riddle, 1999). Wilson *et al.* (1991) report distances of first reaction of up to 1.1 km to aircraft approaching Adélie penguin colonies, while flight behaviour could last until the aircraft was more than 2.8 km away. The potential impact of an aircraft varies with the type of aircraft, speed and altitude, bird species, ambient environment, timing, duration and frequency of exposure (see Harris, 2005). Flight distances at which disturbance is thought to be detrimental have been established via quantification of only overt behavioural reactions, such as percent of population fleeing. However, a disturbance event may additionally interrupt vital behaviour and induce freezing (e.g. Eilam, 2005). Even without any behavioural reaction to a disturbance stimulus additional energy demands can be high solely due to the physiological stress response (Regel and Pütz, 1997).

It is now well recognized that overt behavioural reactions, or lack of them, are a poor guide to the degree of disturbance human proximity or activity may be causing penguins. Thus absence of evidence in the form of alarm behaviours is not evidence of the absence of stress, and researchers have had to probe more deeply.

Physiology

Increased heart rate is part of the stress response to stimuli that are perceived by an animal as being novel, challenging or threatening. Elevated heart rate can occur independently of any overt behavioural reaction to perturbation. The first study to examine changes in penguin heart rate in response to human disturbance was that by Wilson and Culik and colleagues (Culik *et al.*, 1990; Culik and Wilson, 1991; Wilson *et al.*, 1991). During physiological field studies using heart rate recorders implanted in Adélie penguins, increases in heart rate were opportunistically measured in response to human disturbance, and found to occur even though no external signs of stress were evident (Wilson *et al.*, 1991). It was recognized that the process of capture, handling and device implantation had the potential to bias results if there was associative learning that predisposed individual penguins to extreme reactions when sighting humans subsequently (Culik and Wilson, 1995). To get around this potentially serious problem, Nimon *et al.* (1996) developed the use of artificial eggs to record heart rates. Such an egg may be placed within a nest and incubated by the attending adult and its essentially undisturbed mate. Using artificial eggs it was found that heart rate increased by 45–110% in response to the approach of a human to within 1 m of nesting penguins, presumably Gentoo penguins (Nimon *et al.*, 1995). Studies on Magellanic (Ecks, 1996) and African penguins (M. de Villiers, Ushuaia, Argentina, 2003, personal communication) measured similar heart rate responses to human approach. Snares penguins had a lower heart rate response, that quickly dropped to pre-disturbance levels, whereas Yellow-eyed penguins reacted more strongly and needed more time for recovery (Ellenberg, in preparation).

Heart rates were measured in Royal penguins on Macquarie Island in response to experimental approaches to within 5 m, the minimum approach guideline for tourists, and mean heart rate increases of 1.23 times the average resting rate were recorded, greater than that in response to predatory skua overflight (Holmes *et al.*, 2005). Although heart rates returned to normal pre-approach levels within 3 min, cumulative impacts or responses to larger visitor groups are not known and conservative guidelines would require setback distances of 30 m to avoid all physiological responses (Holmes *et al.*, 2005). Humboldt penguins are a recent tourism focus in Chile, where visitor management follows guidelines used for the apparently more disturbance-tolerant Magellanic penguin colonies. However, Humboldt penguins showed elevated heart rate responses to people visible even 150 m away, with a recovery time of up to 30 min and little evidence of habituation potential (Ellenberg *et al.*, 2006).

Elevated heart rate is one manifestation of the vertebrate stress response that is activated by the hypothalamic–pituitary–adrenal axis and mediated by an increase in 'stress hormones' (glucocorticosteroids) from the adrenocortical tissue – the adrenocortical stress response. Short-term increases in circulating levels of glucocorticosteroids enable individuals to escape from or cope with adverse conditions. However, long-term elevation of stress hormones can be physiologically damaging to individuals resulting in higher susceptibility to disease, reduced fertility and lower life expectancy (e.g. Walker *et al.*, 2005). Recent work on quantifying the costs of human disturbance on penguins has used experimental blood sampling protocols to measure changes in corticosterone (the glucocorticosteroid in birds) in response to standardized experimental disturbance. Fowler (1999) found that Magellanic penguins have significantly elevated levels of corticosterone in response to a person visible nearby for only 5 min. In an elegant experimental set up it was demonstrated that Magellanic penguins habituate to human visitation as long as the stimulus is short, intense and consistent (Walker *et al.*, 2006). Corticosterone responses to disturbance are greater in undisturbed adult Magellanic penguins compared to those in tourist areas, with no difference in baseline levels (Fowler, 1999; Walker *et al.*, 2006). However, this difference was found to be due to a decreased capability of the adrenocortical tissue to secrete corticosterone in tourist-visited birds (Walker *et al.*, 2006), and may have disadvantages under different circumstances, e.g. if birds are unable to adequately access stored energy in times of need.

In contrast to Magellanic penguins, Yellow-eyed penguins appear to have been sensitized by human disturbance and show a stronger initial stress response at a breeding site exposed to unregulated tourism compared to an undisturbed area (Ellenberg *et al.*, 2007). In the first study of its kind, Walker *et al.* (2005) examined stress hormone responses of Magellanic penguin chicks, showing that newly hatched chicks in tourist-visited areas had higher corticosterone responses than newly hatched ones in undisturbed areas. So far, it is not known what the longer-term developmental consequences may be of elevated corticosterone responses at hatching (Walker *et al.*, 2005).

Despite the increasing application of sophisticated techniques to quantify physiological responses by penguins to human disturbance, an obvious question remains: so what? So what if heart rate is elevated for certain periods of time? What is the energetic consequence of this? So what if some species of penguin show robust corticosterone responses? Does this carry any fitness costs?

It is generally assumed that the physiological stress response can be correlated to actual energy consumption and thus the metabolic costs of disturbance; however, to date very few studies have attempted the actual calculations. Regel and Pütz (1997) found via an ingested temperature logger that moulting Emperor penguins increased their metabolic rate without showing the slightest behavioural reaction to human approach. Although human presence lasted for less than 5 min, some birds needed several hours to recover. One single disturbance event resulted in an additional energy expenditure of up to 10% of the daily energy demand during moult. As an example they calculated that after one single visit to the resident colony of 6000 pairs the penguins would need a total of 310 kg of krill to compensate for this disturbance, which of course is hardly possible during moult when penguins are confined to land.

The relative significance of a disturbance stimulus will depend on the circumstances: in situations where food is ample and easily obtained, additional energy expenditure may not be a problem; however, the long moult fasts and parts of the breeding cycle require a delicate energy balance. Any additional energy expenditure may lead to starvation or nest desertion. Beyond any individual animal welfare concerns, a key issue is whether human disturbance carries the potential to have population-level consequences that may threaten a species. This is of concern not only for conservation authorities and wildlife managers, but also for tourism operators whose livelihood depends on the maintenance of healthy penguin populations.

Population-level change: distribution, breeding success and abundance

Population-level responses encompass the cumulative effects of individual physiological, behavioural and survival consequences, and include changes in the distribution of breeding colonies through avoidance of human disturbance, declines in breeding success and declines in absolute numbers of breeding birds. The picture to date is far from clear, principally because many other factors apart from human disturbance may affect distribution, breeding success and abundance. Climate, nesting habitat, food availability or natural predation have an influence on the dynamics of individual populations. Extreme weather conditions were the main factor limiting Magellanic penguin breeding success at Carbo Vírgenes, Argentina (Frere et al., 1998), resulting in high nest desertion rates and chick mortality. Although not intended at the time, human modification of Pájaro Niño Island in Central Chile has resulted in higher numbers of nesting Humboldt penguins (Simeone and Bernal, 2000): the construction of a breakwater wall and removal of pine forest provided new nesting habitat while human disturbance was reduced by marina personnel now controlling human access to breeding sites. Gentoo penguin breeding success on Macquarie Island is related to habitat, colony size, presence of other penguin species and the proximity of southern elephant seal harems (Holmes et al., 2006), and the authors suggest that differences related to geographical position of colony sites may be explained by quality of foraging areas around the island. A strong dependence on particular oceanographic conditions for successful breeding has been documented for several penguin species. In Little penguins, for example, 29–47% of the between-year variation in weight and breeding performance is explained by variation in sea surface temperatures; eggs were laid later in the season if sea surface temperatures were higher, presumably when schooling fish were scarcer (Wooler *et al.*, 1991; Dann, 1992; Mickelson *et al.*, 1992). Potentially adverse effects of tourism and research on Adélie and Gentoo penguins may be negligible relative to the effects imposed by long-term changes in other environmental variables (Fraser and Patterson, 1997; Cobley *et al.*, 2000). The increasing disturbance by humans in Hope Bay, Antarctic Peninsula, seems only to have changed the shape of the rookery and the penguins' paths to the sea, but not to have stopped the growth of the rookery (Zale, 1994) which is attributed to locally favourable foraging conditions.

Nevertheless, there is evidence that human disturbance has caused reduced breeding success in several penguin species. Mixed colonies of Humboldt and Magellanic penguins on the Puñihuil Islands in southern Chile have been subject to unregulated tourism activities since 1985, resulting in increased incidence of nesting burrow collapse and possible declines in the numbers of both species since 1991 (Simeone and Schlatter, 1998). Human presence provoked increased predation of Fiordland penguin chicks by Weka, a flightless rail endemic to New Zealand (St Clair and St Clair, 1992), and further south, human disturbance at penguin colonies may be used by skuas or giant petrels to their predatory advantage (Giese, 1996; Descamps *et al.*, 2005). Additionally, the balance between penguin and skua populations appears to be of crucial importance. At Cape Crozier, penguin breeding groups reduced in size through human disturbance were unable to resist skua attacks (Oelke, 1978). However, reduced breeding success may be caused by much more subtle causes than destruction of nesting borrows or facilitation of predation. For example, human passage through low-density breeding areas of African penguins on Jutten Island, South Africa, caused not only egg loss and the exodus of birds but prevented nest-site prospecting (Hockey and Hallinan, 1981). Similarly, human visits may adversely affect the recruitment of pre-breeding birds to Adélie penguin colonies (Woehler *et al.*, 1994). Chronically stressed birds are likely to show suppressed reproductive behaviour (establishment of nesting territories) and reduced fertility (Fowler *et al.*, 1995). Furthermore, ineffective brooding may lead to loss of the clutch or retarded development of the embryos, and greater energy demands on the adults arising due to human disturbance may leave less food for the chicks.

Adélie penguin hatching success was as much as 47% lower and chick survival reduced by 80% in colonies exposed to recreational visits compared to undisturbed sites, whereas investigator disturbance had less impact on breeding success (Giese, 1996). This result was surprising as a person entering a colony for nest-checks was thought to constitute a more intense disturbance event than that of a couple of tourists moving slowly around at 5 m distance from the colony's edge two to four times a day for no more than 10 min per visit.

Although predation by skuas was the main reason for low reproductive success in the disturbed colonies, cooling of eggs or chicks due to ineffective brooding during disturbance events is thought to have caused additional losses. Giese (1996) suggested that frequency of visitation, rather than type of intrusion, might be the critical factor influencing breeding success. Low frequency of disturbance events might explain the lack of difference in the breeding success of Royal and Rockhopper penguin nests that were visited by a careful observer once a week for a breeding biology study, compared with neighbouring nests that remained untouched (Hull and Wilson, 1996). Naturally, the 'type of intrusion' has to be within reasonable limits. People have long debated where to draw the line between 'disturbing to some individuals, but beneficial overall to the population' and 'too disturbing to some or all' (Cheney, 1999). Recently, Wilson and McMahon (2006) drew attention to the fact that the ethics of acceptable practice for scientific intervention is still poorly defined.

Even a short-term disturbance event can have devastating consequences. Approximately 7000 King penguins died by asphyxiation when a stampede occurred on Macquarie Island attributed to an overflight of a Hercules aircraft (Rounsevell and Binns, 1991; Cooper et al., 1994). The deaths resulted from large numbers of fleeing penguins piling up on each other against a natural barrier at one edge of the colony. Three days of helicopter operations caused a 15% decrease in Adélie penguin numbers at 11 breeding sites and 8% of active nest mortality (Wilson et al., 1991).

Over the longer term, disturbance-mediated declines in breeding success may result in reduced number of breeding birds. At Cape Hallet, site of a joint New Zealand–USA Antarctic base, Adélie penguin numbers declined while the base was in operation between 1959 and 1968, increasing again only after the base was mothballed in 1973 (Wilson et al., 1991). Adélie penguin numbers increased by as much as 928% at breeding colonies in Wilkes Land, East Antarctica, with the exception of those at Shirley Island, near the Australian Casey Station (Woehler et al., 1991), where observed changes in distribution and reductions in mean breeding success due to human visitation are believed to have prevented any population increase (Woehler et al., 1994). At Cape Bird, Ross Island, Adélie penguin breeding groups near a field station declined more than 50% over a 20-year period, even though the total population of the colony markedly increased (Young, 1990). The drastic decline in numbers of breeding Adélie penguins at Cape Royds between 1955 and 1963 is attributed to visitor disturbance (Thomson, 1977). Following regulations to control helicopter operations and visitors, as well as the introduction of a caretaker scheme, the colony at Cape Royds recovered, however, a strict cause and effect has not been established. Current management guidelines for aircraft operation are less stringent and less specific than those recommended by the Scientific Committee for Antarctic Research (SCAR) Specialist Group on Birds, and represent a compromise to accommodate operational needs. Harris (2005) argues they should be considered interim measures until new and improved research results appear.

We are still at the very beginning of understanding the far-reaching subtle and thus difficult-to-measure effects of human disturbance. Interspecific differences in response appear to be important. In Magellanic penguins, breeding success was not affected in an area where visitors can walk freely among nests and approach penguins to within a few metres of nest sites, on occasion even touching the birds (Yorio and Boersma, 1992), whereas a Humboldt penguin colony exposed to visitors at close range had virtually no reproductive output (Ellenberg *et al.*, 2006). Even supposedly robust species can be affected by human presence: on Montague Island, Australia, nest site density of Little penguins was positively correlated with distance from footpaths (Weerheim *et al.*, 2003). On the other hand, for sensitive species tourism impacts can be negligible if visitors are managed appropriately. At Port Lockroy, one of the five most-visited sites by tour ships on the Antarctic Peninsula, there was no evidence that tourism was having any impact on Gentoo penguin reproductivity (Cobley and Shears, 1999); and on Cuverville Island visitor presence near apparently well-adapted Gentoo penguin breeding groups (see Fig. 9.3) had no effect on skua predatory behaviour (Crosbie, 1999). At a site on the Otago Peninsula, New Zealand, where

Fig. 9.3. A Gentoo penguin enjoys a rare comfort. After a busy breeding season a Gentoo penguin at Cuverville Island, Antarctic Peninsula, appreciates the luxury of modern insulation. The occupant on the backpack did not even bother to get up while the camera for this shot was carefully reclaimed from a position beneath it. When the rightful owner tried to reclaim his backpack 2 h later, the Gentoo aggressively defended its cosy ground. Cuverville Island has consistently received high numbers of visitors on commercial tours since the mid-1980s (Photograph Rolf Stange).

well-informed visitors watch Yellow-eyed penguins at close range from hides and covered trenches, the rate of food transfer and breeding success was unaffected (Ratz and Thompson, 1999), whereas at a neighbouring site essentially unregulated tourism was associated with reduced breeding success and lower fledgling weights in the same species (Ellenberg et al., 2007).

Summary

The last decade has seen the development of innovative new research techniques and the application of elegant experimental protocols to overcome the problem of separating the confounding effects of observer, manipulation and instrumentation impacts from the measured disturbance response in penguins. There is now a reasonable and growing body of scientific literature exploring the behavioural and physiological responses to human proximity by a number of penguin species. What has become evident is the marked variability in responses. The available evidence suggests that the distance at which a penguin will tolerate the proximity of a human varies with the type of disturbance, according to penguin species, and intraspecifically according to individual, age, condition, current behaviour, stage of breeding and previous experience with humans.

It is clear that any generic guidelines for managing penguin tourism, especially those based on visible responses to human proximity, run the risk of resulting in harmful impacts at both individual and population levels. Even setback distances derived from conservative estimates for one species may trigger significant physiological responses with associated energetic costs in another species. Similarly, disturbance that may be tolerated at certain stages of breeding may stimulate more extreme responses at other stages or sites, and currently little is known about variability in responsiveness between the different life stages. In addition, individuals of the same species and at the same stage of breeding may react differently depending on their character and their previous experience with humans. Habituation to disturbance is not assured, and we still know relatively little about interspecific tendencies to habituation or sensitization, and the form and magnitude of stimuli involved.

Implications

As a consequence it is neither feasible nor recommended to attempt to construct any general 'rules of thumb' to manage the impacts of human disturbance in relation to penguin tourism. As penguin-focused tourism inevitably increases in scale, the potential for serious population-level impacts will increase, i.e. impacts that are additive to all the other factors acting to reduce the viability of penguin populations. It is of concern that nature-based tourism has extended into the most pristine environments, such as Antarctica, and that visitors and tourism operators are turning their gaze to even the most rare and threatened species, such as Fiordland penguins. There will inevitably be pres-

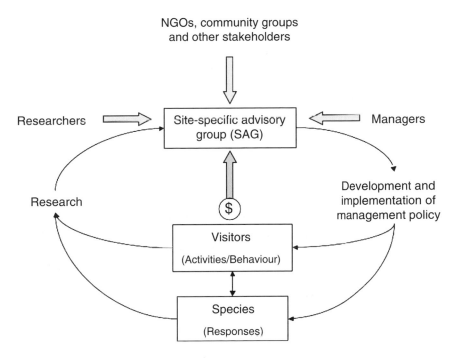

Fig. 9.4. Ideal management approach.

sure from the tourism industry to open up new areas and penguin-viewing opportunities, and from tourists themselves to allow more people closer access to penguin-photo opportunities. It is therefore important to take care in selecting sites that will be promoted for tourism in order to minimize potential impacts from disturbance. Ideally such sites would be chosen prior to tourism development; however, the following site-selection criteria may be applied retrospectively: How important is the site for the focal species (what proportion of the population will be affected by tourist activities)? What other species are present at the site (some of which may be more susceptible to human disturbance than the focal species)? Can the physical setting be used to minimize disturbance effects while enhancing visitor experiences? In the light of increasing tourist pressure improvement of the management of existing sites (where habituation may have occurred or tourist activities may already have selected for less sensitive individuals) should have priority over the opening up of new sites. In addition, key conservation areas should be identified and set aside for low-impact monitoring only. Associated research will enable the recognition of any consequences of human disturbance sufficiently early to allow management changes to minimize negative effects at the population level; such adaptive management would aim thereby to protect the tourism industry itself (Fig. 9.4).

After careful selection of the site to be promoted for tourism a Site-specific Advisory Group (SAG, including representatives of all stakeholder groups) is created. The SAG reviews research findings and has oversight for the development and implementation of management policy. Effectiveness of management tools (education, passive control, licensed guides/warden, enforcement) is regularly assessed and, if necessary, revised. Associated research considers the social as well as the biological effects of visitor–wildlife interactions (lower boxes) and will include monitoring of the type of activity, numbers of visitors involved, group size, frequency of visits, timing (day/season), approach distance, visitor behaviour, etc., as well as their effects on physiology, behaviour, reproduction, and survival of focal and associated species. Such information is vital for understanding both visitor- and species-specific needs and thus forms the basis for anticipatory and adaptive management.

Rigorous research is needed to understand the nature of human disturbance-related impacts on penguins and to derive the hard data necessary to allow conservation managers to formulate appropriate species-specific visitor management guidelines. Such research is resource hungry and it becomes increasingly important that costs be borne by end users. The significant gaps in our knowledge need to be filled urgently, since tourism development and expansion will not necessarily wait for better guidelines. The way forward is via the establishment of mutually beneficial partnerships between the tourism industry, conservation authorities, the scientific community and tourists themselves.

Acknowledgements

This chapter greatly benefited from the insightful comments of Thomas Mattern, Hermann Ellenberg and Yolanda van Heezik.

References

Beale, C. and Monaghan, P. (2004) Behavioural responses to human disturbance: a matter of choice? *Animal Behaviour* 68, 1065–1069.

Boersma, P.D. and Stokes, D.L. (1995) Conservation: threats to penguin populations. In: Williams, T.D. (ed.) *The Penguins*. Oxford University Press, Oxford, pp. 127–139.

Cheney, C. (1999) Ecotourism and penguins: measuring stress effects in Magellanic penguins. *Penguin Conservation* 12, 24–25.

Cobley, N.D. and Shears, J.R. (1999) Breeding performance of Gentoo penguins (*Pygoscelis papua*) at a colony exposed to high levels of human disturbance. *Polar Biology* 21, 355–360.

Cobley, N.D., Shears, J.R. and Downie, R.H. (2000) The impact of tourists on Gentoo penguins at Port Lockroy, Antarctic Peninsula. In: Davison, W., Howard-Williams, C. and Broady, P. (eds) *Antarctic Ecosystems: Models for Wider Ecological Understanding*. New Zealand Natural Sciences, The Caxton Press, Christchurch, pp. 319–323.

Cooper, J., Avenant, N.L. and Lafite, P.W. (1994) Airdrops and king penguins: a potential conservation problem at sub-Antarctic Marion island. *Polar Record* 30, 277–282.

Crosbie, K. (1999) Interactions between skuas *Catharacta* sp. and Gentoo penguins *Pygoscelis papua* in relation to tourist activities at Cuverville Island, Antarctic Peninsula. *Marine Ornithology* 27, 195–197.

Culik, B.M. and Wilson, R.P. (1991) Penguins crowded out? *Nature* 351, 340.

Culik, B.M. and Wilson, R.P. (1995) Penguins disturbed by tourists. *Nature* 376, 301.

Culik, B.M., Adelung, D. and Woakes, A.J. (1990) The effects of disturbance on the heart rate and behaviour of Adélie penguins (*Pygoscelis adeliae*) during the breeding season. In: Kerry,

K.R. and Hempel, G. (eds) *Antarctic Ecosystems. Ecological Change and Conservation.* Springer, pp. 177–182.

Dann, P. (1992) Distribution, population trends and factors influencing the population size of little penguins *Eudyptula minor* on Phillip Island. *Emu* 91, 263–272.

Descamps, S., Gauthier-Clerk, M., Le Bohec, C., Gendner, J.-P., Le Maho, Y. (2005) Impact of predation on king penguin *Aptenodytes patagonicus* in Crozet Archipelago. *Polar Biology* 28, 13–16.

Domjan, M. (2003) *The Principles of Learning and Behaviour*, 5th edn. Wadsworth/Thomson Learning, Belmont, California.

Ecks, M. (1996) Einfluß von Störungen auf die Herzschlagrate brütender Magellanpinguine *Spheniscus magellanicus*. Diplomarbeit, Friedrich-Wilhelms-Universität, Bonn.

Eilam, D. (2005) Die hard: a blend of freezing and fleeing as a dynamic defense – implications for the control of defensive behavior. *Neuroscience and Biobehavioral Reviews* 29, 1181–1191.

Ellenberg, U., Mattern, T., Seddon, P.J. and Luna Jorquera, G. (2006) Physiological and reproductive consequences of human disturbance in Humboldt penguins: the need for species-specific visitor management. *Biological Conservation* 133, 95–106.

Ellenberg, U., Setiawan, A.N., Cree, A., Houston, D.M. and Seddon, P.J. (2007) Elevated hormonal stress response and reduced reproductive output in Yellow-eyed penguins exposed to unregulated tourism. *General and Comparative Endocrinology* 152, 54–63.

Fowler, G., Wingfield, J. and Boersma, P. (1995) Hormonal and reproductive effects of low levels of petroleum fouling in Magellanic penguins. *Auk* 112, 382–389.

Fowler, G.S. (1999) Behavioral and hormonal responses of Magellanic penguins (*Spheniscus magellanicus*) to tourism and nest site visitation. *Biological Conservation* 90, 143–149.

Fraser, W.R. and Patterson, D.L. (1997) Human disturbance and long-term changes in Adélie penguin populations: a natural experiment at Palmer Station, Antarctic Peninsula In: Battaglia, B., Valencia, J. and Walton, D.H. (eds) *Antarctic Communities: Species, Structure and Survival*. Cambridge University Press, Cambridge, pp. 445–446.

Frere, E., Gandini, P. and Boersma, P.D. (1998) The breeding ecology of Magellanic penguins at Cabo Vírgenes, Argentina: what factors determine reproductive success? *Colonial Waterbirds* 21, 205–210.

Giese, M. (1996) Effects of human activity on Adelie penguin *Pycoscelis adeliae* breeding success. *Biological Conservation* 75, 157–164.

Giese, M. and Riddle, M. (1999) Disturbance of Emperor penguin *Aptenodytes fosteri* chicks by helicopters. *Polar Biology* 22, 366–371.

Harris, C. (2005) Aircraft operations near concentrations of birds in Antarctica: the development of practical guidelines. *Biological Conservation* 125, 309–322.

Hockey, P.A.R. and Hallinan, J. (1981) Effect of human disturbance on the breeding behaviour of Jackass penguins *Spheniscus demersus*. *South African Journal of Wildlife Research* 11, 59–62.

Holmes, N., Giese, M. and Kriwoken, L.K. (2005) Testing the minimum approach distance guidelines for incubating Royal penguins *Eudyptes schlegeli*. *Biological Conservation* 126, 339–350.

Holmes, N., Giese, M., Achurch, H., Robinson, S. and Kriwoken, L.K. (2006) Behaviour and breeding success of gentoo penguins *Pygoscelis papua* in areas of low and high human activity. *Polar Biology* 29, 399–412.

Hull, C.L. and Wilson, J. (1996) The effect of investigators on the breeding success of Royal, *Eudyptes schlegeli*, and Rockhopper penguins, *E. chrysocome*, at Macquarie Island. *Polar Biology* 16, 335–337.

Martin, J., de Neve, L., Fargallo, J.A., Polo, V. and Soler, M. (2004) Factors affecting the escape behaviour of juvenile Chinstrap penguins, *Pygoscelis antarctica*, in response to human disturbance. *Polar Biology* 27, 775–781.

Martin, J., de Neve, L., Fargallo, J.A., Polo, V. and Soler, M. (2006) Health-dependent vulnerability to predation affects escape responses of unguarded chinstrap penguin chicks. *Behavioral Ecology and Sociobiology* 60, 778–784.

McClung, M.R., Seddon, P.J., Massaro, M. and Setiawan, A.N. (2004) Nature-based tourism impacts on Yellow-eyed Penguins *Megadyptes antipodes*: does unregulated visitor access affect fledging weight and juvenile survival? *Biological Conservation* 119, 279–285.

Mickelson, M.J., Dann, P. and Cullen, J.M. (1992) Sea temperatures in bass strait and breeding success of little penguins *Eudyptula minor* at Phillip Island, South-eastern Australia. *Emu* 91, 355–368.

Nimon, A.J., Schroter, R.C. and Stonehouse, B. (1995) Heart rate of disturbed penguins. *Nature* 374, 415.

Nimon, A.J., Schroter, R.C. and Oxenham, R.C. (1996) Artificial eggs: measuring heart rate and effects of disturbance in nesting penguins. *Physiology and Behavior* 60, 1019–1022.

Nisbet, I.C.T. (2000) Disturbance, habituation, and management of waterbird colonies. *Waterbirds* 23, 312–332.

Oelke, H. (1978) Natürliche oder anthropogene populationsveränderungen von Adéliepinguinen (*Pygoscelis adeliae*) im Ross-Meer-Sector der Antarktis. *Journal für Ornithologie* 119, 1–13.

Otley, H.M. (2005) Nature-based tourism: experiences at the volunteer point Penguin colony in the Falkland islands. *Marine Ornithology* 33, 181–187.

Petersen, S.L., Branch, G.M., Ainely, D.G., Boersma, P.D., Cooper, J. and Woehler, E.J. (2005) Is flipper banding of penguins a problem? *Marine Ornithology* 33, 75–79.

Ratz, H. and Thompson, C. (1999) Who is watching whom? Checks for impacts of tourists on Yellow-eyed penguins *Megadyptes antipodes*. *Marine Ornithology* 27, 205–210.

Regel, J. and Pütz, K. (1997) Effect of human disturbance on body temperature and energy expenditure in penguins. *Polar Biology* 18, 246–253.

Rounsevell, D. and Binns, D. (1991) Mass deaths of King penguins (*Aptenodytes patagonicus*) at Lusitiana Bay, Macquarie Island. *Aurora* 10, 8–10.

Simeone, A. and Bernal, M. (2000) Effects of habitat modification on breeding seabirds: a case study in central Chile. *Waterbirds* 23, 449–456.

Simeone, A. and Schlatter, R.P. (1998) Threats to a mixed species colony of *Spheniscus* penguins in southern Chile. *Colonial Waterbirds* 21, 418–421.

St Clair, C.C. and St. Clair, R.C. (1992) Weka predation on eggs and chicks of Fiordland crested penguins. *Notornis* 39, 60–63.

Thomson, R.B. (1977) Effects of human disturbance on an Adelie penguin rookery and measures of control. In: Llano, G.A. (ed.) *Adaptations within the Antarctic Ecosystems*. Smithsonian Inst., Washington, DC, pp. 1117–1180.

van Heezik, Y. and Seddon, P.J. (1990) Effect of human disturbance on beach groups of jackass penguins. *South African Journal for Wildlife Research* 20, 89–93.

Walker, B.G., Boersma, P.D. and Wingfield, J.C. (2005) Physiological and behavioral differences in Magellanic penguin chicks in undisturbed and tourist-visited locations of a colony. *Conservation Biology* 19, 1571–1577.

Walker, B.G., Boersma, P.D. and Wingfield, J.C. (2006) Habituation of adult Magellanic Penguins to human visitation as expressed through behavior and corticosterone secretion. *Conservation Biology* 20, 146–154.

Weerheim, M., Klomp, N., Brunsting, A.M.H. and Komdeur, J. (2003) Population size, breeding habitat and nest site distribution of little penguins (*Eudyptula minor*) on Montague Island, New South Wales. *Wildlife Research* 30, 151–157.

Wilson, K.J., Taylor, R.H. and Barton, K.J. (1990) The impact of man on Adélie Penguins at Cape Hallet, Antarctica. In: Kerry, K.R. and Hempel, G. (eds) *Antarctic Ecosystems: Ecological Change and Conservation*. Springer, Berlin, pp. 183–190.

Wilson, R. and McMahon, C. (2006) Measuring devices on wild animals: what constitutes acceptable practice? *Frontiers in Ecology and the Environment* 4, 147–154.

Wilson, R.P., Coria, N.R., Spairani, H.J., Adelung, D. and Culik, B.M. (1989) Human-induced behaviour in Adélie penguins *Pygoscelis adeliae. Polar Biology* 10, 77–80.

Wilson, R.P., Culik, B., Danfeld, R. and Adelung, D. (1991) People in Antarctica – how much do Adélie Penguins (*Pygoscelis adeliae*) care? *Polar Biology* 11, 363–371.

Woehler, E.J., Slip, D.J., Robertson, L.M., Fullagar, P.J. and Burton, H.R. (1991) The distribution, abundance and status of Adélie penguins *Pygoscelis adeliae* at the Windmill Islands, Wilkes Land, Antarctica. *Marine Ornithology* 19, 1–18.

Woehler, E.J., Penney, R.L., Creet, S.M. and Burton, H.R. (1994) Impacts of human visitors on breeding success and long-term population trends in Adélie penguins at Casey, Antarctica. *Polar Biology* 14, 269–274.

Wooler, R.D., Dunlop, J.N., Klomp, N.I., Meathrel, C.E. and Wienecki, B. (1991) Seabird abundance, distribution and breeding patterns in relation to the Leeuwin Current. *Journal of the Royal Society of Western Australia* 74, 129–132.

Wright, M. (1998) Ecotourism on Otago Peninsula: preliminary studies of hoihos (*Megadyptes antipodes*) and Hookers sealion (*Phocartos hookerii*), Science for Conservation Report 68. Department of Conservation, Wellington.

Yorio, P. and Boersma, P.D. (1992) The effects of human disturbance on Magellanic penguin (*Spheniscus magellanicus*) behaviour and breeding success. *Bird Conservation International* 2, 161–173.

Young, E.C. (1990) Long-term stability and human impact in Antarctic skuas and Adélie penguins. In: Kerry, K.R. and Hempel, G. (ed.) *Antarctic Ecosystems: Ecological Change and Conservation*. Springer, Berlin, pp. 231–236.

Zale, R. (1994) Changes in size of the Hope Bay Adélie penguin rookery as inferred from Lake Boeckella sediment. *Ecography* 17, 297–304.

10 Impacts of Tourism on Pinnipeds and Implications for Tourism Management

D. NEWSOME AND K. RODGER

Introduction

Pinnipeds are fin-footed marine mammals with front and hind flippers, such as seals, sea lions and walruses. The behavioural traits of pinnipeds make them appealing for tourism with viewing opportunities ranging from boat cruises, to swim-with interactions to guided onshore tours (Kirkwood *et al.*, 2003). Tourism interest in pinnipeds is increasing in importance and involves a wide range of species utilizing islands and coastlines at various locations around the world. For example, Young (1998) reported 117 boat-based seal-watching operations involving some 500,000 visitors participating in the UK and Ireland in 1997. Kirkwood *et al.* (2003) note that there are some 80 pinniped tourism sites in the southern hemisphere with a yearly economic value of around US$12 million, with the Australian component comprising some 53 operators visiting 23 sites and involving around 400,000 tourists. Other important southern hemisphere locations include the Kaikoura Peninsula, New Zealand (~250,000 tourists per annum); Duiker Island, South Africa (~200,000 tourists per annum); and the Peninsula Valdez, Argentina (~150,000 tourists per annum). Pinniped tourism is also an important activity at locations in North America, the Galápagos Islands and Europe (Table 10.1).

One of the most impressive pinniped breeding sites in the northern hemisphere occurs on San Miguel Island in the Channel Islands National Park and Marine Sanctuary, California, USA. Here, there are approximately 70,000 Californian sea lions (*Zalophus californicus*), 50,000 northern elephant seals (*Mirounga angustirostris*), 5000 northern fur seals (*Callorhinus ursinus*) and 1000 harbour seals (*Phoca vitulina concolor*) (US National Park Service, 2006). Annual tourist numbers run at 60,000 to the Marine Park waters with 30,000 tourist visits to the islands (Channel Islands National Park, 2006).

Pinniped tourism comprises visitation to, and the viewing of, seals and sea lions at breeding colonies and/or haul-out sites. The spectrum of tourism activity comprises both boat-based observation and swim-with tours or land-based watching and/or

Table 10.1. Examples of tourism activities and locations based on the presence of pinnipeds in the wild.

Viewing experience/activity	Species involved	Locations
Use of telescopes and remote video facilities	Australian fur seal	Phillip Island and Seal Rocks Sea Life Centre, Victoria, Australia
Guided tours onshore (boat and/or land-based access)	Australian sea lion	Carnac Island, Western Australia; Seal Bay, Kangaroo Island, South Australia
	Northern elephant seal	Monteray Bay, California, USA
	Walrus	Walrus Islands State Game Sanctuary, Alaska, USA
	Hawaiian monk seal	Hawaii, USA
	Southern elephant seal/ southern sea lion	Peninsula Valdez, Argentina
	Cape fur seal	Cape Cross, Namibia
	New Zealand fur seal	Kaikoura Peninsula, New Zealand
	Weddell seals	Antarctica
Guided tours via boat	Australian fur seal	Montagu Island, New South Wales, Australia
	Common seal	Farne Islands, England
	California sea lions/ Harbour seals/ Northern elephant seal	Channel Islands National Park, California, USA
	Cape fur seal	Duiker Island, South Africa
Sea kayaking	Grey seals	Sweden; Norway
	New Zealand fur seal	Kaikoura Peninsula, New Zealand
Swim-with tours	Australian fur seal	Montagu Island, New South Wales; Port Phillip Bay, Victoria, Australia
	Australian sea lion	Baird Bay, South Australia
	Galápagos fur seal	Galápagos Islands, Ecuador
	New Zealand fur seal	Kaikoura Peninsula, New Zealand
Scuba-diving interactions	Australian fur seal	Montagu Island, New South Wales, Australia
	California sea lions/ Harbour seals/ Northern elephant seal	Channel Islands National Park, California, USA

swimming activities where pinnipeds occur in close proximity to disembarkation points at suitable mainland sites (Table 10.1). In many cases guides are present to facilitate the pinniped-viewing experience, providing interpretation and visitor management.

Although pinniped tourism can be recognized as a discrete subsector of ecotourism-based operations it often forms part of a general marine tourism/sea mammal tourism package. In these situations the focus of attention may also be cetaceans and/or seabird breeding colonies. Pinnipeds may also be incidental to other tourism/recreational activities such as hiking, picnicking and general recreational beach activities that take place in areas that are important for pinnipeds, such as offshore islands that occur in close proximity to urban centres. In such locations, people frequently come into contact with pinnipeds in the absence of guides (Fig. 10.1). This particular form of human intrusion can result in disturbance and interference such as displacement, stampedes, boat strikes and food provisioning (e.g. Lewis, 1987; Constantine, 1999; Shaughnessy, 1999). An example of the problems caused by general recreation is that of the Hawaiian monk seal (*Monachus schauinslandi*) where recreational disturbance by visitors on beaches used by seals as haul-out sites caused a decline in seal numbers. Human use over a period of some 30 years resulted in the hauling-out sites being abandoned. With protection and a reduction in disturbance at Kure Atoll, Hawaii, the seal populations showed recovery over time (Gerrodette and Gilmartin, 1990).

In recent years, concerns have also been raised as to the cumulative threats to the breeding populations of many species of pinniped. These include climate change, over-fishing, pollution, hunting, by-catch in commercial fishing nets, conflicts relating to aquaculture facilities and disturbance due to human intrusion at both

Fig. 10.1. General recreational beach activities taking place on Carnac Island, Western Australia. The area is an important haul-out site for the Australian sea lion. (Photograph Jean-Paul Orsini.)

haul-out areas and breeding sites (e.g. Kirkwood *et al.*, 2003; Forcada *et al.*, 2005; McMahon and Burton, 2005; National Seal Strategy Group, 2005). Given that tourism focuses on many protected populations it is important that tourism activities and influences are understood so that significant habitat, breeding sites and foraging areas are not disturbed to the extent that it results in negative impacts. There is a growing literature on the negative impacts of tourism and recreation on wildlife (e.g. see Newsome *et al.*, 2005) and an emerging database on the effects of tourism on pinnipeds (e.g. Kovacs and Innes, 1990; Shaughnessy *et al.*, 1999; Cassini, 2001; Lelli and Harris, 2001; Orsini and Newsome, 2005; Orsini *et al.*, 2006).

The objective of this chapter, therefore, is to provide an overview of pinniped tourism by addressing three specific questions that relate to its increasing popularity, concerns over impacts and management effectiveness and the overall sustainability of pinniped tourism:

1. What is the importance of gaining knowledge on visitor attitudes and expectations regarding pinniped-viewing activities?
Given that a major part of understanding tourism impacts on wildlife relates to visitor attitudes and behaviour (Newsome *et al.*, 2005), the first question pertains to the visitors themselves.
2. How do pinnipeds respond to the presence of tourists?
In relation to ecological impacts there is an increasing database that indicates that different species respond differently to disturbance and the significance of disturbance will also vary according to whether the focus of tourism is on a breeding site or haul-out area (e.g. Birtles *et al.*, 2001). The second question, therefore, considers this research in order to gain an understanding as to how pinnipeds react to human intrusion and how such data is collected.
3. How effective are management strategies that are designed to mitigate the negative impacts of tourism?
Given that pinniped tourism has increased rapidly in recent years and is expanding in an unregulated fashion (Kirkwood *et al.*, 2003) at many locations it is important to gauge the nature and effectiveness of management. Hence, a third question relating to how human disturbance can be minimized.

In exploring the answers to these questions, this chapter provides some insights into visitor attitudes and expectations and the impacts of tourism at breeding and haul-out sites. The relative significance of boat-based, land-based and swim-with activities is also explored along with current views at to how pinniped tourism should be investigated and managed.

What Is the Importance of Gaining Knowledge on Visitor Attitudes and Expectations Regarding Pinniped-viewing Activities?

Overview

This is an important area of wildlife tourism research because, by understanding the desires of the wildlife tourist, it is possible to anticipate potential problems and design appropriate management accordingly. Interpretation strategies can be better informed if expectations about tourist–wildlife interaction and the risks of impacts are appreciated. Ranger presence and interpretation can also be used to manage inappropriate behaviours that may disrupt pinnipeds. For example, Cassini (2001) found that when visitors ran, shouted and waved (moved) their arms, South American fur seals (*Arctocephalus australis*) reacted more strongly than when approached in a calm manner. Boren (2001) observed human behaviours during seal swims. She recorded the chasing of seals in the water, people approaching to within 10 m, splashing seals with flippers, encircling seals and people engaging in sudden or loud movements towards seals.

Given that some visitors may wish to get close, photograph and touch pinnipeds, it is important that expectations be managed through appropriate wildlife tourism experience marketing strategies. Such marketing plans can be designed to introduce appropriate behaviours before the visitor arrives on site. Illustrations in brochures could show tourists crouching at a viewing experience at reasonable distance. Minimal impact approaches such as these are born out by the work of Kovacs and Innes (1990) who found that breeding harp seals (*Phoca groenlandica*) did not react visibly to tourists who did not approach closely, moved slowly, crouched down and remained calm during the viewing experience.

Investigating the human dimension of pinniped tourism

The human dimensions of wildlife tourism have been increasingly recognized as a necessity for management (Orams, 2000; Lewis and Newsome, 2003; Orsini and Newsome, 2005). Data collected from those interacting with the wildlife can provide useful information to not only assist managers in understanding the nature of the human–wildlife interaction, but also how to help manage the situation. This is of particular concern giving the rising importance of public safety issues due to the increased risk to human safety from visitor–wildlife interactions (e.g. Fig. 10.1). Issues include the protection of visitors from potential aggressive behaviour and the prevention of transmitting diseases between visitors and the animals (National Seal Strategy Group, 2005; Orsini and Newsome, 2005). Yet, despite the increasing recognition for its need, little research on the human dimensions of pinniped tourism has been undertaken.

The two primary methods used to investigate the human dimensions of pinniped tourism include visitor surveys and observation of tourist behaviour (see Barton *et al.*, 1998; Martinez, 2003; Orsini and Newsome, 2005). The utility of data collected from visitor surveys and observation is twofold. It can provide valuable information for managers to assist with conservation as well as insight into the purpose and profile of visitors who engage in wildlife-viewing activities (Orsini, 2004). Outcomes from the research include an understanding of the activities people undertake in the presence of pinnipeds, their expectations including the level and type of interaction they desire and their satisfaction with their experience. Furthermore, investigating human dimensions of pinniped tourism can reveal visitors' perceptions of their impacts upon wildlife, their views on management and their attitudes to the wildlife (Barton *et al.*, 1998; Orsini and Newsome, 2005).

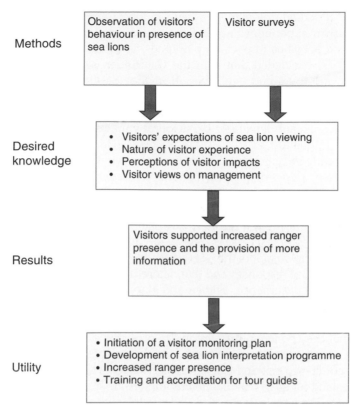

| Methods | Observation of visitors' behaviour in presence of sea lions | Visitor surveys |

Desired knowledge
- Visitors' expectations of sea lion viewing
- Nature of visitor experience
- Perceptions of visitor impacts
- Visitor views on management

Results
Visitors supported increased ranger presence and the provision of more information

Utility
- Initiation of a visitor monitoring plan
- Development of sea lion interpretation programme
- Increased ranger presence
- Training and accreditation for tour guides

Fig. 10.2. Investigating the human dimensions of sea lion tourism on Carnac Island, Western Australia. (Adapted from Orsini and Newsome, 2005.)

The utility of the knowledge gained from these methods is significant. The knowledge provides for management of visitor–wildlife interactions ensuring minimization of impacts on the wildlife and increased visitor satisfaction. A clear example of the importance of human dimensions and their implications for pinniped tourism and management can be seen in the research undertaken by Orsini and Newsome (2005) (Fig. 10.2). This study examined visitor perceptions of hauled-out sea lions on Carnac Island, Western Australia, and provided useful information on visitor perspectives and the likelihood of visitor compliance to particular management actions.

How Do Pinnipeds Respond to the Presence of Tourists?

Overview

As indicated in Table 10.1 access to, and the viewing of, pinnipeds comprise boat trips and tours, land-based pedestrian access, sea kayaking and underwater viewing via snorkelling and scuba-diving excursions. Haul-out sites and the breeding colonies of a wide range of species form the basis of wildlife tourism experiences around the world. The database, as to the extent that these activities impact on pinnipeds, is variable. For example, Tershy *et al.* (1997) concluded that with the Californian sea lions in Mexico, boat-based viewing tours had negligible effect. For Southern elephant seals (*Mirounga leonina*) in the subantarctic region, non-tourism studies suggest that the animals are very tolerant to helicopter noise and human presence on land with no significant detectable behavioural impacts from the presence of researchers (Engelhard *et al.*, 2001a,b, 2002; Burton and van den Hoff, 2002). Similarly at Baird Bay, South Australia, where there is an established tourism industry depending upon boat-based viewing tours and commercial swims with Australian sea lions (*Neophoca cinerea*), no significant impact has been observed apart from increased sea lion vigilance on land due to noise from boats and tourists (Martinez, 2003).

Birtles *et al.* (2001), however, caution that the different species do not necessarily respond the same way to human intrusion because of differing behavioural responses and that different pinnipeds may have differing susceptibilities to disturbance. Suryan and Harvey (1999) also note that variability can exist within a single species. In addition to this, there is evidence to suggest that disturbance to pinnipeds depends upon how people access a viewing area and what they do when they are in the presence of pinnipeds.

Furthermore, recent work has shown that there is variability in response according to sex and age of species. Boren *et al.* (2002) found that with New Zealand fur seals (*Arctocephalus forsteri*) that were approached on foot and in boats, the females were displaced into the sea, while males held their ground and pups attempted to move way and hide from the intrusion. Accordingly, a number of studies have shown that tourism does have the potential to increase levels of vigilance, modify behaviour, alter activity budgets, change foraging habits and potentially bring about site abandonment along with declines in reproductive success (e.g. Allen *et al.*, 1984; Kovacs and Innes, 1990; Lidgard, 1996; Barton *et al.*, 1998; Shaughnessy *et al.*, 1999; Kucey, 2005; Orsini *et al.*, 2006).

Impacts associated with tourism access to areas important for pinnipeds

The use of watercraft such as tour vessels, motorboats and kayaks to access and interact with pinnipeds is increasing worldwide. It has been noted that marine mammal response to watercraft (especially noise) will vary according to type, speed, mode of operation, distance from target species and the intensity and frequency of noise (Richardson and Würsig, 1997; McCauley and Cato, 2003). All of these factors are generally more pronounced in larger vehicles especially when moving rapidly.

In terms of boat approaches, Lelli and Harris (2001) found harbour seals were more easily disturbed by paddled boats than motorboats. They also reported that boat traffic was a strong predictor of the number of seals that were hauled out. Boats were observed to cause the majority of flushing incidents (85 during a 122-day observation period) from haul-out ledges. Allen *et al.*

(1984) had previously reported that harbour seals were less likely to haul out again if disturbance from boat traffic persists.

In terms of the issues associated with aerial sightseeing/fly in access, Born *et al.* (1999) observed that helicopters illicite greater escape responses than fixed-wing aircraft when flying over hauled-out ringed seals (*Phoca hispida*). Salter (1979) found that adult female Atlantic walrus (*Odobenus rosmarus*) respond quicker to aircraft passing over than males do.

In some cases, however, land-based approaches by humans cause the strongest response in hauled-out and breeding pinnipeds. Boren *et al.* (2002) found that New Zealand fur seals were more sensitive to approaches on land than to approaches by Kayak or boats. With Australian fur seals (*Arctocephalus pusillus doriferous*) and New Zealand fur seals on Montague Island, Australia responses to boat approaches included increased alertness and fleeing, and strong responses were detected when the animals were approached by pedestrians (Shaughnessy *et al.*, 1999). Cassini (2001) observed that with South American fur seals, pedestrian approaches of less than 10 m caused much stronger responses than approaches at greater than 10 m.

Impacts associated with tourism activity at haul-out and breeding sites

Tourists on land may approach pinnipeds in order to attain close observation, to photograph the animals, be photographed with the animals, to touch them or to create some other disturbance (prodding or throwing objects) in order to elicit a reaction such as forcing a seal/sea lion to sit up. Such incidents can result in the pinnipeds moving away from the disturbance or stimulate an aggressive response such as charging at people (Constantine, 1999). Visitors who arrive on boats, independent of tour operators, may also disturb pinnipeds in an attempt to get them to swim. Such behaviour has been observed at Seal Island, Shoalwater Bay, Western Australia where Australian sea lions hauled out on a very small beach (40 m long) were subject to hand-feeding attempts and snorkellers were observed to be enticing sea lions to the water. As a result, landing on the beach is no longer allowed at Seal Island (CALM, 1992, unpublished data).

Studies of disturbance at a grey seal (*Halichoerus gryptus*) breeding colony at Donna Brook, England revealed disruption to maternal behaviour in that females in areas of highest disturbance were more vigilant and frequently gave birth later in the season, resulting in a diminished lactation period with possible lowered growth rates for the young (Lidgard, 1996).

Kovacs and Innes (1990) examined the impact of tourism on harp seals in the Gulf of St Lawrence, Canada. Land-based viewing activities caused behavioural disturbance in the form of reduced maternal attendance. Increased vigilance by females also resulted in less time spent nursing pups. The pups themselves showed increased alert times, threat, aggressive and avoidance behaviours. When approached to within 3 m, and when touched, young seals often showed a freeze response. Normal behavioural profiles (absence of disturbance) were evident at around 1 h following the interaction with tourists.

In a study of the impacts of land- and boat-based viewing tours and commercial swims on New Zealand fur seals in Kaikoura (South Island, New Zealand), changes in fur seal behaviour were detected according to the specific site and age of the seals. For example, mother–pup pairs spent less time in 'nuzzling' contact at the Tonga Island (Tasman Bay, New Zealand) tourist site. The seals also gave stronger responses (alertness and fleeing) with pedestrian approaches and seals at undisturbed sites demonstrated higher levels of aggression and avoidance behaviour than at sites with regular human visitation (Boren, 2001; Boren *et al.*, 2002).

Kucey (2005) used single census counts, behavioural observations and detailed analyses of hauling behaviour to investigate human disruption of hauled-out Steller sea lions (*Eumetopias jubatus*) in Alaska. She found that usage of haul-out areas was significantly reduced following disturbance and that full recovery was achieved on average 4.3 days following disturbance at six out of ten sites. None the less, Kucey and Trites (2006) caution that the determination of recovery times is dependent on the criteria used and the need for researchers to take account of temporal variations in hauling-out behaviour.

Van Polanen Petel *et al.* (2001) and van Polanen Petel (2005) have reported that when approached on foot Antarctic Weddell seals (*Leptonychotes weddellii*) became alert, but the distance at which females became alert appears to relate to the numbers and ways people approach, whether the pup was positioned between people and the female and the distance from the water and other Weddell seals. Overall, threat and escape behaviours were not recorded but van Polanen Petel (2005) cautions that individual variation amongst Weddell seals needs to be taken into account when interpreting seal reactions to disturbance.

Investigating the Impact of Tourism on Pinnipeds

There are currently few, if any, long-term studies that investigate the impacts of tourism on pinnipeds (Birtles *et al.*, 2001). Most studies focus on the short-term effects with the majority undertaken several years after the establishment of tourism (Martinez, 2003). Disturbance to pinnipeds through tourism may result in delayed or missed breeding opportunities or injury, stress or mortality to the animal (National Seal Strategy Group, 2005). The biological study of impacts provides essential knowledge for the future management of visitor–wildlife interactions by examining the relationship between the number of visitors and the response of wildlife and determining disturbance thresholds (Barton *et al.*, 1998).

To investigate the impact of tourism on pinnipeds the behavioural responses of wildlife to visitors are examined. Behaviour can be defined as the external expression of an animal's response to stimuli from its internal or external environment (Hinde, 1982). It is the most common response measured in human–wildlife interactions studies. Observations of behaviour in the presence and absence of human activity can reveal what effect this has on an animal and their response to human activity (Toates, 1995). The individual changes in behaviour can reveal how that specific species perceives visitor interactions (van Polanen Petel, 2005).

Behavioural response studies are useful to examine both watercraft- and land-based interactions. Observations of seals' and sea lions' responses to visitor interactions can be investigated, in addition to their reactions to controlled approach experiments and their displacement and return rate to haul outs. To fully understand the response of species to visitors several approaches can be used. For example, to study the impacts of human visitors on Australian sea lions at Carnac Island, Orsini *et al.* (2006) examined the rate of return (haul out) of sea lions between low to moderate and high human visitation, as well as the sea lions' behavioural responses to human visitation and interaction.

Quantitative methods for sampling behaviour have been identified to allow comparisons between studies and to reduce observational bias. For behavioural studies, protocols (how long to observe an animal) and sampling methods (procedures used to sample the behaviour) need to be established (Mann, 1999). For pinniped research, the main sampling techniques (Table 10.2) have included focal animal sampling, scan sampling and incident sampling (Boren, 2001; Martinez, 2003; Orsini, 2004). To research animal populations there are advantages in identifying and following individual animals rather than groups of unidentified animals (Altmann, 1974; Mann, 1999, 2000). This approach is commonly used to gain a baseline inventory of behaviour for specific species at specific sites as it samples occurrences of important biological behaviours (Altmann, 1974). Once this information has been collected the impacts of visitors can then be determined.

A variety of variables are measured to determine species' responses to visitors. These responses are then categorized (e.g. resting, alert, active). However, in many cases different terminology is used for recording observations making comparisons between studies difficult. Although there are many difficulties with observational research, a wider use of quantitative sampling techniques (Table 10.2) will contribute to the management of this fledgling industry.

How Effective Are the Management Strategies that Are Designed to Mitigate Negative Impacts?

Overview

The viewing of pinnipeds comprises accessing and then the viewing of seals and sea lions at breeding colonies and/or haul-out sites, and may include both boat-based observation and/or swim-with tours or land-based watching and/or swimming activities. Many species of pinniped are involved and situated in a diverse range of settings from congregations in harbours and on offshore islands, close to urban centres through to remote locations in the subantarctic region and Antarctica. Human visitation can be casual comprising general recreational visitors and free independent travellers through to highly organized

Table 10.2. Sampling methods to determine behavioural responses of pinnipeds. (From Altmann, 1974; Mann, 1999; Boren, 2001; Martinez, 2003; Orsini, 2004.)

Sampling method	Desired outcome	Variables measured
Focal animal observations: Identification of specific animals that are followed for a specific period of time to understand how long the animal spends in each behavioural state	Gathering of baseline behavioural data	• Duration of interaction • Distance between visitor and animal • Type and size of group • Visitor behaviour (e.g. movement speed, voice level, actions)
Instantaneous scan sampling: Examines behaviour across the whole colony of species	Determines occurrences of particular behaviours for a large group of animals	• Animal response (e.g. resting, alert, move away, flight, aggression, decreased maternal care)
Incident sampling: Observation focused on a predetermined animal when an event takes place	Determines behavioural changes when a particular event such as tourism takes place	• Number of animals displaced from haul out and return rate to haul out after visitor interaction • Animal response (e.g. resting, alert, move away, flight, aggression, decreased maternal care) • Number of animals displaced from haul out and return rate to haul out after visitor interaction

and controlled tourism operations. Depending on the location visits may be seasonal and visitor numbers may vary greatly. An example of the potential scale of organized activities is provided by Constantine (1999) who reported that in Kaikoura, New Zealand, tour operator activities could be divided into: dolphin/seal watch/swim boats (78 trips per week), boat-based seal swimming (119 trips per week), land-based seal swimming (35 trips per week) and land-based seal watching (21 trips per week).

Orams (1999) describes a range of management approaches that can be, and are, utilized in mitigating the negative impacts of pinniped tourism (Table 10.3). Important strategies designed to manage pinniped tourism in a range of geographic situations comprise setting minimum approach distances, designation of sanctuary zones, the use of education/interpretation and site-hardening techniques.

Minimum approach distances

In order to protect pinnipeds from disturbance, managers and researchers have specified minimum distances for human approach. These distances vary

Table 10.3. General aspects of marine tourism management and applicability to pinniped tourism. (Modified from Orams, 1999.)

Management approach and associated issues	Specific components	Applicability to pinniped tourism
Physical		
Site hardening concentrates use but can foster further site development	Boardwalks, viewing areas and fence lines Sacrifice areas	Habitat protection and separation from wildlife Controlled focal point of activity
Educational		
Wide range of approaches for disseminating information to the public	Brochures that promote appropriate behaviour (includes pamphlets for tour operators)	Wording (requests to reduce noise, avoid touching and asking that visitors attempt not to illicit a reaction)
Code of conduct via brochure and signage	Signs at eye level/easy to see plus to convey information to non-English speaking visitors. Need to be encouraging and positive as not to spoil visitor experience	Signs that display information/messages (e.g. approach distances) regarding sea lion interaction. Information about viewing seals to be easily available/use of Internet
	Visitor centre: wide range of facilities and where staff/volunteers talk to tourists	Expensive to set up but can be very effective as a focal point of activities and dissemination of information
	Guided walks/talks. Need good communication/interpretation skills. Requires staffing for personal contact with visitors	Guided walks to supervise set approach distances
Regulatory		
Supervision/enforcement needed	Limit visitor numbers and access via tours. Prohibit unrestricted access, prohibit certain activities	Regulatory signs at all access points
	Zone for specific kinds of use	Exclusion zone for boats and swimmers where pinnipeds are resting
Economic		
Requires enforcement and legislative backing	Fees for group activities Fines for inappropriate behaviour	User pays guided tours

according to species and location in the world. For example, on guided walks to see northern elephant seals at Ano Nuevo, California, USA, tourists are required to stay 8 m away from the seals (Ano Nuevo State Reserve, 2000; Newsome *et al.*, 2005), while Carlson (1996) advises that the recommended approach distance for hauled-out seals in the USA is 9 m. Kirkwood *et al.* (2003) give approach distances of more than 10 m when on land for Australian sea lions in Western Australia and 10 m for land approaches to New Zealand fur seals at Kaikoura Peninsula, New Zealand.

Free, independent travellers and some tour operators who desire very close views of seals and sea lions can ignore guidelines and recommended distances for human approach. Approach distance regulations are much more likely to be effective when there is management presence. For example, visits to Australian sea lion colonies at Seal Bay, Kangaroo Island, Australia are strictly controlled as closer access (6 m approach distance) is by means of a ranger-led guided walk and directed via boardwalks (Harris and Leiper, 1995). Boren (2001) found that New Zealand fur seal response to a guided walk varied according to approach distance and size of group. She observed that the guide reduced the number of avoidance responses by up to 15%. On the other hand, Orsini *et al.* (2006) found that the vigilance of Australian sea lions did not change with varying approach distances of less than 2.5 m to greater than 15 m, or according to the number of humans present. Sea lions, in fact, remained alert at a range of distances that extended beyond the recommended 5–10 m set approach distance.

A number of recent pinniped-tourism interaction research programmes have highlighted problems associated with setting minimum approach distances. Boren (2001) also found that New Zealand fur seals responded at different distances at different sites depending on prior experience with the stimulus. Control site and breeding colony seals showed signs of disturbance at 30–60 m, increasing markedly at a distance of 30 m. Management had previously set a minimum approach distance of 5 m on land with Barton *et al.* (1998) subsequently recommending 20 m. Boren (2001) thus concluded that a minimum land approach distance of 30 m for New Zealand fur seals at non-breeding sites would reduce disturbance but a distance of 30 m at more sensitive breeding sites was unrealistic. Boren (2001) further noted that minimum approach distances should be conservative and precautionary, and added that there is scope for prohibiting land approaches at breeding colonies. Boren (2001) also remarked that the guidelines for approach by boat are also in need of clarification, as the recommended approach distances vary from site to site. Her data indicate that seals at breeding sites were responding to kayaks at a distance of 20 m with the current approach distance guideline set at 10 m for non-breeding sites and 20 m for breeding sites at Kaikoura.

Boren *et al.* (2002) also highlight the issue of what constitutes an acceptable level of disturbance and raise the question of who decides what is acceptable or not. Such a question has also been raised by van Polanen Petel (2005) who found the Australian Antarctic Division (AAD) guidelines for vehicular access to Weddell seals not to be effective because seals were observed to be responding to vehicles well before the set approach distances. Van Polanen Petel (2005) also cautioned that interpretation of disturbance to pinnipeds depends on whether the alert response is deemed to be acceptable or not. If considered acceptable, then the approach distance might be lessened for vehicular access. Additionally, she found that a 20 m separation distance for pedestrian approaches to maternal Weddell seals and pups on their own was inadequate in preventing alert responses. If, as discussed previously, the alert response is considered to be acceptable then approach distances of 15 m for females with young and 5 m for young on their own could be utilized without bringing about 'significant' disturbance.

Van Polanen Petel (2005) also observed that blanket guidelines do not allow for individual sensitivities and reactions to human presence and thus may not be able to reduce disturbance to all individuals during a viewing experience. This issue is supported by the findings of Orsini *et al.* (2006) who observed that, with hauled-out Australian sea lions on Carnac Island, Western Australia, some ignored human presence while others were displaced and left the beach. Nevertheless, if a conservative, precautionary approach is used, by setting large approach distances so that all individuals at a viewing site are not seen to react, visitor satisfaction is potentially reduced.

In a discussion on making allowance for species and site-specific details that account for individual responses when setting approach distances van Polanen Petel (2005) suggests the use of proportional odds regression models in order to generate contour maps that include distance from refuge, a factor that influences an animal's response to disturbance. Therefore, a detailed evaluation of local conditions, such as availability of cover and proximity to water, could allow tourists to approach more closely, while reducing the risk of disturbance. The proportional odds regression model then potentially becomes an important tool where close-up experiences are desired in order to generate higher levels of visitor satisfaction.

It seems, therefore, that the setting of minimum approach distances, although designed to reduce disturbance to pinnipeds if adhered to, may fail to adequately protect the species concerned. In the first instance, approach distances are dependent on compliance, and where serious breaches and harassment of pinnipeds occurs, enforcement through prosecution is required (e.g. Table 10.3). A second and perhaps more difficult aspect is the subjective nature of deciding what constitutes disturbance and, more importantly, significant disturbance. Other reasons that may contribute to the failure in adequately protecting the target species include an absence of pinniped–tourist interaction studies that test the efficacy of minimum approach distances and species, site-specific and tourism operation variability. However, as recognized by Kovacs and Innes (1990), minimum approach distances are important in reducing the negative impacts of tourism but as discussed above should be subject to ongoing review and adaptive management (see Newsome *et al.*, 2005) and used in conjunction with other management strategies.

Education and interpretive strategies

Education is a significant strategy and widely used in recreation/tourism situations (see Table 10.3). Important elements in pinniped tourism include the provision of information designed to mitigate impacts and provide the scope for tourists to learn about pinnipeds, especially hauling-out behaviour and breeding activities. Brochures are commonly used to promote appropriate behaviour (Fig. 10.3). Important focal points for interpretation programmes should be the widespread observation of, and public perception that, some species of pinniped are inquisitive and many are apparently docile in the presence of humans. In terms of sustainable tourism enterprises the success of pinniped tourism can depend on the quality and style of provided information/activities in contributing to visitor satisfaction.

Oaten and Seager (1993) report on the benefits of interpretation from a management perspective at Seal Bay, Kangaroo Island, Australia. They list the following advantages in terms of visitor management:

• Supervision of visitor behaviour;
• Information delivery on desired behaviour;
• Time limits spent on beach;
• Opportunity to educate visitors about biodiversity and conservation.

In addition to such approaches, codes of conduct have been developed at various locations but need to be supported with other educational approaches and management actions (Table 10.4). Codes of conduct or guidelines for visitor behaviour can be described in pamphlets, placed on notice boards, made

Viewing wild seals

In order to assist people wishing to view wild seals, please follow the rules set out below.

1. KEEP A SAFE DISTANCE AT ALL TIMES.

   Sea lions near Perth that are used to people may tolerate you approaching them on land to as close as 5 to 10m. Other sea lions and fur seals are likely to rush to the water, or even attack you if you approach even this close. NEVER TRY TO TOUCH WILD SEALS. The best approach on land is to move slowly, and stop **at least** 5-10m away, or sooner if the animal shows any response to your presence.

   DO NOT LEAVE UNSUPERVISED CHILDREN ANYWHERE NEAR SEALS. ALSO, NEVER WALK BETWEEN A SEAL AND THE WATER; always leave the animal an escape route to the water. Fur seals should not be approached on land, as they will almost always rush to the water, causing great disturbance to the animals. They are best observed quietly from a boat a safe distance from their rocky resting site.

   SEA LIONS AND FUR SEALS ARE VERY DANGEROUS WHEN BREEDING. It is essential to keep away from islands where they breed, and never try to walk among a colony of breeding seals. Your local CALM office can let you know where and when these animals are breeding.

2. DO NOT FEED THE SEALS: LET THE SEA LIONS AND FUR SEALS GET THEIR OWN FOOD.

   People who feed wild seals are endangering the seals and putting themselves and other people at risk. Fed seals quickly lose fear of people. Some will beg for food, behave aggressively towards people to get food, and often become dangerously entangled in hooks, lures and fishing line. On occasion, seals have seriously bitten divers, swimmers and other people, either to get food or perhaps because they see people as competing for food.

Seals accustomed to handouts may try to enter boats or steal fish from fishing nets and may become a nuisance to fishers. Seals that lose their fear of people are at great risk of being attacked by people who do not appreciate their approaches. There are many examples of serious bites being inflicted on well-intentioned people who offer fish to a 'friendly' sea lion.

3. SWIMMING TOO CLOSE TO SEA LIONS AND FUR SEALS CAN BE DANGEROUS.

   Sea lions have inflicted serious injuries to swimmers who were trying to have a close look or imitate their behaviour. Remember to keep **at least** 10m away from seals in the water. If a sea lion or fur seal approaches you closer than 10m in the water, swim slowly away and get out of the water. Seal colonies along the south and west coast are also popular haunts for great white sharks, and swimming in these waters can be dangerous.

4. LEAVE THE DOG AT HOME.

   Sea lions and fur seals do not mix with dogs. An inquisitive dog represents a threat to a seal and the results will invariably be a problem. Serious diseases can be transmitted between dogs and seals.

5. TAKE YOUR RUBBISH HOME WITH YOU.

   Apart from polluting our beautiful coastline, rubbish and debris dropped at sea or on beaches can be dangerous to sea lions and fur seals. Each year seals die a terrible slow death from entanglement or ingestion of plastic and other rubbish.

6. **RESPECT THE ANIMALS AND THEY WILL TOLERATE YOUR PRESENCE.**

   It is very easy to photograph and enjoy sea lions and fur seals in a safe manner, without disturbing them. Use common sense when boating or landing on beaches with seals and give the animals plenty of space. Don't try to get the animals to sit up or enter the water by making noises or throwing anything. Enjoy the privilege of interacting with sea lions and fur seals; if you do it in a responsible manner the opportunity will remain for future generations.

Further Information

For further information or advice, please contact the Department of Conservation and Land Management (CALM) at:

Wildlife Branch
17 Dick Perry Avenue
Technology Park, Western Precinct
KENSINGTON WA 6151
Tel: (08) 9334 0333

Or visit your nearest CALM office and CALM's NatureBase website at http://www.calm.wa.gov.au

Department of Conservation and Land Management

Fig. 10.3. Part of brochure designed to inform the public on how to behave in the presence of sea lions and fur seals in Western Australia. (From CALM, 2003.)

Table 10.4. Generic code of conduct for the observation of seals and sea lions (sea and land access).

Identified aspects for a code of conduct for visitors	Rationale	Potential methods
Respect no access zones	Sanctuary zones where wildlife can be free of disturbance from tourism	Restricted areas off limits to tourists. No approval for tour operator access
Boat/vehicle users keep to recommended approach speed	Vehicle access can disturb and displace pinnipeds	Education. Specified approach distances. Regulations applied to vehicle approach, speed and noise output
Observe with telescopes and binoculars	Reduces disturbance during breeding season. Avoids displacement from haul-out sites	Education. Specified approach distances. Guided tours
Maintain recommended approach distances	Reduces disturbance (vigilance, reduced maternal care, aggressive responses and displacement)	Education. Specified approach distances
Approach seals/sea lions in a slow and quiet fashion	Visitor behaviour such as running, fast movements, shouting and arm waving is likely to lead to greater levels of disturbance	Education, supervision, guided tours
Observe animals for signs of disturbance and/or aggression and if necessary employ a slow, quiet retreat	Visitor safety. Reduces disturbance (vigilance, reduced maternal care, aggressive responses and displacement)	Education (brochures, signage, tour guides, ranger supervision)
Do not interfere with or separate pups from females	Where close approach may occur or pedestrian access is approved at breeding sites. Guidelines for free independent tourists	Education. Specified approach distances, ranger supervision, guided tours
Limit the time spent observing target species	Reduces continuous need for target species to be vigilant. Allows for recovery time following disturbance	Education. Specified approach distances, supervision, guided tours
Do not surround seal/sea lions and leave an escape route	Where close approach may occur or pedestrian access is approved at breeding/haul-out sites. Guidelines for free independent tourists	Education. Specified approach distances, supervision, guided tours. Ranger supervision

Continued

Table 10.4. *Continued*

Identified aspects for a code of conduct for visitors	Rationale	Potential methods
No feeding or touching	Where close approach may occur or pedestrian access is approved at breeding/haul-out sites. Guidelines for free independent tourists	Education. Fines for transgression of regulations. Ranger supervision
No dogs	Potential risk at unsupervised sites. Avoids serious disturbance during breeding season. Avoids displacement from haul-out sites. Reduces risk of disease transmission	Education. Fines for transgression of regulations. Ranger supervision
Learn about seals and sea lions and their way of life	Fosters respect and appropriate behaviours	Brochures, signage, visitor centres, volunteers, ranger presence and education staff

available at tourist centres, be web based and/or be part of an interpretive walk/tour. However, such codes will only be as effective as visitor compliance. Moreover, visitors may not be aware that guidelines for interacting with sea lions exist; this may occur for various reasons such as a lack of effort on the visitor's behalf, a lack of brochures available on site, the absence of regular ranger presence and the absence of a professional tour guide (Fig. 10.4). Educational packages should also extend to the education of the industry in the form of educational displays that disseminate the results of current research, pamphlets and workshops for tour operators that could be based at visitor centres or government offices.

When and How Many Times to Visit?

Arising out of Kovacs and Innes (1990) study on the impacts of tourism on harp seals was the view that access to breeding sites should be restricted during the initial stages of the pupping season. The reasoning behind this recommendation was that if separation of mother and pup occurred before offspring recognition was set then this could result in the pups failing to feed. They indicated that separation and injury to pups could be avoided if the above restrictions and visitor education (should newborns be located during a visit) formed part of a seal interaction plan.

Boren (2001), in her study of New Zealand fur seals, reported that a large number of operators (e.g. whale-watch vessels, land- and boat-based seal swims, kayak tours, whale-watch flights and/or guided walks) were viewing and interacting with seals at various sites around New Zealand. She recommended that the number of operators and frequency of visits be regulated so that saturation does

Fig. 10.4. Recreational visitors encountoring sea lions on Carnac Island, Western Australia. Visitors are present and taking photographs well within the specificd 5–10 m approach distance. (Photograph Jean-Paul Orsini.)

not occur. Boren (2001) compares Kaikoura's 247 possible trips per week to interact with non-breeding and breeding seals with Ano Nuevo (California, USA), which has a seasonal visitation to breeding northern elephant seals of seven supervised trips a week. Possible solutions to heavy pressure situations include the spatial and temporal allocation and rotation of trips and limitations on tourism activity (e.g. number of vessels viewing/people visiting a colony) at specific locations.

In Antarctica, van Polanen Petel (2005) observed apparent short-term rapid habituation in Weddell seals evident when human approach (single person) was frequent occurring over a time period of 2 h or less. Habituation was observed not to occur when approaches were irregular and occurred over a greater time period of up to 3 weeks. In the light of this, van Polanen Petel (2005) suggests two different management options for Antarctic Weddell seals. The first involves allowing only a few visits to all seal colonies over the entire breeding season at irregular intervals. This would serve the purpose of minimizing disturbance to all individuals because it amounts to only a few visits to each seal. The second option involves designating only one colony for human visitation and protecting the remaining colonies from human disturbance. This latter strategy recognizes short-term behavioural impacts, as well as accommodating the short-term habituation findings.

Use of Fence Lines, Boardwalks and Viewing Platforms

Physical site-hardening strategies such as viewing structures and fencing are important in the direction and control of visitor access in wildlife tourism

(Newsome *et al.*, 2005). Cassini *et al.* (2004) conducted one of the first studies to report on the effectiveness of fence lines in reducing disturbance to pinnipeds, assessing how a fence influenced the behavioural responses of South American fur seals to tourist approaches. They tested the fur seals' response 1 year before and then 5 years after the fence was erected. The fence was found to reduce overall seal responses to tourists, reduce intense responses such as threat, attack and leaving the area and reduce responses stimulated by large tourist groups and intrusive visitor behaviour. In conclusion, the fence was found to be effective in reducing the most stressful and aggressive responses of seals to tourists. As tourists who gain close access have the greatest potential to disturb the seals, the fence was important in limiting human encroachment. Cassini *et al.* (2004), however, note that fencing has the potential to confine wildlife, especially where pinniped colonies may be expanding and therefore requiring more space, and thus recommend that fence-line strategies be combined with monitoring so that the position of fences can be adjusted if necessary.

An example of the applicability and use of boardwalks and viewing structures can be seen in the case of the Australian sea lion colony at Seal Bay, Kangaroo Island, Australia. Seal Bay had become a well-known tourist attraction by 1975 with a visitation of around 20,000 visitors per annum. Since then visitor numbers have increased from around 40,000 per annum in 1988 up to 102,000 per annum by 2000 (Seal Bay, 2002). A beach access boardwalk was constructed in the mid-1970s allowing close, but controlled, access to the sea lions. Before this development people were allowed free access to the breeding colony, causing disturbance to the sea lions and trampling of sensitive coastal vegetation. Today access is controlled through formally guided tours and tour guides receive training in crowd control, ways of ensuring consistent group behaviours (e.g. maintaining 6 m minimum approach distances) and how to deal with approaches by sea lions (Harris and Leiper, 1995). Moreover, Seal Bay has been subject to continuous site development and management responses in the face of ever-increasing visitor numbers (Table 10.5). A recent appraisal by Seal Bay Management suggests that the success of this management response can be measured against observations and records that show that sea lion numbers have remained stable, that pups are born in close proximity to visitors and that visitors are achieving close views without apparent undue disturbance to the sea lions (Seal Bay, 2002). Shaughnessy *et al.* (2006), however, claim that sea lion numbers actually decreased over a 13-season period up to 2003; they attribute this decline to fishing activity.

Seal Bay Park Management currently uses a range of management strategies (see Table 10.3) that have evolved over time (Table 10.5) and this has proved to be successful in dealing with an increasing visitor profile. The Seal Bay case study described by Harris and Leiper (1995) demonstrates the need for a combined management approach and the effectiveness of site hardening and control of visitor access via tour guiding in minimizing impacts on the sea lions. Nevertheless, the question remains as to how far management must go in responding to an ever-increasing tourism demand.

Table 10.5. Chronology of visitor management at Seal Bay, Kangaroo Island, South Australia. (Adapted from Harris and Leiper, 1995.)

Date	Management actions
1954	Legal protection for Australian sea lion Regulations prohibiting interference with sea lions
1967	Declaration as a fauna reserve containing two prohibited access zones providing for two secure breeding areas
1970s	Construction of a rest area for visitors and overflow car park at Bales Bay and closure and relocation of car park at Seal Bay Construction of sealed track to the beach. Development of a boardwalk lookout and viewing platform. Ranger presence during peak visitor activity
1980s	Interpretive programmes focusing on ecology and place of humans in the biosphere
1987	Limits of acceptable change planning framework applied. Monitoring of selected conditions, e.g. response of sea lions (aggressive reactions) to visitors and visitor perceptions
1987/88	Introduction of a user pays guided tour programme via restricted access zone
1987 on	Staggered arrival of tour buses in order to avoid crowding
Options for the future	Marketing to promote appropriate on-site behaviour Set upper limit of visitation to 150,000 per annum Extend viewing platform Two-tier viewing option: (i) viewing platform and interpretive centre or (ii) viewing platform and interpretive centre and a guided tour on the beach. Coach tours pre-book for staggered arrival times

Conclusion

Pinniped tourism, an emerging industry, is growing in popularity and demand. The three questions discussed in this chapter highlight the difficulties associated with sustainably managing current and future visitor–wildlife interactions. The first question highlighted the importance and need for knowledge on visitor attitudes and expectations regarding pinniped-viewing activities. By having this knowledge available, current and potential problems can be managed while ensuring visitor satisfaction with their wildlife experiences. The second question examined the impacts of tourism. Different species respond differently to disturbance and the significance of disturbance will also vary (Birtles *et al.*, 2001). To measure disturbance, species' behavioural responses are collected and categorized through a variety of sampling methods. Impacts recorded include changes in behaviour, site abandonment, change in foraging behaviour and declines in reproductive success (Allen *et al.*, 1984; Kovacs and Innes, 1990; Lidgard, 1996; Barton *et al.*, 1998; Shaughnessy *et al.*, 1999; Kucey, 2005). The third question examined strategies to minimize human disturbance. Strategies identified included minimum approach distances, educational and interpretive strategies, controls on the time and duration of visits and the use of fence lines, boardwalks and viewing platforms. Often a combined management approach is required.

A number of issues currently remain unresolved for pinniped tourism. The challenge for management is to prevent or minimize impacts from tourism interactions where close contact takes place while ensuring visitor satisfaction. If this is achieved pinniped tourism could allow for the opportunity to educate the public while conserving the wildlife. Unfortunately, short-term impacts from pinniped–visitor interactions have been identified. What effect this will have in the long term is currently unknown. Past research indicates that pinnipeds' response to visitor interactions varies between sites and species. There is still a great deal of uncertainty as behavioural responses show high individual variations and there is currently a lack of knowledge about the significance of alert behaviours. This makes it difficult to decide on general recommendations. To determine effective management strategies there is a need for research to include both the ecological impacts and the human dimensions (visitor attitudes and expectations). The use of general guidelines may not necessarily be an effective way to minimize disturbance to all pinnipeds whilst maximizing visitor satisfaction (van Polanen Petel, 2005). Until guidelines for management are improved, minimum approach distances need to be conservative and precautionary.

Acknowledgements

We would like to thank Jean-Paul Orsini for initiating tourism-related research on sea lions in Western Australia, for the provision of research articles and for photographic material. We also thank Rebecca Coyle for advice regarding the CALM seals and sea lions educational pamphlet and Peter Shaughnessy for several points of clarification.

References

Allen, S.G., Ainley, D.G., Page, G.W. and Ribic, C.A. (1984) The effect of disturbance on harbour seal haul-out patterns at Balinas Lagoon, California. *Fishery Bulletin* 82, 493–500.
Altmann, J. (1974) Observational study of behaviour sampling methods. *Behaviour* 49, 227–265.
Ano Nuevo State Reserve (2000) *Elephant Seal Walk Information Brochure*. Ano Nuevo State Reserve, California.
Barton, K., Booth, K., Simmons, D.G. and Fairweather, J.R. (1998) *Tourist and New Zealand Fur Seal Interactions Along the Kaikoura Coast*. Education Centre Report No. 9. Lincoln University, Lincoln, New Zealand.
Birtles, A., Valentine, P. and Curnock, M. (2001) *Tourism Based on Free-Ranging Marine Wildlife*. Wildlife Tourism Research Report No. 11. Status Assessment of Wildlife Tourism in Australian Series. STCRC, Gold Coast, Queensland.
Boren, L. (2001) Assessing the impact of tourism on New Zealand fur seals (*Arctocephalus forsteri*). Masters thesis, University of Canterbury, Canterbury, New Zealand.
Boren, L., Gemmell, N.J. and Barton, K.J. (2002) Tourism disturbance on New Zealand fur seals *Arctocephalus forsteri*. *Australian Mammalogy* 24, 85–95.
Born, E.W., Riget, F.F., Dietz, R. and Andriashek, D. (1999) Escape responses of hauled out ringed seals (*Phoca hispida*) to aircraft disturbance. *Polar Biology* 20, 396–403.

Burton, H. and van den Hoff, J. (2002) Humans and southern elephant seals. *Australian Mammalogy* 24, 127–139.

Carlson, C.A. (1996) *A Review of Whale Watching Guidelines and Regulations Around the World*. Report for the International Fund for Animal Welfare, UK.

Cassini, M.H. (2001) Behavioural responses of South American fur seals to approach by tourists – a brief report. *Applied Animal Behaviour Science* 71, 341–346.

Cassini, M.H., Szteren, D. and Fernandez-Juricic, E. (2004) Fence effects on the behavioral responses of South American fur seals to tourist approaches. *Journal of Ethology* 22, 127–133.

Channel Islands National Park (2006) Channel islands National Park. Available at: http://www. channel.islands.national-park-com/info.htm

Constantine, R. (1999) Effects of tourism on marine mammals in New Zealand. Science for Conservation Report No. 106. Department of Conservation, Wellington, New Zealand.

Engelhard, G.H., Creuwels, J.C.S., Broekman, M., Baarspul, A.N.J. and Reijnders, P.J.H. (2001a) Effects of human disturbance on lactation behaviour in the southern elephant seal. In: *Proceedings of the VIII SCAR International Biology Symposium: Antarctic Biology in a Global Context, 27 August–1 September 2001*, Vrije Universteit, Amsterdam, The Netherlands.

Engelhard, G.H., van den Hoff, J., Broekman, M., Baarspul, A.N.J., Field, I., Burton, H.R. and Reijnders, P.J.H. (2001b) Mass of weaned elephant seal pups in areas of low and high human presence. *Polar Biology* 24, 244–251.

Engelhard, G.H., Baarspul, A.N.J., Broekman, M., Creuwels, J.C.S. and Reijnders, P.J.H. (2002) Human disturbance, nursing behaviour, and lactational pup growth in a declining southern elephant seal (*Mirounga leonina*) population. *Canadian Journal of Zoology* 80(11), 1876–1886.

Forcada, J. Trathan, P.N., Reid, K. and Murphy, E.J. (2005) The effects of global climatic variability in pup production of Antarctic fur seals. *Ecology* 86, 2408–2417.

Gerrodette, T. and Gilmartin, W.G. (1990) Demographic consequences of changed pupping and hauling sites of the Hawaiian monk seal. *Conservation Biology* 4, 423–430.

Harris, R. and Leiper, N. (1995) *Sustainable Tourism: An Australian Perspective*. Butterworth-Heinemann, Chatswood, New South Wales, Australia.

Hinde, R.A. (1982) *Ethology: Its Nature and Relations with Other Sciences*. Fontana Press, Glasgow.

Kirkwood, R., Boren, L., Shaughnessy, P., Szteren, D., Mawson, P., Hückstädt, L., Hofmeyr, G., Oosthuizen, H., Schiavini, A., Campagna, C. and Berris, M. (2003) Pinniped-focused tourism in the Southern Hemisphere: a review of the industry. In: Gales, N., Hindell, M. and Kirkwood, R. (eds) *Marine Mammals: Fisheries, Tourism and Management Issues*. CSIRO Publishing, Collingwood, Victoria, pp. 245–264.

Kovacs, K.M. and Innes, S. (1990) The impact of tourism on harp seals (*Phoca groenlandica*) in the Gulf of St Lawrence, Canada. *Applied Animal Behaviour Science* 26(1–2), 15–26.

Kucey, L. (2005) Human disturbance and the hauling out behaviour of Steller sea lions (*Eumetopias jubatus*). MSc thesis, University of British, Columbia, Canada.

Kucey, L. and Trites, A.W. (2006) A review of the potential effects of disturbance on sea lions: assessing response and recovery. In: Trites, A.W., Atkinson, S.K., DeMaster, D.P., Fritz, L.W., Gelatt, T.S., Rea, L.D. and Wynne, K.M. (eds) *Sea Lions of the World*. Lowell Wakefield Fisheries Symposium, USA, pp. 325–352.

Lelli, B. and Harris, D.E. (2001) Human disturbances affect harbour seal haul-out behaviour: can the law protect these seals from boaters? *Macalester Environmental Review* (October), 1–13. Available at: http://www.macalester.edu/~envirost/MacEnvReview/harbor_seal.htm)

Lewis, A. and Newsome, D. (2003) Planning for stingray tourism at Hamelin Bay, Western Australia: the importance of stakeholder perspectives. *International Journal of Tourism Research* 5, 331–346.

Lewis, J.P. (1987) An evaluation of a census-related disturbance of Stellar sea lions. MSc thesis, University of Alaska, Fairbanks, Alaska.

Lidgard, D.C. (1996) The effects of human disturbance on the maternal behaviour and performance of grey seals (*Halichoerus gryptus*) at Donna Brook, Lincolnshire, England. Preliminary Report to the British Ecological Society, UK.

Mann, J. (1999) Behavioral sampling methods for Cetaceans: a review and critique. *Marine Mammal Science* 15(1), 102–122.

Mann, J. (2000) Unraveling the dynamics of social life. In: Mann, J., Connor, R.C., Tyack, P.L. and Whitehead, H. (eds) *Cetacean Societies: Field Studies of Dolphins and Whales.* University of Chicago Press, Chicago, Illinois, pp. 45–64.

Martinez, A. (2003) Swimming with sea lions: friend or foe? Impacts of tourism on Australian sea lions, *Neophoca cinerea*, at Baird Bay, S.A. Honours thesis, Flinders University of South Australia, Adelaide, South Australia.

McCauley, R.D. and Cato, D.H. (2003) Acoustics and marine mammals: introduction, importance, threats and potential as a research tool. In: Gales, N., Hindell, M. and Kirkwood, R. (eds) *Marine Mammals: Fisheries, Tourism and Management Issues.* CSIRO Publishing, Collingwood, Victoria, pp. 245–264.

McMahon, C.R. and Burton, H.R. (2005) Climate change and seal survival: evidence for environmentally mediated changes in elephant seal, *Mirounga leonina*, pup survival. *Proceedings of the Royal Society of London, Series B* 272, 923–928.

National Seal Strategy Group (2005) *Assessment of Interactions between Humans and Seals: Fisheries, Aquaculture and Tourism*, National Seal Strategy Group (final draft 2005). Australian Government Department of Agriculture, Fisheries and Forestry. Canberra, Australia.

Newsome, D., Dowling, R.K. and Moore, S.A. (2005) *Wildlife Tourism*. Channel View Publications, Clevedon, UK.

Oaten, L.R. and Seager, P.G. (1993) Visitor management (by interpretation): Kangaroo Island's seal bay. *Australian Ranger* (Autumn), 35–37.

Orams, M. (1999) *Marine Tourism: Development, Impacts and Management*. Routledge, London.

Orams, M.B. (2000) Tourists getting too close to whales: is it what whale watching is all about? *Tourism Management* 21, 561–569.

Orsini, J.-P. (2004) Human impacts on Australian sea lions, *Neophoca cinerea*, hauled out on Carnac Island (Perth, Western Australia): implications for wildlife and tourism management. Masters thesis, Murdoch University, Perth, Australia.

Orsini, J.-P. and Newsome, D. (2005) Human perceptions of hauled out sea lions (*Neophoca cinerea*) and implications for management: a case study from carnac island, Western Australia. *Tourism in Marine Environments* 2, 23–37.

Orsini, J.-P., Shaughnessy, P.D. and Newsome, D. (2006) Impacts of human visitors on Australian Sea Lions (*Neophoca cinerea*) at Carnac Island, Western Australia: implications for tourism management. *Tourism in Marine Environments* 3(2), 101–116.

Richardson, W.J. and Würsig, B. (1997) Influences of man made noise and other human actions on cetacean behaviour. *Marine and Freshwater Behavioural Physiology* 29, 183–209.

Salter, R.E. (1979) Site utilisation, activity budgets and disturbance responses of Atlantic walruses during terrestrial haul out. *Canadian Journal of Zoology* 57, 1169–1180.

Seal Bay (2002) Seal Bay ParksWeb. Available at: http://www.environment.sa.gov.au/parks/sealbay/park.html

Shaughnessy, P.D. (1999) *The Action Plan for Australian Seals*. Environment Australia, Canberra, Australia.

Shaughnessy, P.D., Nicholls, A.O. and Briggs, S.V. (1999) *Interactions between Tourists and Wildlife at Montague Island: Fur Seals, Little Penguins and Crested Terns*. Report to the South Australian National Parks and Wildlife Service. CSIRO Wildlife and Ecology, Canberra, Australia.

Shaughnessy, P.D., McIntosh, R.R., Goldsworthy, S.D., Dennis, T.E. and Berris, M. (2006) Trends in abundance of Australian sea lions, Neophoca cinerea, at Seal Bay, Kangaroo Island, South Australia. In: Trites, A.W., Atkinson, S.K., DeMaster, D.P., Fritz, L.W., Gelatt, T.S., Rea, L.D. and Wynne, K.M. (eds) *Sea Lions of the World*. Lowell Wakefield Fisheries Symposium, USA, pp. 325–352.

Suryan, R.M. and Harvey, J.T. (1999) Variability in reactions of pacific harbour seals, *Phoca vitulina richardsi*, to disturbance. *Fishery Bulletin* 97, 332–339.

Tershy, B.R., Breese, D. and Croll, D.A. (1997) Human perturbations and conservation strategies for San Pedro Martir Island, Islas del Golfo de California Reserve, Mexico. *Environmental Conservation* 24(3), 261–270.

Toates, F. (1995) *Stress, Conceptual and Biological Aspects*. Wiley, Chichester, UK.

US National Park Service (2006) Channel Islands National Park seal and sea lion viewing. Available at: http://www.nps.gov/chis/planyourvisit/seal-and-sea-lion-viewing.htm

van Polanen Petel, T., Giese, M. and Bryden, M. (2001) Measuring the effects of human activity on Weddell seals (*Leptonychotes weddellii*) in Antarctica. In: *Proceedings of the VIII SCAR International Biology Symposium: Antarctic Biology in a Global Context, 27 August–1 September 2001*. Vrije Universteit, Amsterdam, The Netherlands.

van Polanen Petel, T.D. (2005) Measuring the effect of human activity on Weddell seals (*Leptonychotes weddellii*) in Antarctica. PhD thesis, University of Tasmania, Australia.

Young, K. (1998) *Seal Watching in the UK and Ireland 1997*. International fund for Animal Welfare, UK.

11 Understanding the Impacts of Noise on Marine Mammals

D. Lusseau

Marine Mammals and Tourism

Tourists looking for more diverse experiences are ever increasingly attracted to wildlife viewing as a way to experience a sense of wilderness (Curtin, 2005). Wildlife-oriented tourism has become a major tourism activity generating a revenue of US$47–US$155 billion a year (Rodger *et al.*, 2007). Marine mammal viewing, especially whale watching, has not escaped this trend, and is now worth more than US$1 billion worldwide in total expenditure (Hoyt, 2001). Interactions between tourists and these species can take many forms from swim-with-dolphins activities (Constantine, 2003) to watching polar bears from the safety of tundra vehicles (Dyck and Baydack, 2004). Visitors can experience dolphins, whales, pinnipeds (seals) and otters from a vessel on the water. However, tourism interactions can also take place on land with pinnipeds, otters and polar bears either by foot or using a variety of motorized vehicles from quad bikes to helicopters. Each of these situations presents its own set of challenges and influences on the acoustic behaviour of the targeted species.

We are increasingly becoming aware that noise pollution can influence many aspects of the life of animals such as their reproductive success (Habib *et al.*, 2007), their foraging abilities (Erbe, 2002; Foote *et al.*, 2004), the quality of their habitat (Mace *et al.*, 1999) and fetal development (Chang and Merzenich, 2003). Hearing is the primary sensory mode of cetaceans (whales, dolphins and porpoises) and hence noise pollution has the potential to have drastic impacts on their lives. Marine mammal-oriented tourism, especially whale watching, has the potential to significantly contribute to the noise pollution to which marine mammals are exposed because of the daily interactions between the targeted species and the tourism platforms. These impacts can take several forms from temporarily masking the sound produced by the animals to damaging the hearing sensory organs through chronic exposure (Erbe, 2002).

Tourism platforms in general do not produce sounds with strong intensities. Tour boats will produce sounds which will vary around 140 dB re 1 μPa at 1 m (see Box 11.1), which is approximately equivalent to the sound intensity produced during a rock concert (100 dB re 20 μPa). In contrast, an airgun array used for underwater geological surveys will produce sounds around 260 dB re 1 μPa at 1 m (a sound 100 times stronger than an airplane sonic boom). Yet the repeated exposure to boat traffic can have several consequences.

This chapter overviews the influence of noise on marine mammals with reference to sounds that can be produced by tourism platforms. After presenting the potential impacts of noise pollution on these species, I will discuss the sources of noise from tourism platforms. Finally, I will place noise pollution in an ecosystem

---

Box 11.1. Glossary of terms.

Ambient noise: Background noise of which the source cannot be identified and is not of direct interest during a recording. It can be an integration of many sounds produced close or far.

Audiogram: Information, usually presented as a graph, about the minimum sound level a species can detect at given frequencies.

Critical ratio: The excess intensity necessary for a sound to be heard above the ambient noise. If the difference between a sound level and the ambient noise level is too small, that sound cannot be detected.

Decibel (dB): Unit of sound level (0.1 Bel) measured by comparing a sound pressure (P) to a reference pressure (P_{ref}, smallest sound pressure audible to humans: 1 μPa for underwater sound reference; in air, the reference sound pressure is usually 20 μPa). Decibels are on a logarithmic scale (usually sound level (dB) = 20 $\log(P/P_{ref})$). Hence an increase in 20 dB re 1 μPa (with a reference pressure of 1 μPa) represents a sound 10 times stronger. A reference to the distance at which a sound is measured is also given with the sound pressure level. For example, a sound may be 120 dB re 1 μPa at 1 m, which means that it was 120 dB with a reference pressure of 1 μPa when measured 1 m away from the source of the sound.

Evoked potential: Electrical signal that is emitted in the nervous system in response to a stimulus such as a sound.

Frequency: A measure of the pitch of a sound. It is measured in Hertz (cycles per second) because sounds are acoustic waves propagating through a medium and hence a sound frequency measures the rate of oscillation of the sound and is related to its wavelength.

Permanent threshold shift (PTS): Permanent hearing impairment that leads to an increased hearing threshold at given frequencies. Sounds should then have higher intensity to be detected (for an evoked potential to be produced). This natural ageing process can be accelerated by prolonged and/or repeated exposure to high levels of sounds. It can also occur abruptly for exceptionally high sound levels such as explosions.

Temporary threshold shift (TTS): Temporary hearing impairment that leads to an increased hearing threshold at given frequencies. Sounds should then have higher intensity to be detected. These shifts can last from minutes to days depending on many factors after exposure to high sound intensity.

context and assess how it can contribute to overall habitat degradation and the consequences of such degradation on the lives of marine mammals.

Marine Mammals and Noise

The importance of sound to cetaceans and other marine mammals

Cetaceans use sound to navigate, find food and interact with conspecifics, while other marine mammals have a use of their hearing sense more similar to other mammals. Therefore, noise pollution has the potential to have greater implications for cetaceans than for pinnipeds, polar bears and otters. This discrepancy in biological relevance of noise impacts means that we have much more information available about the impact of sound generated by tourism operations for cetaceans than for other marine mammals (Richardson *et al.*, 1995). It also means that it is easier to tease apart for these species the influence of the tourism platform noise from other platform factors which may elicit responses from marine mammals (Patenaude *et al.*, 2002; Dyck and Baydack, 2004). In addition, there is a difference in how much the hearing range of these species overlaps with sounds produced by tourism platforms. Humans can hear sounds ranging from 20 Hz to 20 kHz with greatest sensitivity, therefore smallest critical ratio (Box 11.1), to sounds between 200 Hz and 9 kHz. Typically whale-watching platforms will produce sounds predominantly around 800 Hz–3 kHz. In comparison, most odontocetes (toothed whales, dolphins and porpoises) can hear sounds that are between 100 Hz and 200 kHz with the greatest sensitivity in their audiogram (Box 11.1) in frequencies ranging from 2–5 to 100 kHz. Pinnipeds have a more restricted underwater audiogram (~200 Hz to ~60 kHz, smallest critical ratio: 1–30 kHz). Little is known about the hearing of mysticetes (baleen whales), but given the frequency range of their vocalizations and observed reactions to strong sounds (Richardson *et al.*, 1995) their upper limit is probably 10 kHz and a lower limit which could be as low as 5 Hz for some species. Most mysticete species have calls with dominant frequencies below 1 kHz and for fin and blue whales (the two largest species) as low as 20–50 Hz. There is therefore less overlap between tourism noise and the vocalizations of these latter species.

Marine mammals use vocalizations for communication with conspecifics. In addition, odontocetes alone have evolved an echolocation sense, producing echolocation clicks, which allows them to navigate and discriminate objects such as prey items. Echolocation patterns vary widely across species but tend to be produced at higher frequencies than vocalizations (high-pitched sounds). Altering the production and reception of echolocation sounds can therefore have direct impacts on the energetic budget of individuals by preventing them from orienting themselves and finding their food. Disruptions in vocalization patterns can also impact the lives of marine mammals but tend to only temporarily isolate individuals from their conspecifics. This effect can still have important consequences in group-living species because it can increase the likelihood that an individual is predated. In the case of mother–calf pairs this isolation can lead to suckling disruptions, which affect the survival probability of calves.

Temporary disturbances induced by noise and their biological relevance

As in other extremely vocal species, such as birds (Slabbekoorn and Peet, 2003; Leonard and Horn, 2005), cetacean acoustic responses to boat-related noise exposure are to increase the intensity of vocalizations they produce and repeat those signals more often. Noise pollution during tourism interactions is perceived by the animals as an increase in ambient noise. Therefore, for vocalizations to be detected they need to match or exceed the critical ratio of the species (Box 11.1). To achieve this, cetaceans tend to produce more intense calls and repeat them to increase the likelihood that the signal will be detected by the potential receivers. Whistles are used as contact calls for group cohesion in dolphins (Janik, 2000; McCowan and Reiss, 2001). These species change the shape of their whistles and the frequency at which they are produced in response to ambient noise (Morisaka *et al.*, 2005). They produce simpler, i.e. less modulated, whistles at lower frequencies when ambient noise is stronger (Boisseau, 2004; Dawson *et al.*, 2004; Morisaka *et al.*, 2005). Such whistles are more likely to be detected by conspecifics in a noisy environment. Increase of whistling rate in dolphins is prevalent during boat interactions (Buckstaff, 2004) and is especially noticeable during mother–calf pair acoustic separation due to boat traffic noise (Scarpaci *et al.*, 2001; Van Parijs and Corkeron, 2001). Adaptations to increased noise exposure occur over long periods as well, indicating that anthropogenic noise can act as a selection pressure on cetaceans (Foote *et al.*, 2004). Killer whales increase the duration of their calls in the presence of whale-watching boats, but this increase only took place after a sharp increase in the number of interactions between tour boats and whales (Foote *et al.*, 2004).

It is important to note that these adaptations to boat noise exposure are very much situation-specific, with an overarching principle to decrease the influence of the added noise on the transmission of the acoustic signals. For example, beluga whales in the presence of boat noise will not only increase their call rate, but also produce higher pitched vocalizations than when boats are absent (Lesage *et al.*, 1999).

While cetaceans are highly vocal species, they also have a highly developed hearing sense. They use passive listening for locating objects such as prey by listening for their vocalizations or movement. Boat interactions with sperm whales seem to cause individuals to start echolocating sooner after starting a dive (Richter *et al.*, 2006). This shows that boat noise alters the efficiency of passive listening at least at the beginning of the dive (when they are close to the surface), forcing individuals to echolocate sooner to find food or topographic marks.

There is little evidence supporting the notion that baleen whales (mysticetes) echolocate. Most of the calls produced by these whales are lower pitched than boat noises and hence few opportunities exist for overlap between the two sounds (Au and Green, 2000). This explains why tour boats do not seem to have acoustic influences on mysticetes. However, whale-watching activities still affect other factors in the life of these species, which can lead to long-term consequences such as site abandonment (Reeves, 1977). In addition, baleen whale calls exhibit similar adaptive responses to other types of sounds that overlap their vocalizations and can therefore jeopardize their reception (Miller *et al.*, 2000). The dominant frequency at which a boat emits its sound is related to how fast its propeller rotates and hence to its size: the larger the boat, the lower the frequency. Therefore, there may be ways for tourism to influence mysticete calls in the future if platforms become larger or if cruise ships become favoured as whale-watching platforms.

Temporary physical damage

Acute or chronic exposure to intense sounds can also result in temporary physical damage (temporary threshold shift or TTS, Box 11.1). Only one study to date has documented the potential for tourism platforms to cause TTS in cetaceans (Erbe, 2002). Killer whales can have their critical ratio increased by 5 dB after being exposed to one fast-moving boat within 450 m for 30–50 min (Erbe, 2002). A boat travelling at typical speed used during encounters, would have to be within 20 m of the whale for 30–50 min to elicit TTS (Erbe, 2002). However, whale schools from the population Erbe (2002) studied, the southern resident killer whales living off Vancouver Island, Canada, are rarely exposed to only one tour boat, and usually interact with on average 21 different vessels in a 1 km² zone centred on the whales (Otis and Osborne, 2001). Tourism traffic is so elevated that they can interact with up to 100 tourism platforms at a time (Otis and Osborne, 2001; Erbe, 2002). It is therefore highly likely that killer whales in this area constantly face TTS caused by tourism activities. While the occurrence of TTS from tourism interactions has not been studied in other populations, it can be expected that it represents a threat to many odontocete populations exposed to high level of tourism interactions.

Laboratory tests with bottlenose dolphins show that TTS typically last 1–2 h (Nachtigall *et al.*, 2003; Nachtigall *et al.*, 2004). That would be in a case where the animals would not be exposed to other tourism interactions within that period, an unlikely situation in many tourism destinations. Hence, the impact of tourism interactions lasts quite some time after the interactions are over. Temporary threshold shifts decrease the likelihood that an individual will hear vocalizations from its conspecifics over the ambient noise. It impairs the hearing capability of this individual for a while and can thus jeopardize its echolocation abilities if the frequency range over which TTS occurred overlaps the frequencies at which that species echolocates. In this way, TTS accentuates the impact of masking due to noise during encounters with tourism platforms, and also carries this isolating effect beyond the encounter itself, increasing the period over which tourism impacts the targeted animals. In many instances, the repeated exposure to TTS sound level could mean that during peak tourism season animals are constantly in a TTS state, which could abruptly or chronically lead to permanent threshold shift (PTS, Box 11.1).

Permanent physical damage

Little is known about the factors influencing the emergence of PTS in marine mammals and indeed in many mammals (Erbe, 2002). Much of it has to be inferred from studies on humans, which have been regularly extended to the study of other mammals (Richardson *et al.*, 1995). Tourism platforms do not produce sounds intense enough to acutely elicit PTS (i.e. after only one exposure). However, chronic exposure, such as is the situation for odontocete populations residing in an area heavily used for tourism, could lead to PTS. For PTS to be elicited by exposure to one boat, it would take 50 years of exposure, 5 days/week, 8 h/day, at 50 m for a 2–5 dB PTS (Erbe, 2002). However, exposure to multiple boats could more readily lead to PTS. For example, a whale with the same duration of exposure but to five boats within 400 m would have a 2–5 dB PTS (Erbe, 2002). These killer whales are exposed to an average of 21 boats (maximum of 100 m) for 8–10 h/day, 7 days/week from May to September within 500 m from them (Otis and Osborne, 2001; Erbe, 2002). Such exposure levels can raise concerns for the emergence of PTS within a whale generation, which, combined with an increase in ambient noise due to anthropogenic activities such as tourism, will highly jeopardize the use killer whales can make of this habitat.

The energetic cost of masking

Masking is one particular case of temporary impact which deserves more attention because of its consequences for the survival of individuals. Although the influence of masking on individual's sensory isolation was covered in the previous section, masking can also have a more pervasive influence on the behaviour of certain species. Some odontocetes echolocate at low frequencies (e.g. killer whales: 12–25 kHz) and therefore, an anthropogenic increase in ambient noise can lead to a decrease in the 'active space' of individuals. The active space can be defined as the zone within which the whale can detect food and communicate with conspecifics, hence a decrease in that area will result in a decrease in foraging efficiency and prey capture (Bain *et al.*, 2002). Current estimates with killer whales show that exposure to whale watching can decrease the active space of these whales approximately threefold (Bain *et al.*, 2002).

It is worth noting the pervasiveness of boat sounds. For example, tour boats travelling at cruising speed can be heard by killer whales up to 16 km away, and mask their calls up to 14 km (Erbe, 2002). That is, the active space of a killer whale can be reduced by a boat which passes at a distance of 14 km. If boats are travelling at slow speed, similar to the speed used during encounters with killer whales, it will mask whale calls up to 1 km away (Erbe, 2002). This masking impact then becomes highly relevant for the survival of killer whale populations in situations, such as in southern resident killer whale population, where whales are followed during summer months from sunrise to sunset by tens of boats within 500 m of the whale schools. A recent study reinforced this proposed masking mechanism by showing that killer whales were significantly less likely to be foraging in the presence of tour boats (Bain *et al.*, submitted). In addition, the distance to the closest boat influenced the effect size: the closer the boats, the less likely whales were to be foraging (Bain *et al.*, submitted). Indeed in the case of this endangered population, masking impact may have led tourism to play a key role in the present decline in population abundance (Bain *et al.*, 2002).

Noise and Tourism

Factors influencing the noise produced by a tourism platform

There are several sources of noise on tourism platforms (Box 11.2) that contribute differently to the overall sound characteristics of the vehicle depending on the type of platform. For example, the noise coming from a non-powered vessel, such as a kayak, will be dominated by the behaviour of the paddlers (splashing, shouting, etc.). In the case of a sailboat under sail, some noise can still be emitted by electric generators, winches, etc. These sounds are not interacting with the acoustics of marine mammals, yet they can still elicit flight responses (Cassini *et al.*, 2004). Powerboats have noisier outputs than other types of vessels and the dominant frequency of their sound, along with its intensity, is usually driven by the size of the vessel and its mode of propulsion. The energy emitted by an engine on a powerboat can be translated in movement in two different ways. It can be used to rotate a propeller that pushes the boat, or it can be used to rotate an impeller that drives a pump and creates a pressurized jet of water that effectively pushes the boat. While jet-powered vessels are quite noisy in air, the underwater component of the sound they emit is smaller than the one of propeller vessels (Kipple and Gabriele, 2003). This explains why jet propulsion is now often favoured for wildlife-viewing vessels.

Jet propulsion is quiet because it does not result in as much cavitation (Box 11.2). The cavitation of a water jet unit is limited to the impeller, which is located inside the

---

Box 11.2. Sources of noise from marine vessels.

Tourists: voice and physical interactions with the boats are conducted underwater by the boat hull.

Machinery: rotating shafts, friction, generators (importance of this source varies with the state of maintenance of the vessel).

Propeller singing: resonant vibration of propeller blade when turning.

Propeller cavitation: bubble forming and collapsing due to the vortex created by the rotating propeller (primary source of noise). The size of those bubbles is related to the rotation speed and dictates the noise's dominan frequency.

Water hitting the hull can be a significant source of noise in rougher conditions.

vessel, and therefore less of this cavitation sound is propagated outside the vessel. Conversely, the rotation of a propeller results in the creation of a vortex outside the vessel. This vortex lowers the water pressure in the vicinity of the propeller blades, resulting in the water changing from its liquid phase to a vapour phase, forming bubbles. As these bubbles move away from the blades and are once again exposed to normal pressure they implode. Cavitation noise, therefore, originates from a series of implosions, hence the high intensity of this sound. The rotation speed of the propeller dictates the size of the bubbles which in turn determine the cavitation sound frequency. Larger vessels tend to have lower dominant frequencies because their propeller turns slower. Many other factors such as engine size, number of engines, hull design, number of blades composing the propeller, etc. influence the dominant frequency of a vessel. To give examples, a 5m zodiac was recorded to have a dominant frequency of 6.3kHz, a fishing vessel (12m) 0.25–1kHz, a supply ship (25m) 100Hz, a freighter (135m) 41Hz and finally a supertanker (340m) 6.8Hz (Richardson *et al.*, 1995). Vessels that are typically used to watch wildlife will have dominant frequencies ranging from 0.1 to 1kHz for small ships to 2 to 10kHz for outboard boats (Richardson *et al.*, 1995; Au and Green, 2000; Erbe, 2002; Dawson *et al.*, 2004).

The intensity of these sounds will vary greatly depending on the size and speed of the boats. For example, a jet ski (Personal Watercraft, PWC) underway will emit less strong sound underwater, 90 dB re 1μPa (Evans *et al.*, 1993), than a similarly powered powerboat, 120 dB re 1μPa (Richardson *et al.*, 1995). A zodiac (with a twin 175 horsepower Evinrude outboard engine) travelling at 10 km/h will emit an underwater source level of 147 dB re 1μPa at 1m, but this sound will be close to 10 times louder when it travels at 55 km/h (163 dB re 1μPa at 1m).

This information can help in deciding on the best type of vessel to be used with certain species. For example, larger vessels will have less overlap with the audiogram of dolphins. They can also carry more people at once and therefore reduce the number of interactions needed to satisfy a certain visitor volume. On the other hand, small outboard boats may be more useful to work with mysticetes for the same reason, that they barely overlap with the inferred audiogram of these species and their calls. However, other factors have to be taken into consideration when deciding the type of vessel to use in an area. For example, larger vessels may result in lower satisfaction level of visitors because of a potential feeling of crowdedness that diminishes the wilderness value of the trip (Higham, 1998). While single jet-powered vessels are much quieter than their propeller-powered counterparts, they are also harder to manoeuvre at slow speed. This may increase the unpredictability of the tourism platform during interactions with marine mammals and hence increase its potential impact (Lusseau, 2003). Therefore, information about the acoustic interactions between the tourism platform and the targeted species in a specific area needs to be put into the relative context of the ecological socio-economic situation of the site to find the best trade-off.

Is it all noise? Other factors influencing the quality of interactions

The platform's behaviour can significantly influence the behaviour of marine mammals (Lusseau, 2003). While many of the elicited responses observed can be linked to sound production (e.g. boats getting in and out of gear or fast speed around the animals), platforms producing very little noise such as kayaks can also elicit these responses (Lusseau, 2003). Indeed, in some instances the lack of noise can elicit stronger avoidance response in some species (Kovacs and Innes, 1990; Suryan and Harvey, 1999). Harbour seals are, for example, more likely to be flushed from a haul-out site into the water by a kayak approaching silently than by a powerboat approaching at a regular speed (Suryan and Harvey, 1999; Johnson and Acevedo-Gutiérrez, 2007). That is because stealthy vessels can get much closer to these animals without being detected and stealth movement is associated with predatory behaviour (Frid and Dill, 2002).

Flushing from haul-out sites can have significant biological effects on individuals. It alters their resting pattern, increasing their energy expenditure and diminishing suckling behaviour opportunities and mothers' attendance to their pups (Kovacs and Innes, 1990). Return to haul-out sites can be significantly delayed after an interaction (Kovacs and Innes, 1990; Suryan and Harvey, 1999), increasing the biological impact of interactions.

Noise as Habitat Degradation

One biologically significant impact of tourism-induced noise pollution is the reduction of the benefits of visiting the concerned area for marine mammals (Box 11.3). The previous sections of this chapter show that tourism-related noise can impair the way marine mammals use their habitats and, indeed, tourism-induced displace-

---

Box 11.3. What is habitat degradation?

Animals use their home range in a way that minimizes the costs of being present in a given area and maximizes the benefits of occupying that site. Costs and benefits relate to survival and reproduction and therefore encompass decisions about predation pressure, foraging opportunities and access to mates.

A site becomes degraded for an animal population, or an individual, if its cost/benefit ratio is worsened. When the new cost/benefit ratio departs significantly from its original trade-off, it is no longer beneficial for a population to visit a site. This can occur when foraging opportunities are worsened, for example, due to a decrease in acoustic active space in cetaceans. It can also happen when predation pressure increases in cases where individuals are more likely to lose contact with their conspecifics due to increased noise.

Four tour boats approach dolphins in Milford Sound, New Zealand. Resident bottlenose dolphins leave this fjord altogether during tourism peak times because the cost of interacting with so many vessels outweighs the benefits of being in this fjord (Lusseau, 2005). These dolphins then lose one-seventh of their home range because of tourism-linked habitat degradation.

ment has been shown for dolphins (Lusseau, 2005) and whales (Reeves, 1977; Bryant *et al.*, 1984). However, the cost/benefit trade-offs are different depending on the type of population using the tourism site. Resident populations, i.e. populations that use the site continuously throughout the year, will have higher cumulative exposure rates to tourism-related noise impacts than populations that only temporarily visit this site. They will therefore incur a higher cumulative 'tourism cost' than transiting individuals at the site. Hence, while the cost of interactions may be the same for resident and transient populations, the biological impact will vary and so will the habitat degradation perceived by the two types of populations.

In contrast, if a resident population does not have the option to extend its home range to cope with degradation of part of its existing range, it will have to utilize the degraded habitat regardless of the cost. Since part of the home range is degraded, it cannot carry as many individuals as previously; therefore, the carrying capacity of the population is decreased and the population abundance declines. Individuals that are more likely to cope with the disturbance, in our case increased noise levels, will be more likely to survive. This mechanism explains how human disturbance, such as tourism-related noise, can act as a selective force on the evolution of marine mammals.

The same principle applies to intrapopulation variation with some components of the population, e.g. social communities, spending more time in one area than others. There is such a case in the population of bottlenose dolphins residing in the Moray Firth (Wilson *et al.*, 2004; Lusseau *et al.*, 2006). This population is composed of two social communities: the inner community restricted to the inner Moray Firth and the outer community whose home range is wider and encompass coastal waters beyond the Moray Firth (Lusseau *et al.*, 2006). The outer community is spending less and less time in the inner Moray Firth, which has been considered as degrading over the past decades (Wilson *et al.*, 2004). However, the inner community is not changing its residency pattern in the inner Moray Firth. There are indications that this latter section of the population may be faring less well than the outer Moray Firth social community, potentially as a result of adaptation of its carrying capacity to its new environment (Thompson *et al.*, 2004). The same principles may hold true for migrating species, which may have several options for foraging and breeding grounds and therefore the abandonment of degraded sites may not be as costly. Indeed, there are indications that grey whales have returned to some previously abandoned wintering grounds after their quality was improved (Bryant *et al.*, 1984). However, further work is needed to understand and demonstrate the mechanisms linking habitat selection and habitat ensonification.

Conclusion

Tourism contributes to the increasing noise pollution of the oceans. The chronic exposure of marine mammals to tourism activities can lead the noise from this industry to have biologically significant influences on the lives of these animals. Exposure to tourism platforms can temporarily damage their primary sense, degrade their habitat and impair their foraging ability and efficiency. Wildlife-oriented tourism strives to be a non-consumptive industry and as such should minimize its ecological footprint. There is much scope to mitigate the noise influence of this industry in the marine environment thanks to new technology and a better understanding of the needs of the targeted species. Tourism therefore has the possibility to act as a flagship of good practice for other marine industries and pave the way to quieter oceans.

References

Au, W.L. and Green, M. (2000) Acoustic interaction of humpack whales and whale-watching boats. *Marine Environmental Research* 49, 469–481.

Bain, D.E., Lusseau, D., Williams, R. and Smith, J.C. (submitted) Effects of vessels on activity states of southern resident killer whales (*Orcinus* sp.). *Canadian Journal of Zoology*.

Bain, D.E., Williams, R.J. and Trites, A.W. (2002) *A Model Linking Energetic Effects of Whale Watching to Killer Whale* (Orcinus orca) *Endangered Species Research*. Friday Harbor Labs, University of Washington, Friday Harbor, Washington.

Boisseau, O.J. (2004) *The Acoustic Behaviour of Resident Bottlenose Dolphins in Fiordland, New Zealand*. Department of Marine Sciences, University of Otago, Dunedin.

Bryant, P.J., Lafferty, C.M. and Lafferty, S.K. (1984) Reoccupation of Laguna Guerrero Negro, Baja California, Mexico, by gray whales. In: Jones, M.L., Swartz, S.L. and Leatherwood, S. (eds) *The Gray Whale* Eschrichtius robustus. Academic Press, New York, pp. 375–387.

Buckstaff, K.C. (2004) Effects of watercraft noise on the acoustic behavior of bottlenose dolphins, *Tursiops truncatus*, in Sarasota Bay, Florida. *Marine Mammal Science* 20, 709–725.

Cassini, M.H., Szteren, D. and Fernàndez-Juricic, E. (2004) Fence effects on the behavioural responses of South American fur seals to tourist approaches. *Journal of Ethology* 22, 127–133.

Chang, E.F. and Merzenich, M.M. (2003) Environmental noise retards auditory cortical development. *Science* 300, 498–502.

Constantine, R. (2003) Increased avoidance of swimmers by wild bottlenose dolphins (*Tursiops truncatus*) due to long-term exposure to swim-with-dolphin tourism. *Marine Mammal Science* 17, 689–702.

Curtin, S. (2005) Nature, wild animals and tourism: an experiential view. *Journal of Ecotourism* 4, 1–15.

Dawson, S.M., Boisseau, O.J., Rayment, W. and Lusseau, D. (2004) *A Quantitative Acoustic Study of the Fiordland Underwater Environment*. Department of Conservation Southland Conservancy Commercial Marine Mammal Viewing Research Strategy, Invercargill, New Zealand, p. 61.

Dyck, M.G. and Baydack, R.K. (2004) Vigilance behaviour of polar bears (*Ursus maritimus*) in the context of wildlife-viewing activities at Churchill, Manitoba, Canada. *Biological Conservation* 116, 343–350.

Erbe, C. (2002) Underwater noise of whale-watching boats and potential effects on killer whales (Orcinus orca), based on an acoustic impact model. *Marine Mammal Science* 18, 394–418.

Evans, P.G.H., Canwell, P.J. and Lewis, E. (1993) An experimental study of the effects of pleasure craft noise upon bottlenosed dolphins in Cardigan Bay, west Wales. European Cetacean Research, pp. 43–46.

Foote, A.D., Osborne, R.W. and Hoelzel, R.A. (2004) Whale call response to masking boat noise. *Nature* 428, 910.

Frid, A. and Dill, L.M. (2002) Human-caused disturbance stimuli as a form of predation risk. *Conservation Ecology* 6, 11.

Habib, L., Bayne, E.M. and Boutin, S. (2007) Chronic industrial noise affects pairing success and age structure of ovenbirds. *Seiurus aurocapilla*. *Journal of Applied Ecology* 44, 176–184.

Higham, J.E.S. (1998) Sustaining the physical and social dimensions of wilderness tourism: the perceptual approach to wilderness management in New Zealand. *Journal of Sustainable Tourism* 6, 26–51.

Hoyt, E. (2001) *Whale Watching 2001*. International Fund for Animal Welfare, London.

Janik, V.M. (2000) Whistle matching in wild bottlenose dolphins (*Tursiops truncatus*). *Science* 289, 1355–1357.

Johnson, A. and Acevedo-Gutiérrez, A. (2007) Regulation compliance by vessels and disturbance of harbour seals (*Phoca vitulina*). *Canadian Journal of Zoology* 85, 290–294.

Kipple, B. and Gabriele, C. (2003) *Glacier Bay Watercraft Noise*. Naval Surface Warfare Center Technical Report NSWCCD-71-TR-2003/522, Anchorage, Alaska.

Kovacs, K.M. and Innes, S. (1990) The impact of tourism on harp seals (*Phoca groenlandica*) in the Gulf of St. Lawrence, Canada. *Applied Animal Behaviour Science* 26, 15–26.

Leonard, M.L. and Horn, A.G. (2005) Ambient noise and the design of begging signals. *Proceedings of the Royal Society of London Series B* 272, 651–656.

Lesage, V., Barrette, C., Kingsley, M.C.S. and Sjare, B. (1999) The effect of vessel noise on the vocal behavior of belugas in the St. Lawrence river estuary, Canada. *Marine Mammal Science* 15, 65–84.

Lusseau, D. (2003) Male and female bottlenose dolphins *Tursiops* sp. have different strategies to avoid interactions with tour boats in Doubtful Sound, New Zealand. *Marine Ecology-Progress Series* 257, 267–274.

Lusseau, D. (2005) The residency pattern of bottlenose dolphins (*Tursiops* spp.) in Milford Sound, New Zealand, is related to boat traffic. *Marine Ecology Progress Series* 295, 265–272.

Lusseau, D., Wilson, B., Hammond, P.S., Grellier, K., Durban, J.W., Parsons, K.M., Barton, T.R. and Thompson, P.M. (2006) Quantifying the influence of sociality on population structure in bottlenose dolphins. *Journal of Animal Ecology* 75, 14–24.

Mace, R.D., Waller, J.S., Manley, T.L., Ake, K. and Wittinger, W.T. (1999) Landscape evaluation of grizzly bear habitat in Western Montana. *Conservation Biology* 13, 367–377.

McCowan, B. and Reiss, D. (2001) The fallacy of 'signature whistles' in bottlenose dolphins: a comparative perspective of 'signature information' in animal vocalizations. *Animal Behaviour* 62, 1151–1162.

Miller, P.J.O., Biassoni, N., Samuels, A. and Tyack, P.L. (2000) Whale songs lengthen in response to sonar. *Nature* 405, 903.

Morisaka, T., Shinohara, M., Nakahara, F. and Akamatsu, T. (2005) Effects of ambient noise on the whistles of Indo-Pacific bottlenose dolphin populations. *Journal of Mammalogy* 86, 541–546.

Nachtigall, P.E., Pawloski, J.L. and Au, W.L. (2003) Temporary threshold shifts and recovery following noise exposure in the Atlantic bottlenose dolphin (*Tursiops truncatus*). *Journal of the Acoustical Society of America* 113, 3425–3429.

Nachtigall, P.E., Supin, A.Y., Pawloski, J.L. and Au, W.L. (2004) Temporary threshold shifts after noise exposure in the bottlenose dolphin (*Tursiops truncatus*) measured using evoked auditory potentials. *Marine Mammal Science* 20, 673–687.

Otis, R.E. and Osborne, R.W. (2001) Historical trends in vessel-based killer whale watching in Haro Strait along the boundary of British Columbia and Washington State: 1976–2001. *Society for Marine Mammalogy Conference*. Society for Marine Mammalogy, Vancouver, British Columbia.

Patenaude, N.J., Richardson, W.J., Smultea, M.A., Koski, W.R., Miller, G.W., Würsig, B. and Greene, C.R. (2002) Aircraft sound and disturbance to bowhead and beluga whales during spring migration in the Alaskan Beaufort sea. *Marine Mammal Science* 18, 309–335.

Reeves, R.R. (1977) *The Problem of Gray Whale (*Eschrichtius robustus*) Harassment: At the Breeding Lagoons and During Migration*. US Marine Mammal Commission MMC-76/06, Washington, DC.

Richardson, W.J., Greene, C.R., Malme, C.I. and Thomson, D.H. (1995) *Marine Mammals and Noise*. Academic Press, San Diego, California.

Richter, C., Dawson, S.M. and Slooten, E. (2006) Impacts of commercial whale watching on male sperm whales at Kaikoura, New Zealand. *Marine Mammal Science* 22, 46–63.

Rodger, K., Moore, S.A. and Newsome, D. (2007) Wildlife tours in Australia: characteristics, the place of science and sustainable futures. *Journal of Sustainable Tourism* 15, 160–179.

Scarpaci, C., Bigger, S.W., Corkeron, P.J. and Nugegoda, D. (2001) Bottlenose dolphins (*Tursiops truncatus*) increase whistling in the presence of 'swim-with-dolphin' tour operations. *Journal of Cetacean Research and Management* 2, 183–185.

Slabbekoorn, H. and Peet, M. (2003) Birds sing at a higher pitch in urban noise. *Nature* 424, 267–268.

Suryan, R.M. and Harvey, J.T. (1999) Variability in reactions of pacific harbor seals, *Phoca vitulina richardsi*, to disturbance. *Fishery Bulletin* 97, 332–339.

Thompson, P.M., Lusseau, D., Corkrey, R. and Hammond, P.S. (2004) *Moray Firth Bottlenose Dolphin Monitoring Strategy Options*. Scottish Natural Heritage Commissioned Report No. 079 (ROAME No. F02AA409).

Van Parijs, S. and Corkeron, P.J. (2001) Boat traffic affects the acoustic behaviour of Pacific humpback dolphins, *Sousa chinensis*. *Journal of the Marine Biological Association of the United Kingdom* 81, 533–538.

Wilson, B., Reid, R.J., Grellier, K., Thompson, P.M. and Hammond, P.S. (2004) Considering the temporal when managing the spatial: a population range expansion impacts protected areas-based management for bottlenose dolphins. *Animal Conservation* 7, 331–338.

12 Shooting Fish in a Barrel: Tourists as Easy Targets

E.J. SHELTON AND B. MCKINLAY

Both authors are situated within, and in their daily lives contribute to, the conversation that revolves around appropriate management of Yellow-eyed penguins (YEPs) (*Megadyptes antipodes*). Our intention is to describe some aspects of the relationship between wildlife, in this case sea lions and penguins, and visitors at a physically and socially dynamic site. We wish also to explore the potential application of the precautionary principle to some management issues that emerge from this consideration. We have elected to couch this particular contribution to the conversation in non-technical language in order to engage with as large a readership as possible, though, perhaps at the cost of failing to mention more abstract theory that informs our position. We hope that the gains in readability outweigh any losses incurred by this lack of elaboration.

There would be little disagreement with the claim that, over the short history of the species, the activities of *Homo sapiens* (humans) are thought to have, directly and indirectly, significantly affected the prevalence and distribution of many other species. The collection of human-induced processes directly affecting other species, the atmosphere, oceans and landforms commonly is discussed in the vernacular as *interference with nature*. Such a formulation, that nature should be left to take its course, can be taken to imply that there is some underlying optimal state or set of processes that, left alone, would provide the appropriate outcome for planet Earth. We need not develop this environmental philosophical debate here (see O'Riordan (1976) for an early introduction to the philosophical issues involved in environmentalism). For our purpose, it is sufficient to accept that humans can and do exist within ecosystems that they affect. This process of affecting the ecosystems within which we live can usefully be described as human disturbance, and often results in a decrease in indigenous biodiversity, a development that is widely regarded as being undesirable. There also exists *non-human disturbance*, a term reserved for those processes that are instigated and seen through to completion without the direct involvement of human activity. Increasingly, though, these disturbances also reflect some indirect human influence. The restoration of an ecosystem to an approximation of what it was like before human activity simplified it, an increasingly popular activity in some parts of the world, will be disruptive to plants and animals that are benefiting from the new steady state, but can be argued as justifiable if certain specific management outcomes or agreed objectives, like a sustainable increase in indigenous biodiversity, are produced. It is beyond the scope of this chapter to pursue this discussion further. Many journals, for example, *Conservation Biology* and *Biological Conservation*, are devoted to doing exactly that (see Seddon *et al.* (2007) for an elaboration of the science of reintroduction). Rather, here we focus on one aspect of late-capitalist production and consumption, nature-based tourism, in one setting, and ask: Does this activity disturb the wildlife, in this case YEPs at Sandfly Bay, Otago Peninsula, New Zealand? If so, how and, most importantly, with what effect?

Beach settings are intriguing places for both wildlife and tourists. To the naïve visitor the beach seems to provide plenty of scope for wildlife to avoid human interactions if they so wish; the sea is available for escape at any time. Sandfly Bay is open and windswept and subject to severe storm surges which sometimes penetrate the foredunes. The beach hosts native gulls, oystercatchers and occasional bar-tailed godwits and Australasian harriers, as well as a variety of introduced passerines. Of the many activities undertaken at the site, including surfing and sand sledding, only some involve the resident wildlife. This beach also enables one of the most rewarding experiences possible for the nature tourist, the sight of YEPs heading away to sea in the morning or coming ashore in the evening. These birds inhabit the east coast of Southern New Zealand, although their main populations are found on the subantarctic islands to the south, and are a major tourist attraction, particularly on Otago Peninsula and on the Catlins coast further south. Viewing penguins in beach settings on Otago Peninsula has been reported as having emotional benefits for visitors (Schänzel and McIntosh, 2000). Wildlife may not view beaches in quite the same way. Perhaps as recently as 300 years ago, a giant bird of prey, Haast's eagle (*Harpagornis moorei*), may have swooped on YEPs as they traversed the beach. Indigenous hunting to extinction of *Harpagornis*'s main prey, the moa and a native goose, consigned the eagle to extinction as well (Hutching, 1998). By contrast, and after a period of 400 years of local extinction due to indigenous hunting, New Zealand sea lions (*Phocarctos hookerii*) are back, breeding and beginning to make previously empty penguin-friendly beaches seem distinctly crowded. The birds can be observed either not proceeding or changing their route to and from the sea in response to the presence of sea lions on the beach. These changes are specifically in response to sea lions and are not elicited by the presence of New Zealand fur seals. One interesting situation that has arisen, though not at this beach, is the predation of YEPs, which are a fully protected species, by New Zealand sea lions, another fully protected species. One individual sea lion has developed such an appetite for penguins that the zoologists most intimately involved with monitoring the interactions between the two species have suggested that her being removed from the wild population and taken into captivity should seriously be considered (Lalas *et al.*, 2007). Previous human activity, specifically indigenous hunting, led to the removal of these two penguin predators, one of which has returned under protection and may now need to be actively managed for the benefit of the penguin. We know very little about the effect of indigenous hunting directly of the penguin, but it is unlikely to have increased their numbers. Clearly, the nature tourist must be situated within this matrix of ongoing human and non-human predator–prey relationships. This background material is presented to demonstrate that YEPs probably have always been, and still are, subjected to chronic and acute disturbance from other species with which they share the coastal environment. Even without nature tourists the Sandfly Bay penguins face major challenges throughout their life course; ferocious storms erode their habitat, sharks and orca hunt them at sea, introduced predators such as the stoat (*Mustela erminea*) raid their nests and kill their chicks, cyclic weather patterns reduce the abundance of preferred prey and devastating bacterial disease has on occasion almost wiped out a whole season's chicks. The penguins are also the subjects of non-tourist human activities; bottom trawlers carve up the sea floor where the birds feed and inshore fishing nets kill a significant number of birds as by-catch (Darby and Dawson, 2000). Scientists routinely catch, weigh and measure the birds and take blood samples (e.g. Ellenberg *et al.*, 2007). The staff of the Department of Conservation collect birds who are found moulting or exhausted in unsuitable places and move them to safely. Surfers share with the birds the waters close to the beach. Clearly, the effects of the nature tourist must also be integrated into this other existing matrix of potentially disturbing, and sometimes fatal, interactions.

Any consideration of nature tourists' disturbance of wildlife must engage with the issue of scale that pervades all of ecology (Levin, 1992) and ecotourism (Hall, 2007) since any disturbance will, by definition, affect ecological processes and relationships at various levels of organization. If humans are to be treated as just another species that exists inside nature, and if tourists are then just humans behaving in certain ways, then our first consideration could be: At what level of organization should tourism be considered? At the level of the individual visitor, the total number of visitors to a site over a specified time, the industry sector directly involved in delivering the tourism product, the whole industry or that condition of existing within postmodernity for which tourism is purported to serve as a metaphor (Dann, 2002)? Analyses that fail to treat the problem of scale seriously are at risk of being poorly formulated, with the result that any insights gained may not be useful. The musings on tourism as a metaphor for everyday postmodern life by Dann's contributors may not generate obviously useful suggestions for dealing with problems emanating from the behaviour of individual visitors to Sandfly Bay, but they may be useful for suggesting the values that the tourists may bring with them to the visit. The putative relationships between tourists' values and behaviours are central to the development of visitor management strategies (e.g. Orams, 1996) but are beyond the scope of this chapter. Second, what level of penguin organization will provide the most useful insights into the nature and effects of human interaction? Should the individual bird be the focus of the analysis, or the family group, the colony, the mainland population or the total population, which includes those birds who live on the subantarctic Campbell and Auckland

Islands? Just as the scale of analysis of human behaviour may be moved upward, so too may the scale of penguin organization being considered, to the point where we might consider as human disturbance the contribution to global warming, and consequent loss of habitat, made by a long-haul-jetting tourist who has never heard of penguins, upon a population of penguins who have never yet seen a human. To some extent the problem of scale is alluded to in the slogan *think globally, act locally*, although this injunction, of uncertain origin, does not specify the optimal scale of global or local analysis and action.

The challenge is to decide upon the scale of analysis that best illustrates the nature of human disturbance that could, and does, occur at Sandfly Bay. It is suggested (Ellenberg *et al.*, 2007), but has not yet been demonstrated, that individual birds prospecting to find a breeding site may be put off choosing Sandfly Bay owing to the frequent intrusive presence of people. This interruption to recruitment of new birds may eventually threaten the viability of the site as a breeding area. Other sites also, though without significant human disturbance, or any other identifiable disturbance, have experienced falling numbers and even abandonment. At Tavora, a site extensively restored by the Yellow-eyed Penguin Trust, the number of birds nesting at the site has dropped since restoration began. Richdale's 'Colony Z' (Director, 1986), the setting for much early observational material (Richdale, 1944), was abandoned for unknown reasons while other sites have experienced increasing numbers of breeding pairs (Moore, 2001). If we take the site as our scale of analysis then there is again no evidence that tourists have ever brought about abandonment (Moore, 2001). If the Sandfly Bay site is, in fact, abandoned in the future then it will be important to postulate the aspects uniquely of tourist behaviour that have proved most disruptive to the penguins' site fidelity. Historically, birds have had to accommodate major changes in sea level, and thus must have had a history of mobility in site selection. The South Island population of penguins does not seem to be replenished by individuals moving from the subantarctic (Triggs and Darby, 1989), unlike the New Zealand sea lions that routinely commute, and it remains unclear how plentiful the birds have ever been on the mainland. Historical population dynamics are important when considering human and non-human disturbance since such disturbances are best superimposed on to long-term population trends, and not on to short-term fluctuations.

Perhaps our consideration of the case of Sandfly Bay could be informed by more general accounts of nature-tourist–wildlife interactions. There have been attempts to produce a balance sheet of positive and negative effects of nature-based tourism. The report of Higginbottom *et al.* (2001), *Positive Effects of Wildlife Tourism on Wildlife,* is a distressingly slim volume. Its companion, Green and Higginbottom's (2001) *The Negative Effects of Wildlife Tourism on Wildlife* is more substantial, with over twice as many pages. Of course, the comparative sizes of the reports cannot be taken as analogues of any *true* state of the benefit/cost ratio of such tourism. Different effects deserve to be weighted according to perceived seriousness, and need to be formulated site-by-site and species-by-species. This is not to deny that certain generalizations may be possible. Establishing such relationships, where causality is strongly implied, fits within the class of challenges known as *wicked problems* (Ludwig, 2001), where apparently simple questions have a habit of being devilishly difficult to answer, and any solution put forward by one investigator by definition fails to satisfy the conditions demanded by some other investigator who invokes different values within which to situate both cost and benefit. Also, to the extent that the starting conditions of the problem rapidly become irrelevant through the number and complexity of interactions, consideration must be given to whether or not chaos is operating (Morris, 1990). Our question of what effect, if any, tourism has on a specified colony of YEPs may be a good candidate for being considered a wicked problem. For a moment, though, let us pretend that we have never come across the concept of wicked problems. How to approach this question of impacts then seems obvious: define what we mean by tourism and tourists, specify what is meant by effects or impacts, devise an appropriate set of measures and apply them. Using this common sense approach it should be possible to *do the science first*, that is, to discover everything there is to know about penguin ecology, and then to factor in the characteristics of this site and then consider the advent of tourism. In the best of all possible worlds the results of the measuring will be unequivocal. The conclusions derived from the data collected then should clearly inform site management. Is it probable, though, that such a clear-cut approach will ever be available? Is this how science works? This issue, whether science is the gradual, or sometimes sudden, uncovering of the truth about the world, or a socio-cultural activity engaged in constructing limited truths about particular small aspects of the world, has interested philosophers of science for some time (Kuhn, 1996/1970). We mention this here since the position we take is that it is not possible to do the science first, that is, to find out everything there is to know about a situation and then to study the effects of one newly introduced variable. Particularly in applied settings, measuring, a key aspect of doing the science, constitutes an intervention, often a non-trivial one. For example, repeated handling of penguins by scientists taking samples of various kinds, more or less intrusively, to demonstrate the effects of nature tourism, may confound the attribution of any such disturbance effects discovered to nature tourists. Admittedly there are ways to attempt to compensate for such disturbances, for example, by providing control sites where tourists never go (Ellenberg *et al.*, 2007, p. 40), but it is very difficult to provide such sites that are identical to the visited sites in every important way, apart from visitation. For example, one of the two strongest claims to date involving Sandfly Bay, that the 'general trend from this study is that two breeding areas with very different levels of human visitation show a difference in YEP chick fledging weight, in the year of study' (McClung *et al.*, 2004), relies for its explanatory power on visitation being the only salient difference between sites. These authors, though, identify other possibly explanatory differences between the sites, for example, the amount of shade available. The results of this study, as is so often the case, are frustratingly tentative. Ellenberg *et al.* (2007), however, is categorical in stating that 'breeding success (at Sandfly Bay) was significantly reduced with only about half the number of chicks fledged per pair . . . compared with undisturbed Green island'(p. 60).

A brief historical account may make it clear that any data gathered about penguins at Sandfly Bay are likely to be equivocal, and any conclusion tentative. On the mainland, the penguins have been the focus of close attention since the mid-1980s, when a census identified that the species seemed to be in sharp decline and without intervention would become locally extinct (Lalas, 1985). This realization prompted a conservation effort that is still in operation. An environmental non-governmental organization (NGO), the Yellow-eyed Penguin Trust, was formed with the mission statement: 'To work towards an increase in the number of YEPs on a self-sustaining basis within their natural coastal ecosystem' (Yellow-eyed Penguin Trust Board of Trustees, 1987). In addition, the bird was the subject of a revised species recovery plan developed by the New Zealand Department of Conservation *Te Papa Atawhai* (McKinlay, 2001). This plan specified a collection of interventions, including habitat restoration and predator control, and was intended to reverse the drop in numbers and lessen the threat of extinction. Shortly after this conservation effort was launched, and designed to be consistent with it, the YEP was identified as a species able to be developed as the focus of tourism (Tisdell, 1988). Thus, for this species, conservation and tourism have been entwined for the last 20 years, resulting in the development of an interesting literature; some focused on penguin biology, some on tourism and some on the interaction of the two. Some ecotourism ventures involving YEPs have been established on private land. Two such are Penguin Place and Elm Wildlife Tours. These ventures employ zoologists, actively restore habitat, carry out intensive predator control, closely monitor breeding success and play an active part in species recovery. At these sites, previously farmland but now transformed into suitable penguin habitat, visitor behaviour is closely managed through the involvement of professional guides. At other sites public access is unregulated. Sandfly Bay is one such site. The contested nature of Sandfly Bay, involving multiple perceptions of both the penguins and the site, has previously been described (Shelton and Lübcke, 2005). Specifically, visitors to this site bring with them a variety of values and behaviours not all of which match the values of the site manager, in this case the New Zealand Department of Conservation. Certain facts about Sandfly Bay are uncontested. In the absence of hard data the observations of concessionaires who visit daily over the summer, and several times per week during the rest of the year, are given credence. The number of unregulated visitors has been on the rise for well over 20 years. Guides report that, frequently, there may be 50 visitors to the site over a fine summer evening. Attempts to formally quantify this rise have been sporadic; a pressure-activated counter that had been in place for several years broke down and was not quickly replaced. The only well-structured attempt to estimate total annual number of visitors, their activities and distribution over the site (Seddon *et al.*, 2004) has yet to be replicated and elaborated to facilitate the development of a spatial model. Everyone associated with the site agrees that, especially over the last 5 years, the number of nest sites near to the public hide, and the number of penguins frequenting that area, has decreased. Guides report that, although 7 years ago it was commonplace to see 25 birds in the vicinity of the hide, now the number fluctuates between two and ten. Does this situation constitute a clear case then, for nature-tourist-induced disturbance, at an individual and breeding colony scale, of the Sandfly Bay YEPs?

Before implicating solely tourism, an understanding of the YEP breeding population is necessary. The Department of Conservation organizes an annual nest count along the whole of the southern coast. Nest monitoring of YEPs at Sandfly Bay began in 1990 with three nests monitored. Since then the largest number of nests has been 22 in 1996/97 and 1997/98. Since those seasons there has been a decline to around ten nests for the period 2000–2005 (Fig. 12.1).

Since 1990, there have been some 235 nesting attempts at Sandfly Bay. These attempts have involved 417 banded birds and 45 unbanded or unidentified birds. In this sample 26 birds identified as females made 78 nesting attempts and 30 males made 80 nesting attempts. Total numbers of nesting attempts for banded males vary between one and 12 and for females between one and ten. Nest site fidelity is reflected by the proportion of nesting individuals who returned to Sandfly Bay from one season to the next. These results are presented below (Fig. 12.2). Of the 235 nesting attempts 124 involved males who were present the previous year and 87 involved females and 82 were where both parents were present the previous year. In 1993, over 90% of males returned to nest at Sandfly Bay. Since 2000, the per cent returning has decreased, reflecting, or reflected by, the overall decrease in nest numbers. Compared to the males the proportion of females who return to Sandfly Bay has been lower. A point to note in this is that the total number of banded birds in the population is relatively low.

Clearly, there has been a reduction in nest fidelity since the late 1990s. Has this been in response to chronic human interaction? Current data are inadequate to answer this question. None the less, site managers must make

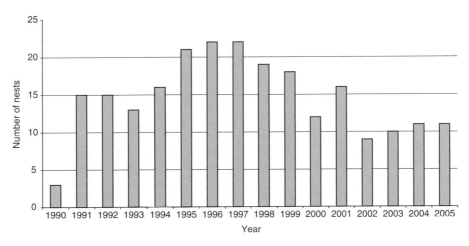

Fig. 12.1. Recorded number of Yellow-eyed penguin nests at Sandfly Bay, Otago Peninsula, New Zealand, 1990–2005. (Unpublished data, Yellow-eyed Penguin Database).

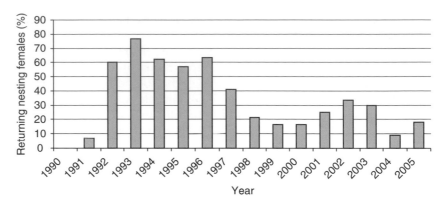

Fig. 12.2. Percentage of nesting female Yellow-eyed penguins (YEPs) at Sandfly Bay who had nested there the previous season.

responsible decisions about the nature and amount of visitor access to the breeding colony.

One response to such incomplete data, and the subsequent uncertainty they generate, is then to take a principled approach to management and apply *the precautionary principle* (O'Riordan and Cameron, 1994; O'Riordan *et al.*, 2001). This approach is situated within scientific uncertainty and addresses two problems: 'First, uncertainty creates an evidentiary problem: what must we know before taking measures to protect the environment, and with what degree of certainty? Second, uncertainty raises the management question: how should we respond to uncertain risks?' (Bodansky, 1994). The precautionary principle has been applied to many different types of questions involving many different academic disciplines (see *The Precautionary Principle Project* (Author, 2003, 2006) for a description of a sophisticated example of applying this approach to the question of biodiversity). How this principle could be integrated into tourism planning and practice has also been opened up for investigation (Fennell and Ebert, 2004). In the case of Sandfly Bay, application of the Precautionary Principle should reduce Type 2 error (Anderson *et al.*, 2000), that is, mistakenly accepting that there is no negative visitor effect on penguins when in fact there is such an effect. Of course there is an opportunity cost involved in this Type 2 error reduction. There is the possibility of there being both positive and negative visitor effects, with a net positive effect. Establishing that this is the case requires that more be learned about the system. Active adaptive management provides an approach for judging 'the benefit of management actions that accelerate information gain', in this case more scientific enquiry, 'relative to the benefit of making the best management decisions' (McCarthy and Possingham, 2007). Active adaptive management is an optimization strategy and therefore does not privilege Type 1 or Type 2 error reduction, but, as a Bayesian procedure, does require more previous knowledge of the system than commonly is available (McCarthy and Possingham, 2007). Thus there is a tension between applying the precautionary principle and developing an optimal site management strategy.

The second principled approach of interest is *the cumulative or additive impacts principle*. This principle, for our purposes here, means simply that if a species is under pressure, through predation from other species or habitat or climatic factors, then why add human disturbance to the list? The additive impacts approach assumes that encounters with tourists will comprise a disturbance in addition to that provided by non-human agents. Our point here is that the application of these two principles, the precautionary principle and the additive or cumulative principle, to a wicked problem does not generate an optimal solution but does support certain management actions over others.

One precautionary approach could involve stopping all visitations. This would almost certainly increase visitation to adjacent sites that are currently either unvisited or lightly visited. An alternative approach would be to attempt to change the nature of tourist access from unregulated to regulated. Such a change would inevitably require the allocation of increased resources to the management of the site. To allocate these scarce resources most effectively it would be sensible to target those visitor behaviours for which the manager has evidence of disruption potential. What would such a hierarchy of tourist-induced disturbance look like? Population Viability Analysis (Keedwell, 2004) allows the use of sensitivity analysis to establish the importance of events, like permanently removing from a population breeding adults early in their reproductive career, on species viability. Tourist behaviour at this site does not constitute such a threat. For penguins in their early reproductive phase, being the victim of fisheries by catch would be the most serious current human-induced threat other than being run over by a vehicle. Getting commercial fishermen even to report instances of penguin by-catch is difficult. Intervening to reduce the occurrences is still somewhere in the distant future. There is no risk from motor vehicles at Sandfly Bay, though at another site judicious fencing has redirected birds away from crossing a busy tourist road at Nugget Point, thus avoiding inevitable deaths. Disruption of diurnal and seasonal routines is the most likely effect of unregulated tourist presence at Sandfly Bay (McClung *et al.*, 2004; Ellenberg *et al.*, 2007) and was formally monitored over summer 2007. Disruption, therefore, is the target of current site management planning. One conceivable approach would be to use habituation as an active management tool (Shelton *et al.*, 2004) although it is difficult to demonstrate unequivocal habituation to human presence in this simultaneously timid and inquisitive species (Shelton and Higham, 2007). YEPs have been described as habituating to human presence under controlled conditions (Ratz and Thompson, 1999). Such controlled conditions are much more easily obtained by commercial operators on private land than by the Department of Conservation on a public beach. Previously suggested differences in level of habituation between birds near the hide at Sandfly Bay and those further along the beach (Shelton and Lübcke, 2005) may have been due simply to the birds near the hide needing to pause longer in order to cool down before their steep climb to their nesting sites and thus demonstrating reduced avoidance behaviour erroneously interpreted as habituation. If we dismiss the notion of getting the birds more used to human presence, then the

obvious option is to change the ways that visitors behave. To that end, the emphasis of the signage at the site is to be changed during the 2007 low season to iconi-
cally illustrate problem behaviours and suggest appropriate visitor behaviour more
directly. This change in signage is one element of a wider intervention, based on
design principles coupled with the concept of bringing visitor behaviour under
stimulus control, which has been under consideration for some time (Shelton and
Abbott, 2004; Abbott, 2006). This more ambitious intervention was modified in
response to practical considerations such as storm surges washing away key way-
finder sites, budgetary cycles and constraints and the availability of volunteer labour
at specific times of the year. Another intervention is planned to be implemented
alongside the modification of signage. Volunteers are being recruited to staff the
site and provide visitors directly with advice on how best to interact with the wildlife. This approach has worked well at a Little Blue penguin (*Eudyptula minor*)
site at Pilot's Beach on Otago Peninsula. Sandfly Bay is less easily accessible for
volunteers but once the scheme is fully operational for the 2007/08 November–
March high season it will be able to be evaluated.

This discussion of human impact on YEPs at Sandfly Bay began by presenting some big questions, for example: How does science work? Can we rely solely
on science for day-to-day site management decisions? It ended with a description
of two modest interventions designed to modify tourists' behaviour. Why is there
such a contrast between theoretically ideal approaches to conservation decision
making and implementation and what actually happens in practical site management? One implication of problems being wicked problems is that not all stakeholders share an epistemological framework. What seems like good science to
one may be perceived as *paralysis by analysis* by another. The urge to protective
action by a wide range of interested parties, in this case those individuals and
groups committed to a sustainable breeding population of penguins at a Sandfly
Bay, brings about a socially negotiated, culturally mediated political intervention,
rather than a theoretically optimal solution backed by hard scientific data. In
response to observed tourist behaviour that is incompatible with even weakly supported notions of best site management practice, the most cost-effective intervention in this case is to attempt to modify tourists' behaviour. The disadvantage of
such an intervention is that, irrespective of the outcome with respect to recruitment and retention of pairs of breeding birds, the true effect of nature tourists'
activities can never be known. Such is the price of principled management. There
is a proverb which is apposite: 'an ounce of prevention is worth a pound of cure'
(Fennell and Ebert, 2004). In the case of the YEPs of Sandfly Bay and the effects
of their visitors on them, we may rephrase this folk wisdom as: in the absence of
hard evidence, but in the face of consistent anecdote and suggestive research findings, prudence dictates that we modify tourists' behaviour as a precaution. The
allocation of scarce resources to such an intervention, which is in accordance with
the wishes of multiple stakeholders, but the effectiveness of which can never be
known, is the reality of everyday conservation site management.

Disclaimer
The views expressed here are those of Bruce McKinlay as an individual and do not
necessarily reflect those of the New Zealand Department of Conservation.

References

Abbott, M. (2006) Beyond the visitor in New Zealand's public conservation estate. Paper presented at the ATLAS Asia-Pacific Conference Tourism after Oil 3–5 December, Dunedin, New Zealand.

Anderson, D., Burnham, K. and Thompson, W. (2000) Null hypothesis testing: problems, prevalence and an alternative. *The Journal of Wildlife Management* 64(4), 912–923.

Author (2003, 2006) *The Precautionary Principle Project: Sustainable development, natural resource management and biodiversity conservation.* Available at http://www.pprinciple.net/

Bodansky, D. (1994) The precautionary principle in US environmental law. In: O'Riordan T. and Cameron, J. (eds) *Interpreting the Precautionary Principle.* Earthscan Publications, London, pp. 203–228.

Dann, G. (2002) *The tourist as a Metaphor of the Social World.* CAB International, Wallingford, UK.

Darby, J. and Dawson, S. (2000) Bycatch of yellow-eyed penguins (*Megadyptes antipodes*) in gillnets in New Zealand waters 1979–1997. *Biological Conservation* 93(3), 327–332.

Director (1986) *Colony Z.* Available at: http://www.nhnz.tv/cat/colonyz.html

Ellenberg, U., Setiawan, A., Cree, A., Houston, D. and Seddon, P. (2007) Elevated hormonal stress response and reduced reproductive output in Yellow-eyed penguins exposed to unregulated tourism. *General and Comparative Endocrinology* 152, 54–63.

Fennell, D.A. and Ebert, K. (2004) Tourism and the precautionary principle. *Journal of Sustainable Tourism* 12(6), 461–479.

Green, R. and Higgenbottom, K. (2001) The negative effects of wildlife tourism on wildlife. *CRC for Sustainable Tourism,* Gold Coast, Queensland.

Hall, C.M. (2007) Scaling ecotourism: the role of scale in understanding the impacts of ecotourism. In: Higham, J.E.S. (ed.) *Critical Issues in Ecotourism: Understanding a Complex Tourism Phenomenon* Butterworth-Heinemann, Oxford, pp. 243–255.

Higgenbottom, K., Green, R. and Northrope, C. (2001) Positive effects of wildlife tourism on wildlife. *CRC for Sustainable Tourism* Gold Coast, Queensland.

Hutching, G. (1998) *The Natural World of New Zealand: An Encyclopaedia of New Zealand's Natural Heritage.* Viking, Auckland.

Keedwell, R.J. (2004) *Use of Population Viability Analysis in Conservation Management in New Zealand.* Department of Conservation, Wellington, New Zealand.

Kuhn, T. (1996/1970) *The Structure of Scientific Revolutions,* 3rd edn. University of Chicago Press, Chicago, Illinois.

Lalas, C. (1985) *Management Strategy for the Conservation of Yellow-eyed Penguins in Otago Reserves: Draft Report.* Department of Lands and Survey, Dunedin, New Zealand.

Lalas, C., Ratz, H., McEwan, K. and McConkey, S. (2007) Predation by New Zealand sea lions (*Phocarctos hookeri*) of Yellow-eyed penguins (*Megadyptes antipodes*) at Otago Peninsula, New Zealand. *Biological Conservation* 135, 235–246.

Levin, S. (1992) The problem of pattern and scale in ecology. *Ecology* 73(6), 1943–1967.

Ludwig, D. (2001) The era of management is over. *Ecosystems* 4, 758–764.

McCarthy, M. and Possingham, H. (2007) Active adaptive management for conservation. *Conservation Biology 2007*(16 April) (online early articles).

McClung, M., Seddon, P., Massaro, M. and Setiawan, A. (2004) Nature-based tourism impacts on yellow-eyed penguins *Megadyptes antipodes*: does unregulated visitor access affect fledging weight and juvenile survival. *Biological Conservation* 119(2), 279–285.

McKinlay, B. (2001) *Hoiho* (Megadyptes antipodes) *Recovery Plan.* New Zealand Department of Conservation *Te Papa Atawhai*, Wellington, New Zealand.

Moore, P.J. (2001) Historical records of yellow-eyed penguin (*Megadyptes antipodes*) in southern new Zealand. *Notornis* 48, 145–156.

Morris, W.F. (1990) Problems in detecting chaotic behaviour in natural populations by fitting simple discrete models. *Ecology* 71(5), 1849–1862.

Orams, M.B. (1996) Using interpretation to manage nature-based tourism. *Journal of Sustainable Tourism* 4(2), 81–94.

O'Riordan, T. (1976) *Environmentalism*. Pion, London.

O'Riordan, T. and Cameron, J. (eds) (1994) *Interpreting the Precautionary Principle*. Earthscan Publications, London.

O'Riordan, T., Cameron, J. and Jordan, A. (eds) (2001) *Reinterpreting the Precautionary Principle*. Cameron May, London.

Ratz, H. and Thompson, C. (1999) Who is watching whom? Checks for impacts of tourism on Yellow-eyed penguins *Megadyptes antipodes*. *Marine Ornithology* 27, 205–210.

Richdale, L. (1944) *Camera Studies of New Zealand Birds: Series B, Number 6*. Available at: http://divcom.otago.ac.nz/tourism/staff/pdfs/RichdaleCameraStudies.pdf

Schänzel, H. and McIntosh, A. (2000) An insight into the personal and emotive context of wildlife viewing at the Penguin Place, Otago Peninsula, New Zealand. *Journal of Sustainable Tourism* 8(1), 36–52.

Seddon, P., Armstrong, D. and Maloney, R. (2007) Developing the science of reintroduction biology. *Conservation Biology* 21(2), 303–312.

Seddon, P., Smith, P., Dunlop, E. and Mathieu, R. (2004) Touirist visitor attitudes, activities and impacts at a Yellow-eyed penguin breeding site on the Otago Peninsula, Dunedin, New Zealand. *New Zealand Journal of Zoology* 31, 119.

Shelton, E.J. and Abbott, M. (2004) Understanding NO! People, persuasive communication and penguins. Paper presented at the NZ Tourism and Hospitality Research Conference 6–10 December, Wellington, New Zealand.

Shelton, E.J. and Higham, J.E.S. (2007) Ecotourism and wildlife habituation. In: Higham, J.E.S. (ed.) *Critical Issues in Ecotourism: Understanding a Complex Tourism Phenomenon*. Butterworth-Heinemann, Oxford, pp. 271–286.

Shelton, E.J. and Lübcke, H. (2005) Penguins as sights, penguins as sites: the problematics of contestation. In: Hall, C.M. and Boyd, S. (eds) *Nature-based Tourism in Peripheral Areas: Development or Disaster?* Channel View Publications, Clevedon, pp. 218–230.

Shelton, E.J., Higham, J.E.S. and Seddon, P. (2004) Habituation, penguin research and ecotourism: Some thoughts from left field. *New Zealand Journal of Zoology* 31, 19.

Tisdell, C. (1988) *Economic Potential of Wildlife on the Otago Peninsula, Especially the Yellow-eyed Penguin*. University of Otago Economic Discussion paper 8818.

Triggs, S. and Darby, J. (1989) *Genetics and Conservation of Yellow-eyed Penguin: An Interim Report*. New Zealand Department of Conservation, Wellington, New Zealand.

Yellow-eyed Penguin Trust Board of Trustees (1987) *Mission Statement*. Available at: http://www.yellow-eyedpenguin.org.nz/

III The Legislative and Ethical Contexts

13 Marine Wildlife Tourism Management: Mandates and Protected Area Challenges

M.L. MILLER

This is a national park. If you want to find wilderness of the komodo dragon and other wildlife, you should stay longer.

(Sign at the entrance to Komodo National Park, Komodo Island, Indonesia)

Introduction

Marine wildlife tourism broadly denotes tourism that is designed by tourism brokers and tourists themselves for the optimal aesthetic (and some might say religious) appreciation of the structure and wonder of marine nature rather than its utilization. It is in wildlife tourism – whether marine or not – that flora and fauna and other life forms have what Hargrove (1989, pp. 10–11) terms 'anthropocentric intrinsic' as opposed to 'anthropocentric instrumental' value. Orams (2002) notes that the wildlife tourism experience takes place along a continuum in which wildlife is: (i) held captive in an environment completely constructed by humans (e.g. zoos, aquariums, oceanariums and aviaries; see, Davis, 1997); (ii) held semi-captive in environments partially constructed by humans (e.g. wildlife parks, rehabilitation centres, gardens, sea pens) (see Fig. 13.1); (iii) fed by tourists in natural environments (e.g. shark and reef fish feeding, bird feeding); or (iv) observed by tourists in the wild (e.g. national parks and protected areas, undeveloped land and sea areas) (see Fig. 13.2). Marine wildlife in this touristic context is something to be enjoyed for its natural essence and diversity, not something to be hunted or harvested and treated as natural resources (see, e.g. Masters, 1998; Orams, 1999; Warburton, 1999; Hall and Boyd, 2005).

In a world where the impulse to travel is ever increasing in potency and in which the coastal zone is becoming ever more congested, marine wildlife tourism competes not only with industrial and residential uses of nature, but also with a host of other variants of marine recreation and tourism (see, e.g. Edwards, 1988; Miller and Auyong, 1991a, 1998a; Miller, 1993b; Conlin and

Fig. 13.1. Tourists engage dolphins in Cozumel, Mexico. In this instance, tourists pay substantial fees for opportunities to see, touch and ride dolphins. Though memorable to the tourist, such tourism is controversial for the domestication of the semi-captive dolphins.

Baum, 1995; Hall and Page, 1996; Lockhart and Drakakis-Smith, 1997; Apostolopoulos and Gayle, 2002; Miller *et al.*, 2002a; Boissevain and Selwyn, 2004; Pattullo, 2005). With the assumption that marine wildlife tourism can have merit and should be promoted, this chapter examines the theme of marine wildlife tourism management, focusing on mandates for regulation and the creation of marine protected areas (MPAs).

Defining marine wildlife tourism

For purposes here, marine wildlife tourism is taken as a special case of nature tourism which takes place in regions encompassing the open ocean and coastlines along with all manner of associated bays, harbours, inlets and estuaries. Nature tourism is interpreted as tourism in which both biotic and abiotic elements of nature are regarded by the tourist as primary touristic amenities. Wildlife tourism is that component of nature tourism in which non-human life is visited, witnessed, appreciated and revered – and in some instances, more formally studied and restored – with minimal violence to the integrity of ecological systems. The spectrum of wildlife animals that have attracted tourism has included insects, fish and invertebrates, birds, reptiles and mammals (see, e.g. Newsome *et al.*, 2005). Plants of touristic value are equally diverse. Marine wildlife tourism, then, is wildlife tourism

Fig. 13.2. Marine wildlife tourism in Suncheon Bay on the southern coast of Korea. Tourism here takes place at the first Korean coastal wetland site to be included on the Ramsar List of wetlands, which followed from the Convention on Wetlands signed in Ramsar, Iran, in 1971. This tourism is carefully planned to minimize degradation of nature while allowing access to the sensitive tidal flat ecosystem. Tourists spend most of their time on a boardwalk designed for viewing wildlife and habitat by a non-profit environmental educational facility.

where the wildlife at issue has natural connections to ocean and coastal ecologies.

This definition of marine wildlife tourism fits with the vocabulary of the tourist, which is to say it excludes engagements of tourists with nature emphasizing forms of human hunting (e.g. the shooting or capture of birds and game, fishing) and it also excludes touristic endeavours that focus first on collecting, gathering or other techniques of removal. Marine wildlife tourism denotes activities that tourists themselves would call marine wildlife tourism.

Marine wildlife tourism understood in this way as an essentially non-consumptive activity overlaps with ecotourism (Kusler, 1991; Kaae and Miller, 1993; Miller, 1993a; Liu, 1994; Shackley, 1996; Fennell, 1999; Honey, 1999), geotourism, wildlife tourism (Higgenbottom, 2004; Newsome *et al.*, 2005), environmental tourism, nature tourism (Whelan, 1991; Wallace *et al.*, 1995; Deng *et al.*, 2002; Waitt *et al.*, 2003; Högmander and Leivo, 2004; Nyaupane *et al.*, 2004; Hall and Boyd, 2005; UNEP/CMS, 2006), polar tourism (Hall and Johnston, 1995; Bauer 2001), alternative tourism (Smith and Eadington, 1992), science tourism (e.g. in which scientist tourists are assisted in their work by tourists who are not scientists), volunteer tourism (e.g. in which tourists respond to

wildlife crises) and activism tourism (e.g. in which tourists make a political statement by objecting with their presence and behaviour to the harvesting of, say,
seals or whales), among other touristic activities when these take place in the
vicinity of oceans.

Marine wildlife tourism takes many shapes and particular types vary according
to the kind of platform used by the tourist, the way special technologies and tools
facilitate the experience, the extent to which the tourist is physically active (or passive) and the way in which the sociology of experience is structured to involve (or
not involve) other tourists and tourist providers such as guides, interpreters and
other experts (see Figs 13.3 and 13.4). Marine wildlife tourism is exemplified in the
watching of marine mammals, fish and seabirds and in the inspection of coastal
cliffs, mountains and land forms. It may involve snorkelling or scuba gear, boats of
diverse kinds, binoculars and telescopes, helicopters and fixed-wing aircraft.

It should be remarked that there are times when marine wildlife tourism is
not the principal motivating or explanatory concept in the mind of the tourist.
From the point of view of the tourist, many marine wildlife tourism moments
are not explicitly described as such because they are embedded in other experiences with other names. None the less, it is commonplace that a surfer, sailor,
beachcomber or sunbather may – in the course of participating in what they
see as a sport or a kind of beach leisure – relate to marine wildlife and the environment in profound ways.

Fig. 13.3. Tourists in the Galápagos Islands examine shoreline wildlife. In this
instance, neither the tourists nor nature benefit from the oversight and education
of trained interpreters or guides.

Fig. 13.4. Tourism on the northern coast of Mallorca. Tourists in the Mediterranean use a variety of forms of small watercraft to visit island coastlines and destinations for a variety of purposes. Marine wildlife is for many tourists an incidental, but important, amenity. In this example, tourist brokers provide a technology of glass-bottomed vessels.

Conceptual framework

Marine wildlife tourism systems are first and foremost tourism systems. As such they involve interactions between people and place in destinations that are situated in marine and coastal environments. Drawing from the work of Miller and colleagues (Miller and Auyong, 1991b,c, 1998b; Miller *et al.*, 2002b; Miller and Hadley, 2005) marine wildlife tourism systems can be seen to have sociologies with three kinds of actors: *tourism brokers*, *tourism locals* and *tourists*.

A 'broker–local–tourist' (BLT) model of marine wildlife tourism is displayed in Fig. 13.5. Tourism brokers comprise persons who in one way or another manage, design or otherwise seek in their occupational work to control tourism outcomes. In Fig. 13.5, on-site brokers are shown to be part of the community because they reside in the touristic region. (Off-site brokers perform the same functions as on-site brokers, but they may have their residences in locations very distant from tourist destinations.)

A first main category of private sector brokers includes individuals and firms that are part of the tourism industry. Marine wildlife tourism examples include guides and tour operators, ecotourism businesses and scuba instructors (Andersen and Miller, 2006).

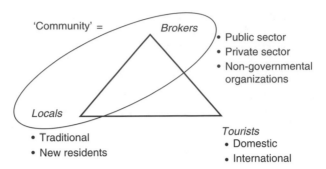

Fig. 13.5. Broker–local–tourist (BLT) model of Marine Wildlife Tourism System. (Adapted from Miller and Auyong, 1991, p. 75.)

A second category of public sector brokers are those employed in government who attend to tourism regulations, policies and enforcement. Marine wildlife tourism examples include many specialists who work for national parks and MPA entities.

A third category of brokers includes a range of civil society organizations, non-governmental organizations (NGOs), non-profit organizations that have programmes or initiatives that address tourism issues as well as the behaviour of other kinds of brokers. Examples of environmental NGOs with marine wildlife tourism agendas include the Nature Conservancy, the World Wild Fund for Nature (WWF), the Palau Conservation Society, Amigos de Sian Ka'an (Quintana Roo, Mexico) and the Coastal Conservation and Education Foundation (Cebu City, Philippines).

Tourism locals consist of persons who reside in the general vicinity of a tourism destination, but who do not depend on tourism for an income or seek in any organized way to control tourism. Examples in the marine wildlife tourism context would include agriculturalists and fishermen who live in the Galápagos and along the coastlines of South America.

Finally, tourists consist of persons who are motivated to visit a tourist destination and who subsequently return home. Needless to say, tourists with marine wildlife interests come from virtually all walks of life.

It is common in the literature of tourism for the non-tourists who inhabit a region visited by tourists to be referred to 'the community', 'the residents' or, in the marine context, 'the islanders'. Unfortunately, this convention masks the fact that brokers (especially those who work in the tourism industry) may have fundamentally different opinions than the locals regarding the desirability of tourism.

Dynamics

Interactions between brokers, locals and tourists are influenced by a mixture of economic, cultural, political and demographic processes. As Butler (1980; see also Pearce, 1989) has shown, tourism destinations – and those with marine wildlife amenities are no exception – have natural histories. Butler's destination life cycle tracks destinations as they are first discovered, and then evolve with a logistic trajectory (and through phases of *exploration, involvement, development* and *consolidation*) to a point at which the carrying capacity of tourists (or of the environment) is reached. It is here that destinations continue on one of the several paths. Thus, some destinations remain in a *consolidation* stage, some begin a *decline* stage and others reinvent themselves (by modifying the quality of experience and kinds of amenities offered to tourists) in a *rejuvenation* stage.

As marine wildlife tourism destinations go through life-cycle phases, the relative absolute numbers and proportions of brokers, locals and tourists change. Such population dynamics can be studied with variations of Lotka-Volterra predator–prey models.

From a sociological point of view, human population dynamics in marine wildlife tourism systems determine the cultural quality of destinations. There are many examples worldwide of places where locals have diversified occupationally to become private sector entrepreneurs and workers. This is illustrated, for example, by small-scale subsistence and commercial fishermen in Palau who have decided to enter the tourist industry as ecotourism guides. There are also many instances where wildlife tourists have been so intrigued by a destination that they elected to remain permanently in the capacity of a local or that of a broker (e.g. by starting an ecotourism business or by finding work with an environmental NGO in the area). There are also many other examples of one kind of broker transforming to become another kind of broker. This is illustrated when government tourist officials leave the public sector to work with private sector developers, or when tourism industry workers leave the private sector to form new NGOs.

Power

Power in marine wildlife tourism systems can be negative or positive. When power is exerted in counterproductive ways, a victim is created. When power is utilized in productive ways (as, e.g. when education rather than propaganda is transmitted) people's lives are enhanced.

It is typically presumed that power in tourism systems is more heavily distributed across tourists than locals or brokers. Such a presumption is based on the facts that tourists do make conscious decisions to travel and often have more discretionary income and general life advantages than the people they visit. Furthermore, tourists have certainly been seen to behave badly towards people and the environment while on vacation (see, e.g. Butcher, 2003).

Although tourists do have leverage, it should be noted that they are also vulnerable by design. In discussing touristic power with a Foucauldian framework, Cheong and Miller (2000) have argued that tourists have less power and that brokers (and sometimes locals) have more power than is generally thought. In wildlife tourism, tourism brokers (e.g. guides and transportation providers) have great influence over where tourists are permitted (or instructed) to go, what they are encouraged to do (or discouraged from doing) and what they learn.

A larger implication of the fact that power in tourism systems is to be found in the hands of brokers is that broker–broker relations are critical to both positive and negative touristic results. The lesson for the tourism analyst is that tourism failures (or successes) follow from broker–broker conflicts (or cooperation).

Pursuit of Touristic Contrast: Serious Leisure, Duty and Beauty

Many marine wildlife tourists exhibit an intense personal dedication to the correct practice of wildlife tourism. Such tourists can (and do) discuss the standards they observe when engaging wildlife and they also can (and do) specify what standards and best practices they demand in tourism brokers providing products and services. In addition, many marine wildlife tourists have considerable scientific knowledge regarding the species of most interest. Marine wildlife tourists often show great passion and expertise when they go to the field, and they can be extra-attentive to their own feelings and environmental conduct. Such touristic experiences are signals of what Stebbins (1992) has termed 'serious leisure'.

Given that marine wildlife tourists are so serious about their leisure it is fair to ask what the special features of marine wildlife are that so attract visitors. The basic motive for all tourism is the pursuit of contrast (Miller and Ditton, 1986, p. 11). In the case of marine wildlife tourism, two kinds of contrast seem particularly prominent.

The first kind of contrast has to do with *duty*. A great many marine wildlife tourists (including those who at times describe themselves as 'ecotourists', 'conservationists' and 'preservationists') are concerned with problems of environmental degradation. At a time in which the anthropogenic causes of environmental decline are increasingly established scientifically and acknowledged generally, many tourists organize their travel to view endangered and threatened species, and also to volunteer in efforts to rebuild populations and restore habitats. Examples of tourism motivated by conservationistic duty are found in Costa Rican turtle tourism and in seabird and wetlands tourism in Korea.

The second kind of contrast important to marine wildlife tourists concerns *beauty*. The earliest full-fledged inquiry into questions of beauty is seen to be that of Plato (Beardsley, 1966). In the 1st century AD, Longinus expounded that the sublime in rhetoric (e.g. in Homer's *Odyssey*) is to be known through its effect, which 'not only persuades, but even throws an audience into transport' (cited in Monk, 1960, p. 12).

It is in 18th-century England, however, that the lexicon of aesthetic evaluation was developed. Aesthetic debates during this 'Century of Taste' centred on fundamental and subtle differences in the meaning of such terms as 'the beautiful', 'the interesting', 'the sublime' and 'the picturesque', as these were descriptors of prospects and nature. In a series of essays in *The Spectator*, which appeared in 1712, Joseph Addison distinguished the 'beautiful' from the 'strange' (or uncommon) and the 'great' (or sublime). The great as illustrated by 'huge heaps of mountains' and 'wide expanses of waters' inspires 'delight' in the irregular as well as in the chaotic and the terrible. In 1757, Edmund Burke agreed that the sublime complemented the beautiful: 'sublime objects are vast in their dimensions, beautiful ones comparatively small; beauty should be smooth and polished; the great, rugged and negligent' (1990, p. 113).

In elaboration, Burke pointed out that the 'whatever is qualified to cause terror is a foundation capable of the sublime' (1990, p. 119; see also Monk, 1960). Writing of the sublime in nature, Burke says:

> The passion caused by the great and sublime in *nature* . . . is Astonishment; and astonishment is that state of the soul, in which all its motions are suspended, with some degree of horror. . . . Astonishment . . . is the effect of the sublime in its highest degree; the inferior effects are admiration, reverence and respect.
>
> (1990, p. 53)

During the Romantic Era between 1750 and 1850, tourists on the Grand Tour began to expand their sensitivities beyond an interest in antiquity (as, e.g. represented by the ruins of Ancient Rome) to an interest in the nature. The 19th century was also the time in which visits to zoos, amateur botanizing and the collection of natural objects became fashionable in England (see, e.g. Blunt, 1976; Barber, 1980; Allen, 1994). Throughout this period, a sign of a properly educated English gentleman was the ability to discuss and evaluate natural beauty using the terminology of Addison and Burke (see, e.g. Nicolson, 1997).

Immanuel Kant took interest in aesthetic issues in the second half of the 18th century. In his first formulations, Kant (1991, pp. 47–48) contrasted the 'feeling of the beautiful' with the 'feeling of the sublime':

> The sublime *moves*, the beautiful *charms* . . . [The sublime feeling] is sometimes accompanied with a certain dread, or melancholy; in some cases merely with quiet wonder; and in still others with a beauty completely pervading a sublime plan. The first [kind of sublime] I shall call the *terrifying sublime*, the second the *noble*, and the third the *splendid*.

Today, many tourists trek to see marine wildlife, which they anticipate will generate in them feelings of awe, beauty and encounters with the sublime. Examples of sublime spectacles are found in the totality of the Great Barrier Reef, sea lion rookeries in the Bering Sea, penguin congregations along the Argentinean coast and great white and other large sharks off the coasts of South Africa and Australia. Abiotic touristic amenities which enthrall marine wildlife tourists include glaciers calving in Alaska, huge waves in Hawaii and 20 miles of tidal flats exposed daily by retreating tides on the western coast of Korea.

Marine wildlife tourism management systems

Marine wildlife tourism systems can be analysed as natural resource management systems (see Miller *et al.*, 1986a). Sociologically, the human component of a *marine wildlife tourism management system* (MWTMS) is composed of a *management element* (typically, a public sector executive agency) and a *constituency element* of diverse stakeholders (typically composed of private sector or industry interests, organized civil society and NGOs and general public). Although the idea of marine wildlife tourism management systems has not resulted in the creation of government agencies or geographic jurisdictions with exactly a 'marine wildlife tourism' name, de facto examples of such systems are found worldwide. In the USA, for example, many policies that are designed and implemented by the National Park Service, the US Fish and Wildlife Service and the National Marine Fisheries Service promote both the conservation of marine wildlife and tourism (for introductions to some of these agencies, see Brockman, 1959; Runte, 1979; Everhart, 1983; Reed and Drabelle, 1984; Miller *et al.*, 1986a; Knight and Bates, 1995; Knight and Gutzwiller, 1995). The Great Barrier Reef Marine Park (GBRMP) and many marine reserves, sanctuaries and protected areas provide additional examples.

Management decisions

Marine wildlife tourism management may be defined as an activity (legitimated by federal statutes and cultural mandates) by which an authority designs and enforces policies on behalf of its constituency. Policies and regulations developed by the management element of marine wildlife tourism systems are informed not only by special-interest group preferences, but also by multidisciplinary science (involving, e.g. biology, ecology, sociology, economics, cultural anthropology and archaeology).

In theory, the policies of managers should sustain marine wildlife tourism systems and should foster proper relationship between humankind and nature. Practically, management decisions are related to two overarching questions of quality:

1. *The impact question*: What is the *quality* (or significance, or value, or meaning, or importance) of marine wildlife tourism for the marine or coastal ecosystem and its components and the citizenry of the area and humankind at large?

The impact question – which may be framed as concerning 'Optimum Visitation' – is answered with an analytical agenda that resembles that of quantitative fishery and tourism management. It is also informed by multidisciplinary human dimensions research, education and advice.

2. *The design question:* What is the best way to create a *quality* biological, social and technological environment that fosters personal growth and responsible environmental conduct through marine wildlife tourism?

The design question concerns notions of 'Optimum Experience' and quality of landscape, restoration, architecture and education. This question is answered with an analytical agenda that requires a combination of natural science and human dimensions expertise.

Management mandates

In many nations throughout the world, the federal (or national) government explicitly addresses tourism matters. In some instances, tourism is the sole focus of management and development agencies; in other cases, tourism is treated together with fisheries, cultural and other topics. To illustrate, South Africa has a Department of Environmental Affairs and Tourism; Antigua and Barbuda are served by a Ministry of Tourism and Environment; New Zealand has a Ministry for Tourism and a Tourist Board; the Solomon Islands have a Visitor Bureau; Fiji has a Ministry of Tourism; and Indonesia has a Ministry of Culture and Tourism. Interestingly, the USA is anomalous for not having a major executive agency or authority with tourism responsibilities. However, as will be seen below, recreation and tourism functions are assigned to a range on natural resource management agencies.

Policy decisions in the realm of marine wildlife tourism management are more than technical or scientific fixes. Management policies hinge on value judgements. They are inherently controversial precisely because allocation decisions (regarding which marine wildlife tourism species are to be targeted, and also what kind of tourists shall have access to these) reflect conceptualizations of social and environmental justice. With this situation, marine wildlife tourism managers are guided in fulfilling their responsibilities by two kinds of instructions or mandates reflecting the values of society.

PHILOSOPHICAL MANDATES   The first and most powerful of these – the *philosophical mandate* – is expressed informally in the norms of society. This type of mandate mirrors a cultural worldview or perspective in which a constellation of (sometimes competing) standards and values structure human conduct. Thus, a philosophical mandate may feature a scientific value, a political or ideological value, an aesthetic value or some combination of these.

Marine wildlife tourism management decisions can be considered as they fit (or fail to fit) with a philosophical mandate. In his paradigmatic-setting volume, *A Sand County Almanac*, Aldo Leopold (1968, p. 214) initiates a discussion of environmental responsibilities by pointing out that: '[w]e can be ethical only in relation to something we can see, feel, understand, love, or otherwise have faith in'. He then proposes his classic three-part test for a *land ethic*:

> A thing is right when it tends to preserve the *integrity*, *stability*, and *beauty* of the biotic environment. It is wrong when it tends otherwise. (emphasis added)
>
> (1968, pp. 224–225)

Extending Leopold's framework to marine wildlife tourism policy making it is suggested here that marine wildlife tourism regulations (ethical decisions) are 'right' if, and only if, they are based on scientific knowledge (epistemological conclusions) and formal kinds of aesthetic intercourse with nature (appreciations of beauty). In language that casts this argument in terms of what we need to know about to make 'right' marine wildlife tourism policies, this amounts to the idea that scientific knowledge coupled with knowledge of environmental aesthetics positions us to have confidence in outcomes concerning environmental ethics.

Marine wildlife tourism activities raise questions of three kinds for managers:

1. *Scientific questions*: What is scientifically *true*? (Science informs how the world behaves, or how it might behave. Science cannot provide normative advice except insofar as societal values are taken as given. In such cases, the specification of value assumptions must be made explicit.) Scientific questions include those that pertain to the Leopoldean *integrity* and *stability* of a marine wildlife tourism system and its parts.

2. *Aesthetic questions*: What is *beautiful*? (Here, beautiful is a cover term that encompasses the tasteful, the sublime, the picturesque and the interesting.

Importantly, evaluations of beauty are educational, but not normative.) Aesthetic questions include those that concern the Leopoldean *beauty* of a marine wild-life tourism system and its constituent parts.

3. *Ethical questions*: What is *right*? (Ethical questions can be seen to overlap with philosophical questions, religious questions and ideological questions. Questions of the ethical kind lead to recommendations about what one *should* do.) Answers to ethical questions are arrived at with considerations of Leopoldean *stability, integrity* and *beauty*.

Taken together, these scientific, aesthetic and ethical questions pose a philosophical challenge for marine wildlife tourism managers. The three questions can be asked not only of the human and natural components of the system, but also of the total system and the relationship linking the two. To study and worry about people instead of nature (or the reverse) is not the point. The idea is not to sustain tourists or marine wildlife separately (although in some instances this may be an acceptable goal), but to sustain the relationship.

INSTITUTIONAL MANDATES   The second type of instruction to marine wildlife managers is found in its *institutional mandates*. Institutional mandates are embodiments of cultural norms and philosophies, but they are, relatively speaking, more codified and formalized. Institutional mandates are expressed most obviously in laws, conventions, treaties, decrees, presidential orders and other documents and acts of government. They are also expressed somewhat less formally in quasi-institutional initiatives, and agendas that represent the changing priorities of society.

Contemporary institutional mandates for marine wildlife tourism management have their foundations in a basic concern for sustainability. Three variants of a conservation ethic reflect this orientation (see Fig. 13.6).

The origins of the sustainability ideal trace to the emergence of the conservation movement in the USA during the Progressive Era at the end of the 19th century. At that time – and in response to private sector natural resource monopolies and

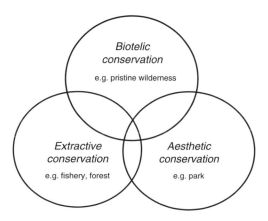

Fig. 13.6. Three conservation ethics fostering sustainability.

short-term horizons – the federal government began to create laws and executive agencies attuned to the theme or mandate of sustainability. As this occurred, two variants of a conservation ethic emerged. The first of these, *extractive conservation*, endorsed the goal of sustainable yield and is famously associated with the US Forest Service (established in 1905) and the forceful personality of Gifford Pinchot. The second variant, *aesthetic conservation*, promoted the goal of sustainable experience and has its roots in the transcendental philosophy of John Muir and the National Park Service (established in 1916) (Miller *et al.*, 1986b) (see Table 13.1). From the Progressive Era until the present, natural resource management systems in the USA

Table 13.1. Selected legislation and events providing a foundation for marine wildlife tourism management (US focus).

1916 – The National Park Service Organic Act (USA)
1956 – Fish and Wildlife Act (USA)
1962 – First World Conference on National Parks (1st World Parks Congress, Seattle, USA) considers need for protection of coastal and marine areas
1966 – National Wildlife Refuge System Administration Act (USA)
1969 – National Environmental Policy Act (USA)
1972 – Scoond World Conference on National Parks (2nd World Parks Congress, Yellowstone, USA)
1972 – National Marine Sanctuaries Act (USA)
1972 – UNESCO adopts the Convention Concerning the Protection of the World Cultural and Natural Heritage
1975 – Great Barrier Reef Marine Park Act (Australia)
1975 – IUCN (now the World Conservation Union) conducts conference on MPAs (Tokyo)
1982 – Third World National Parks Congress (3rd World Parks Congress, Bali, Indonesia)
1982 – IUCN Commission on National Parks and Protected Areas conducts workshops for marine and coastal protected areas (held as part of the 3rd World Parks Congress, Bali, Indonesia)
1986 – 1990 – IUCN Commission on National Parks and Protected Areas (now World Commission on Protected Areas) promotes global system of MPAs
1988 – Resolution 17.38 of the IUCN General Assembly (reaffirmed in Resolution 19.46 in 1994) defines MPA and specifies goals
1990 – Congress on Coastal and Marine Tourism (CMT 1990, Honolulu, USA)
1992 – Fourth World Congress on National Parks and Protected Areas (4th World Parks Congress, Caracas, Venezuela)
1992 – United Nations Conference on Environment and Development (World Summit on Sustainable Development or Earth Summit, Rio de Janeiro, Brazil) generates the Rio Declaration on Environment and Development and also *Agenda 21*
1996 – World Congress on Coastal and Marine Tourism (CMT, 1996, Honolulu, USA)
1997 – National Wildlife Refuge System Improvement Act
1999 – International Symposium on Coastal and Marine Tourism (CMT, 1999, Vancouver, Canada)
2000 – President Clinton's Executive Order 13158 calls for national integrated system of MPAs (USA)
2002 – World Summit on Sustainable Development (WSSD, Johannesburg, South Africa)
2003 – Fifth IUCN World Parks Congress (5th World Parks Congress, Durban, South Africa)
2005 – Fourth Coastal and Marine Tourism Congress (CMT, 2005, Çesme, Turkey)
2007 – Fifth International Coastal and Marine Tourism Congress (CMT, 2007, Auckland, New Zealand)

have implemented policies that provide for sustainability through either extractive or aesthetic conservation.

In the 1960s, a third variant of the conservation ethic, *biotelic (bio* = life, *télos* = purpose (Greek)) *conservation*, emerged as the non-materialistic or intrinsic value of nature and habitat was seen to be at risk due to human activities. This resulted in the passage of a wide array of environmental statutes, for example, the National Environmental Policy Act of 1969, the Marine Mammal Protection Act of 1972, the Coastal Zone Management Act of 1972 and the Endangered Species Act of 1973 (see also Table 13.1). With this, natural resource management agencies such as the National Marine Fisheries Service and the National Park Service were obligated to adjust their policies so that the sustainability of life forms without (immediate or even future) commercial and aesthetic value was guaranteed.

The ideal of sustainability has also become a global theme. Beginning at the end of World War II academics and practitioners in developed nations endeavoured to share their technologies and ideologies with lesser-developed nations under the banner of 'development'. From the 1960s, this theme has generally been displaced by that of 'sustainable development' in recognition of the unacceptable environmental costs of the first strategy. Perhaps the best-known formulation of sustainable development is found in *Our Common Future* (see also Lele, 1991):

> [Sustainable development is] the ability to make development sustainable—to ensure that it meets the needs of the present without compromising the ability of future generations to meet their own needs.
>
> (WCED, 1987, p. 8)

The theme of sustainable development began to be even more widely diffused at the 1992 United Nations Conference on Environment and Development – also known as the World Summit on Sustainable Development (WSSD) or the Earth Summit – held in Rio de Janeiro. At that time, delegates crafted the Rio Declaration on Environment and Development and also accepted the 900+ page *Agenda 21*, which specified a global approach to a wide range of issues including poverty, environmental protection of fragile ecosystems and the integration of environment and development in decision-making processes. In 2002, ideas and programmes introduced at the Rio de Janeiro WSSD were further advanced at a Johannesburg WSSD (see Table 13.1). (For a selection of papers and volumes that examine the potential for sustainable tourism, see Edwards, 1988; Kaae and Miller, 1993; France, 1997; Honey, 1999; Hall and Richards, 2000; Reid, 2003; Edgell, 2005.)

Although some critics continue to insist that sustainable development is an oxymoron, it seems itself to be sustainable as an ideal. To wit, Kates *et al.* (2005, p. 20) point out that:

> [There is] near universal agreement that sustainability is a worthwhile value and goal—a powerful feature in diverse and conflicted social contexts.

In combination, the ideals of sustainable development and the several variants of conservation provide an overarching institutional mandate for marine wildlife tourism management. In addition, a number of pivotal international meetings and events have provided direction to managers, and to all citizens, regarding what is important to consider when the future of systems linking humanity to nature is at issue (see Table 13.1).

The aforementioned institutional mandates pertain to direction channelled to marine wildlife tourism managers as opposed to their constituencies. Another kind of institutional mandate seeks to promote sustainability and conservation beyond management bureaucracies by appealing directly to stakeholders in general, and marine wildlife private sector brokers and tourists in particular.

The United Nations World Tourism (UNWTO) is the leading global organization concerned with tourism policy and practices. The general assembly of UNWTO consists of 150 member states, seven associate members and 350 affiliate members. UNWTO is especially concerned with tourism in coastal areas and small-island states, and has created special sustainable development and tourism ethics committees. UNWTO's Global Code of Ethics for Tourism is based on principles of sustainability, with a special emphasis on local communities and monitoring development. The Code consists of ten principles covering such topics as obligations of stakeholders in tourism development, rights of workers and entrepreneurs in the tourism industry, the right to tourism and the use of tourism as a beneficial activity for host countries and communities.

The ethical mandate of the UNWTO is echoed in a variety of ecotourism codes attuned to responsible travel and conservation developed by organizations such as The International Ecotourism Society (TIES). A mandate not only for the protection of environmental and cultural entities of special value protection (and exemplified, e.g. by the Great Barrier Reef) but also for the practice of responsible visitation is implicit in the IUCN World Heritage Mission. As still another example, the National Geographic Society's Centre for Sustainable Destinations strives to 'preserve all the world's distinctive places through wisely managed tourism and enlightened destination management' (NCSD, 2007). (For other marine wildlife watching codes, see Scottish Natural Heritage (2004).)

Protected areas as marine wildlife tourism management tools

Mark Orams (1999, pp. 71–93) has identified four main strategies for the management of marine tourism: (i) *physical* strategies having to do with facility design and placement, provisions for sacrifice areas which permit intensive use in some areas so that others may remain pristine, and other manipulations of place; (ii) *regulatory* strategies with limiting access and the number of visitors, and separate activities (while prohibiting some uses); (iii) *economic* strategies that promote desired behaviours and change through the use of (dis)incentives; and (iv) *educational* strategies that effect desired change (including personal growth) *via* interpretative centres, signage and printed materials, guided walks and personal contact. To this, could be added: (v) *planning* strategies commonplace in fields of urban and tourism planning (see, e.g. Grenier *et al.*, 1993 and Miller and Hadley, 2005). While all of these approaches have a future in marine wildlife tourism management, this section concentrates on a regulatory strategy of zoning.

Of all regulatory possibilities, zoning is perhaps the easiest to conceptualize, implement and enforce. This is, of course, conditional on the extent to which the management authority and its institutional mandate are accepted by stakeholders as legitimate.

From a zoning perspective, marine wildlife tourism can be sustainable to the extent that wildlife conservation and visitor enjoyment goals are balanced. The most influential framework for marine wildlife tourism zoning – one which promotes the design of protected areas – does not come from the literature and experience of tourism management, but instead from experiences of the World Conservation Union (IUCN).

Protected areas are of many types, but generally function as hedges against unconstrained development. The IUCN Program on Protected Areas (PPA) supports the work of the IUCN World Commission on Protected Areas (WCPA). Both IUCN entities agree on the basic definition:

> Protected Area: An area of land and/or sea especially dedicated to the protection and maintenance of biological diversity, and of natural and associated cultural resources, and managed through legal or other effective means.
>
> (IUCN, 2007a)

PPA and WCPA also share a mission:

> To promote the establishment and effective management of a worldwide, representative network of terrestrial and marine protected areas as an integral contribution to the IUCN mission.
>
> (IUCN, 2007b)

PPA activities are conducted in 16 regions. Importantly, regional authorities are encouraged to develop their own protected area regimes tailored to local conditions. Table 13.2 displays the six categories of IUCN protected areas. Marine wildlife amenities of touristic value are of course found in all types of protected areas. Indeed, marine wildlife can also be found well outside of any protected area and in even the most urban and industrial of settings. However, marine wildlife tourism is most encouraged in Category II (National park: protected area managed mainly for ecosystem protection and recreation), Category III (Natural monument: protected area managed mainly for conservation of specific natural features), Category V (Protected Landscape/Seascape: protected area managed mainly for landscape/seascape conservation or recreation) and Category VI (Managed Resource Protected Area: protected area managed mainly for the sustainable use of natural resources).

With IUCN encouragement, many nations now strive to develop domestic protected area systems (see, e.g. McNeely and Miller, 1984, Committee on the Evaluation, Design and Monitoring of Marine Reserves and Protected Areas in the USA; Ocean Studies Board; Commission on Geosciences; Environment and Resources; National Research Council, 2001; Eagles *et al.*, 2002; and The International Bank for Reconstruction and Development/The World Bank, 2006). Such progress is illustrated in the Mediterranean by the Italian System of MPAs, which includes 29 MPAs, two underwater archaeological sites, one cetacean sanctuary (jointly managed with France and Monaco) and two terrestrial

Table 13.2. IUCN Protected Area Management Categories. (From IUCN Information Sheet No. 3. Available at: www.iucn.org/themes/wcpa/wpc2003/pdfs/outputs/pascat/pascatrev_info3.pdf)

Category Ia: Strict nature reserve/wilderness protection area managed mainly for science or wilderness protection – an area of land and/or sea possessing some outstanding or representative ecosystems, geological or physiological features and/or species, available primarily for scientific research and/or environmental monitoring.

Category Ib: Wilderness area: protected area managed mainly for wilderness protection – large area of unmodified or slightly modified land and/or sea, retaining its natural characteristics and influence, without permanent or significant habitation, which is protected to manage and preserve its natural condition.

Category II: National park: protected area managed mainly for ecosystem protection and recreation – natural of land and/or sea designated to a) protect the ecological integrity of one or more ecosystems for present and future generations, b) exclude exploitation or occupation inimical to the purposes of designation of the area and c) provide a foundation for spiritual, scientific, educational, recreational, and visitor opportunities, all of which must be environmentally and culturally compatible.

Category III: Natural monument: protected area managed mainly for conservation of specific natural features – area containing specific natural or natural/cultural feature(s) of outstanding or unique value because of their inherent rarity, representativeness or aesthetic qualities or cultural significance.

Category IV: Habitat/Species Management area: protected area managed mainly for conservation through management intervention – area of land and/or sea subject to active intervention for management purposes so as to ensure the maintenance of habitats to meet the requirements of particular species.

Category V: Protected Landscape/Seascape: protected area managed mainly for landscape/seascape conservation or recreation – protected area managed mainly for landscape/seascape conservation or recreation – area of land, with coast or sea as appropriate, where the interaction of people and nature over time has produced an area of distinct character with significant aesthetic, ecological and or cultural value, and often with high biological diversity. Safeguarding the integrity of this traditional interaction is vital to the protection, maintenance and evolution of such an area.

Category VI: Managed Resource Protected Area: protected area managed mainly for the sustainable use of natural resources – area containing predominantly unmodified natural systems, managed to ensure long-term protection and maintenance of biological diversity, while also providing a sustainable flow of natural products and services to meet community needs.

---

national parks with jurisdiction over marine resources (Cosentino, 2005). (For discussion of reasons for protected areas, tourism guidelines and principles, tour operator contributions, management and social science and human dimensions agendas, see also Salm and Clarke, 1989; Buckley, 2002; Christie *et al.*, 2003; Harmon, 2004; Högmander and Leivo, 2004; Oles *et al.*, 2006; Tour Operators' Initiative Secretariat, no date.)

In the USA, institutional momentum for MPAs traces to the 2000 Executive Order 13158 of President Clinton calling for an integrated system of MPAs. In response to this Order, the National Marine Protected Areas Center (NMPAC, 2005) in the USA has recently developed a classification scheme for MPAs based on six fundamental design characteristics: primary conservation focus, level of protection afforded, permanence of protection, constancy of protection, ecological scale of protection and restrictions on extraction.

Marine Protected Area Goals

It is especially important for marine wildlife tourism managers and constituencies to keep in mind that MPAs are not *ends* in themselves, but *means* to ends. There can be no a priori best specification of the *primary purpose* of MPA. In practice, the cover term 'Marine Protected Area' refers to 'protected areas', 'sanctuaries', 'reserves', 'parks' and a host of similar terms which – depending on the circumstance – emphasize one conservation value over others. Moreover, these terms are not used with any consistency. Thus, a sanctuary in one area may permit only visitation of a scientific brand, while a sanctuary in another may heavily promote wildlife or nature tourism.

The three variants of a conservation ethic in Fig. 13.2 are useful for understanding the intended purpose of MPAs. MPAs that would fit into a marine wildlife tourism management agenda would be those that would fall in the intersection of *biotelic conservation* (which emphasizes protection of species and habitat) and *aesthetic conservation* (which emphasizes visitation for aesthetic appreciation).

The Great Barrier Reef is sometimes described as the largest natural feature on Earth stretching some 2300 km along the north-east coast of Australia (see Fig. 13.7). In 1975, the GBRMP Act created a complex MPA with management

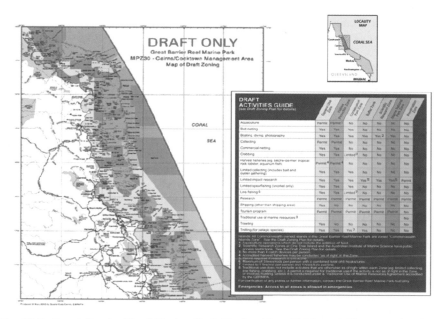

Fig. 13.7. Great Barrier Reef Marine Park Zoning Plan 2003: activities permitted in Cairns/Cooktown Management Area. (From GBRMP. Available at: http://www.gbrmpa.gov.au/data/assets/pdf file/0020/8282/mpz 30.pdf)

responsibilities vested in a GBRMP Authority. The primary mandate of the authority is:

> [T]o provide for the protection, wise use, understanding and enjoyment of the Great Barrier Reef in perpetuity through the care and development of the Great Barrier Reef Marine Park.
>
> (GBRMP, 2007)

Selected subordinate goals require the authority:

- 'to protect the natural qualities of the Great Barrier Reef, while providing for reasonable use of the Reef Region,
- to provide for economic development consistent with meeting the goal and other aims of the Authority,
- to provide recognition of Aboriginal and Torres Strait Islander traditional affiliations and rights in management of the Marine Park and
- to adapt actively the Marine Park and the operations of the Authority to changing circumstances.'

> (GBRMP, 2007)

The GBRMP qualifies as the single most impressive example of an MPA. Although labelled a 'park', GBRMP is much more; it is a multizoned regime that allows a wide variety of uses (and non-uses). This complexity of balancing multiple uses is apparent in Fig. 13.7, which shows a draft-zoning plan for the Cairns/Cooktown Management Area.

Although the GBRMP may be the most widely known example of MPA management that supports both marine wildlife protection and marine tourism, it is by no means the only success. A cursory consideration of other MPAs shows that there is a growing global record of marine wildlife tourism results. MPAs where marine tourism and wildlife protection are coexisting – and in more than a few instances co-flourishing – are found, for example, in reef preserves and protected areas in Bermuda, Folkestone Park and Marine Reserve in Barbados, Hol Chan and Shark Ray Alley Marine Reserves in Belize, West End/Sandy Bay Marine Reserve on the island of Roatán in Honduras, Goat Island Marine Reserve in New Zealand, Managama Marine Conservation Area in the Northern Marianas and Suncheon Bay and other Wetlands Conservation Areas in Korea. (IUCN has published a variety of directories of protected areas, many of which have to do with wildlife in island, coastal and marine settings.)

Discussion and Conclusion

This chapter has introduced a conceptual framework for understanding how issues of marine tourism development and marine wildlife protection can be raised and resolved jointly and sustainably under the name of marine wildlife tourism management. While successes in this regard are certainly to be found in the agendas of the World Conservation Union, the GBRMP Authority and many other MPAs, as many challenges as opportunities remain for managers.

Looking ahead, four recommendations detail the areas to be addressed for a sustainable marine wildlife tourism management.

Marine wildlife tourism managers must take greater advantage of professionals and academics whose expertise lies outside the natural and social sciences. In this regard, the fields of environmental philosophy, environmental history, tourism law, conflict resolution and mitigation have much to offer.

Integration across academic disciplines must be augmented with integration across sectors. Marine wildlife tourism managers need to be aware of ideas and approaches in such fields as fisheries management, coastal zone management, waste management, ecosystem management and the like.

Marine wildlife tourism policy lessons learned by managers in government must be exchanged for lessons learned about marine wildlife tourism by private sector tourism brokers who organize their efforts to a profit or project process and by NGOs and environmental organizations that respond to the rhythms of the campaign process.

Finally, marine wildlife tourism brokers (whether these are housed in the private sector, the public sector or the NGO sector) must develop more ways of communicating to constituencies with techniques of environmental educational and outreach (see, e.g. Orams, 1993). In particular, marine wildlife tourists should not be underestimated as easily satisfied.

Marine wildlife tourism can unfold in a 'win–win' way for wildlife and for tourists, but this will never be automatic. Innovation in management, however, has the potential to foster sustainability.

References

Allen, D.E. (1994) *The Naturalist in Britain: A Social History*, 2nd edn. Princeton University Press, Princeton, New Jersey.

Andersen, M.S. and Miller, M.L. (2006) Onboard marine environmental education: whale watching in the San Juan Island, Washington. *Tourism in Marine Environments* 2(2), 111–118.

Apostolopoulos, Y. and Gayle, D.J. (eds) (2002) *Island Tourism and Sustainable Development: Caribbean, Pacific, and Mediterranean Experiences*. Praeger, London.

Barber, L. (1980) *The Heyday of Natural History 1820–1870*. Doubleday and Company, Garden City, New York.

Bauer, T.G. (2001) *Tourism in the Antarctic: Opportunities, Constraints, and Future Prospects*. The Hayworth Press, New York.

Beardsley, M.C. (1966) *Aesthetics from Classical Greece to the Present: A Short History*. University of Alabama Press, Tuscaloosa, Alabama.

Blunt, W. (1976) *The Ark in the Park: The Zoo in the Ninteenth Century*. Hamish Hamilton in association with The Tryon Gallery, London.

Boissevain, J. and Selwyn, T. (eds) (2004) *Contesting the Foreshore: Tourism, Society, and Politics on the Coast*. Amsterdam University Press, Amsterdam, The Netherlands.

Brockman, C.F. (1959) *Recreational Use of Wild Lands*. McGraw-Hill, New York.

Buckley, R. (2002) Draft principles for tourism in protected areas. *Journal of Ecotourism* 1(1), 77–80.

Burke, E. (1990 [1757]) *A Philosophical Enquiry into the Origin of Our Ideas of the Sublime and the Beautiful*. Oxford University Press, New York.

Butcher, J. (2003) *The Moralization of Tourism: Sun, Sand . . . and Saving the World?* Routledge, New York.

Butler, R.W. (1980) The concept of a tourist area cycle of evolution: implications for management of resources. *Canadian Geographer* 23(1), 5–16.

Cheong, S.M. and Miller, M.L. (2000) Power and tourism: a Foucauldian observation. *Annals of Tourism Research* 27(2), 371–390.

Christie, P., McCay, B.J., Miller, M.L., Lowe, C., White, A.T., Stoffle, R., Fluharty, D.F., McManus, L. T., Chuenpagdee, R., Pomeroy, C., Suman, D.O., Blount, B.G., Huppert, D., Eisma, R.-L. V., Oracion, E., Lowry, K. and Pollnac, R.B. (2003) Toward developing a complete understanding: a social science research agenda for marine protected areas. *Fisheries* 28(12), 22–26.

Committee on the Evaluation, Design, and Monitoring of Marine Reserves and Protected Areas in the USA, Ocean Studies Board, Commission on Geosciences, Environment and Resources, National Research Council (2001) *Marine Protected Areas: Tools for Sustaining Ocean Ecosystems.* National Academy Press, Washington, DC.

Conlin, M.V. and Baum, T. (eds) (1995) *Island Tourism: Management Principles and Practice.* Wiley, New York.

Cosentino, A. (2005) *The Italian System of Marine and Coastal Protected Areas.* Global Forum on Oceans, Coasts and Islands, Lisbon, Portugal (10–14 October 2005). Available at: www.globaloceans.org/tops2005/pdf/AldoCosentino.pdf

Davis, S.G. (1997) *Spectacular Nature: Corporate Culture and the Sea World Experience.* University of California Press, Berkeley, California.

Deng, J., King, B. and Bauer, T. (2002) Evaluating natural attractions for tourism. *Annals of Tourism Research* 29(2), 422–438.

Eagles, F.J., McCool, S.F. and Haynes, C.D. (2002) *Sustainable Tourism in Protected areas: Guidelines for Planning and Management.* IUCN, Gland, Switzerland.

Edgell, D.I. Sr (2005) *Sustainable Tourism as an Economic Development Strategy along Coastlines.* [Report] Institute for Tourism (Department of Nutrition and Hospitality Management), East Carolina University, Greenville, North Carolina.

Edwards, F. (1988) *Environmentally Sound Tourism in the Caribbean.* University of Calgary Press, Calgary, Alberta, Canada.

Everhart, W.C. (1983) *The National Park Service,* 2nd edn. Westview Press, Boulder, Colorado.

Fennell, D. (1999) *Ecotourism: An Introduction.* Routledege, London.

France, L. (ed.) (1997) *The Earthscan Reader in Sustainable Tourism.* Earthscan Publications, London.

Great Barrier Reef Marine Park (GBRMP) (2007) Goals and Aims. Available at: http://www.gbrmpa.gov.au/corp_site/about_us/goals_aims

Grenier, D., Kaae, B.C., Miller, M.L. and Mobsley, R.W. (1993) Ecotourism, landscape architecture and urban planning. *Landscape and Urban Planning* 25, 1–16.

Hall, C.M. and Boyd, S. (eds) (2005) *Nature-based Tourism in Peripheral Areas: Development or Disaster?* Channel View Publications, Buffalo, New York.

Hall, C.M. and Johnston, M.E. (eds) (1995) *Polar Tourism: Tourism in the Arctic and Antarctic Regions.* Wiley, New York.

Hall, C.M. and Page, S.J. (eds) (1996) *Tourism in the Pacific: Issues and Cases,* International Thomson Business Press, London.

Hall, D. and Richards, G. (eds) (2000) *Tourism and Sustainable Community Development.* Routledge, New York.

Hargrove, E.C. (1989) *Foundations of Environmental Ethics.* Environmental Ethics Books, Denton, Texas.

Harmon, D. (2004) Intangible values of protected areas: What are they? Why do they matter? *The George Wright FORUM* 21(2), 9–22.

Higgenbottom, K. (ed.) (2004) *Wildlife Tourism: Impacts, Management, and Planning.* Sustainable Tourism Cooperative Research Centre, Australia. Available at: http://www. crctourism.com.au/CRCServer/page.aspx?page_id=42

Högmander, J. and Leivo, A. (2004) General principles for sustainable nature tourism in protected areas administered by Metsähallitus, Finland. Working papers of the Finnish Forest Research Institute 2. Available at: www.metla.fi/julkaisut/workingpapers/2004/mwp002. htm

Honey, M. (1999) *Ecotourism and Sustainable Development: Who Owns Paradise?* Island Press, Washington, DC.

IUCN World Commission on Protected Areas (WCPA) (2007a) The IUCN Protected Area Management Categories. Available at: www.iucn.org/themes/wcpa/wpc2003/pdfs/ outputs/pascat/pascatrev_info3.pdf

IUCN World Commission on Protected Areas (WCPA) (2007b) IUCN WCPA Protected Areas Programme. Available at: www.iucn.org/themes/wcpa/ppa/programme.htm

Kaae, B.C. and Miller, M.L. (1993) Coastal and marine ecotourism: a formula for sustainable development? *TRENDS* 30(2), 35–41.

Kant, I. (1991 [1764]) *Observations on the Feeling of the Beautiful and Sublime.* (Translated by J.T. Goldthwait.) University of California Press, Berkeley, California.

Kates, R.W., Parris, T.M. and Leiserowitz, A.A. (2005) What is sustainable development? Goals, indicators, values, and practice. *Environment: Science and Policy for Sustainable Development* 47(3), 8–21.

Knight, R.L. and Bates, S.F. (eds) (1995) *A New Century for Natural Resources Management.* Island Press, Washington, DC.

Knight, R.L. and Gutzwiller, K.J. (eds) (1995) *Wildlife and Recreationalists: Coexistence Through Management and Research.* Island Press, Washington, DC.

Kusler, J.A. (ed.) (1991) *Ecotourism and Resource Conservation: A Collection of Papers,* 2 Vols. State Wetlands Managers, Berne, New York.

Liu, J.C. (1994) *Pacific Islands Ecotourism: A Public Policy and Planning Guide (The Ecotourism Planning Kit).* The Pacific Business Center Program, University of Hawaii, Honolulu, Hawaii.

Lockhart, D.G. and Drakakis-Smith (eds) (1997) *Island Tourism: Trends and Prospects.* Pinter, London.

Leopold, A. (1968 [1949]) *A Sand County Almanac: And Sketches Here and There.* Oxford University Press, New York.

Lele, S.M. (1991) Sustainable development: a critical review. *World Development* 19(6), 607–621.

Masters, D. (1998) *Marine Wildlife Tourism: Developing a Quality Approach in the Highlands and Islands.* [Report] Tourism and Environment Initiative and Scottish Natural Heritage, Inverness, Scotland.

McNeely, J.A. and Miller, K.R. (eds) (1984) *National Parks, Conservation and Development: The Role of Protected Areas in Sustaining Society.* Proceedings of the World Congress on National Parks, Bali, Indonesia, October 1982. Smithsonian Institution Press, Washington, DC.

Miller, M.L. (ed.) (1993a) Ecotourism in marine and coastal areas. *Ocean and Coastal Management* 20(special issue).

Miller, M.L. (1993b) The rise of coastal and marine tourism. *Ocean and Coastal Management* 20, 181–199.

Miller, M.L. and Auyong, J. (eds) (1991a) *Proceedings of the 1990 Congress on Coastal and Marine Tourism: A Symposium and Workshop on Balancing Conservation and Economic Development,* 2 Vols. National Coastal Resources and Development Institute, Newport, Oregon.

Miller, M.L. and Auyong, J. (1991b) Tourism in the coastal zone: Portents, Problems, and Possibilities. In: Miller, M.L. and Auyong, J. (eds) *Proceedings of the 1990 Congress on Coastal and Marine Tourism: A Symposium and Workshop on Balancing Conservation and Economic Development*, 2 Vols. National Coastal Resources and Development Institute, Newport, Oregon, pp. 1–8.

Miller, M.L. and Auyong, J. (1991c) Coastal zone tourism: a potent force affecting environment and society. *Marine Policy* 15, 75–99.

Miller, M.L. and Auyong, J. (eds) (1998a) *Proceedings of the 1996 World Congress on Coastal and Marine Tourism: Experiences in Management and Development*. Washington Sea Grant Program and the School of Marine Affairs, University of Washington and Oregon Sea Grant Program, Oregon State University, Oregon.

Miller, M.L. and Auyong, J. (1998b) Remarks on tourism terminologies: anti-tourism, mass tourism, and alternative tourism. In: Miller, M.L. and Auyong, J. (eds) *Proceedings of the 1996 World Congress on Coastal and Marine Tourism: Experiences in Management and Development*. Washington Sea Grant Program and the School of Marine Affairs, University of Washington and Oregon Sea Grant Program, Oregon State University, Oregon, pp. 1–24.

Miller, M.L. and Ditton, R. (1986) Travel, tourism, and marine affairs. *Costal Zone Management Journal* 14(1/2), 1–19.

Miller, M.L. and Hadley, N.H. (2005) Tourism and coastal development. In: Schwartz, M.L. (ed.) *Encyclopedia of Coastal Science*. Springer, Dordrecht, pp. 1002–1009.

Miller, M.L., Gale, R.P. and Brown, P.J. (eds) (1986a) *Social Science in Natural Resource Management Systems*. Westview Press, Boulder, Colorado.

Miller, M.L., Gale, R.P. and Brown, P.J. (1986b) Natural resource management systems. In: Miller, M.L., Gale, R.P. and Brown, P.J. (eds) *Social Science in Natural Resource Management Systems*. Westview Press, Boulder, Colorado.

Miller, M.L., Auyong, J. and Hadley, N.P. (eds) (2002a) *Proceedings of the 1999 International Symposium on Coastal and Marine Tourism: Balancing Tourism and Conservation*. Washington Sea Grant Program and School of Marine Affairs, University of Washington; Oregon Sea Grant College Program, Oregon State University; and Oceans Blue Foundation. Seattle, Washington.

Miller, M.L., Auyong, J. and Hadley, N.P. (2002b) Sustainable coastal tourism. In: Miller, M.L., Auyong, J. and Hadley, N.A. (eds) *Proceedings of the 1999 International Symposium on Coastal and Marine Tourism: Balancing Tourism and Conservation*. Washington Sea Grant Program and School of Marine Affairs, University of Washington; Oregon Sea Grant College Program, Oregon State University and Oceans Blue Foundation, Seattle, Washington, pp. 3–20.

Monk, S.H. (1960 [1935]) *The Sublime: A Study of Critical Theories in XVIII-century England*. University of Michigan Press, Ann Arbor, Michigan.

National Geographic Center for Sustainable Destinations (NCSD) (2007) Available at: www. nationalgeographic.com/travel/sustainable/

National Marine Protected Areas Center (NMPAC) (2005) A functional classification system for US marine protected areas; an objective tool for understanding the purpose and effects of MPAs. National Marine Protected Areas Center, Santa Cruz and Monterey, California.

Newsome, D., Dowling, R.K. and Moore, S.A. (2005) *Wildlife Tourism*. Channel View Publications, Buffalo, New York.

Nicolson, M.H. (1997) *Mountain Gloom and Mountain Glory: The Development of the Aesthetics of the Infinite*. (With a Foreward by W. Cronon.) University of Washington Press, Seattle, Washington.

Nyaupane, G.P., Morais, D.B. and Graefe, A.R. (2004) Nature tourism constraints: a cross-activity comparison. *Annals of Tourism Research* 31(3), 540–555.

Oles, B., Wahle, C.M., Fischer, S., Miller, M.L. and Christie, P. (2006) Using regional workshops to understand the human dimension of MPAs. *MPA News* I7(10), 7.

Orams, M.B. (1993) The role of education in managing marine wildlife-tourist interaction. 1993 National MESA Conference Papers, pp. 101–106. Available at: http://www.whalewatchcruises.com.au/dolphinweb/research/papers/paper1.pdf

Orams, M.B. (1999) *Marine Tourism: Development and Management*. Routledge, New York.

Orams, M.B. (2002) Feeding wildlife as a tourist attraction: a review of issues and impacts. *Tourism Management* 23, 281–193.

Pattullo, P. (2005) *Last Resorts: The Cost of Tourism in the Caribbean*, 2nd edn. Latin American Bureau, London.

Pearce, D. (1989) *Tourism Development*, 2nd edn. Wiley, New York.

Reed, N.P. and Drabelle, D. (1984) *The United States Fish and Wildlife Service*. Westview Press, Boulder, Colorado.

Reid, D.G. (2003) *Tourism, Globalization and Development: Responsible Tourism Planning*. Pluto Press, London.

Runte, A. (1979) *National Parks: The American Experience*. University of Nebraska Press, Lincoln, Nebraska.

Salm, R.V. and Clarke, J.R. (eds) (1989) *Marine and Coastal Areas: A Guide for Planners and Managers*. IUCN and Natural Resources, Gland, Switzerland.

Scottish Natural Heritage (2004) *Scottish Marine Wildlife Watching Code: A Guide for Best Practice for Watching Marine Wildlife*. Scottish Natural Heritage, Inverness, Scotland, UK. Available at: www.marinecode.org/documents/Scottish-Marine-Code-web.pdf

Shackley, M. (1996) *Wildlife Tourism*. International Thomson Business Press, London.

Smith, V.L. and Eadington, W.R. (eds) (1992) *Tourism Alternatives: Potentials and Problems in the Development of Tourism*. University of Pennsylvania Press, Philadelphia, Pennsylvania.

Stebbins, R.A. (1992) *Amateurs,Professionals, and Serious Leisure*. McGill-Queen's University Press, Montreal.

The International Bank for Reconstruction and Development/The World Bank (2006) *Scaling Up Marine Management: The Role of Marine Protected Areas*. The World Bank, Washington, DC.

Tour Operator's Initiative Secretariat (no date) *Tour Operators' Contributions to Sustainable Tourism in Protected Areas*. c/o United Nations Environment Programme (Division of Technology, Industry and Economics. Available at: www.world-tourism.org/tour/destinations/FS_protectedareas.pdf)

United Nations Environment Programme (UNEP) and the Secretariat of the Convention on the Conservation of Migratory Species (CMS) (2006) *Wildlife Watching and Tourism: A Study on the Benefits and Risks of a Growing Tourism Activity and Its Impacts on Species*. UNEP/CMP Secretariat, Bonn.

Waitt, G., Lane, R. and Head, L. (2003) The boundaries of nature tourism. *Annals of Tourism Research* 30(3), 523–545.

Wallace, D.R., Holing, D. and Methvin, S. (1995) *Nature Travel: A Nature Company Guide*. Time-Life Books, New York.

Warburton, C.A. (1999) *Marine Wildlife Tourism and Whale-watching on the Island of Mull, West Scotland*. [Report] The Hebridean Whale and Dolphin Trust, Mull, UK.

Whelan, T. (ed.) (1991) *Nature Tourism: Managing for the Environment*. Island Press, Washington, DC.

World Commission on Environment and Development (WCED) (1987) *Our Common Future* (also known as 'The Brundtland Report'). Oxford University Press, New York.

14 Marine Wildlife Tourism and Ethics

B. GARROD

Wildlife tourism markets have grown considerably in recent years, both in volume and in value (Reynolds and Braithwaite, 2001; Curtin and Wilkes, 2005), and marine wildlife tourism has proven no exception to this trend. Thus, for example, Garrod and Wilson (2004), calculate that between 1990 and 1999, commercial whale watching grew at 12% per annum in terms of volume and 18.6% per annum at current prices in terms of value (based on Hoyt, 2000). This compares to an overall growth in international tourism receipts of 7.3% per annum at current prices (WTO, 2001). Meanwhile, the nominal rate of world economic growth over the last quarter of the 20th century was around 3.5–4% per annum (Larsen, 1999).

In spite of the spectacular growth in wildlife tourism markets, the ethical issues prompted by the human–animal interactions implicit in such activities have rarely troubled academics. Thus, while a handful of books and journal papers have dealt with the ethics of international tourism (Hultsman, 1995; Payne and Dimanche, 1996; Fennell and Malloy, 1999; Butcher, 2003; Holden, 2003; Fennell, 2006) and of ecotourism more specifically (Sirakaya, 1997; Honey, 1999; Fennell, 2003; Buckley, 2005), there has been relatively little academic focus on the ethics of wildlife tourism. Marine wildlife tourism, in particular, has received scant attention.

The kinds of ethical issue raised in examining marine wildlife tourism are essentially no different to those involved in terrestrial wildlife tourism. It can be argued, however, that the special nature of marine wildlife tourism raises the profile and hence the significance of such issues. Indeed, marine wildlife tourism is fundamentally different to terrestrial wildlife tourism in a number of important respects. Perhaps the most important is that interacting with marine wildlife generally requires tourists to leave their own natural surroundings and to enter an environment that is essentially alien to them (Cater and Cater, 2001). The effect is to make the marine wildlife tourism experience a much more embodied and visceral one; an experience that often involves physical exertion, exhilaration, mental challenge and the sharpening of the senses (Cater, 2007). Marine wildlife tourism experiences – and the ethical considerations that underlie the choices tourists encounter in undertaking them – can somehow seem more 'real' and consequential as a result.

The fact that the tourists are outside of their normal environment also raises the profile of ethical considerations. In the marine environment, tourists are often less certain how they should behave. When faced with an ethical dilemma – for example, whether they should touch a dolphin that is coming close to their tour boat – tourists often have little prior experience on which to base their decision making. Having a close-up encounter with any species of wild animal is no longer an everyday occurrence for many people, let alone a large marine mammal such as a dolphin. Furthermore, the tourist may well have less knowledge about such unfamiliar species of marine wildlife to help guide their decision making.

The nature of tourist engagement with wildlife (whether marine or terrestrial) also tends to make ethical decision making more complex in comparison to other forms of tourism. The human–wildlife interactions that are implicit in both marine and terrestrial wildlife tourism may be indirect, for example, in the form of watching, observing, studying, photographing or listening to the wildlife concerned, or they may be direct, for example, the touching, feeding, riding on or even capturing and killing of wildlife. Each of these forms of engagement raises particular ethical issues. To complicate matters, wildlife tours often involve more than one type of human–wildlife interactions. They may also share access to the wildlife with other users (e.g. commercial harvesting or informal recreational use). The ethical considerations involved can consequently become overlapping and compounded (Buckley, 2005).

The purpose of this chapter is to explore the ethical dimension of marine wildlife tourism. The focus will be on investigating a number of prominent ethical issues that are bound up in the provision of marine wildlife tourism experiences. These focus respectively on the acceptability of 'consumptive' forms of marine wildlife tourism, such as marine angling; on holding marine animals, such as dolphins, captive as the focus of a tourism attraction; and on the feeding of wildlife as part of the tourism experience. In so doing, the paper will also touch briefly on the ethical issues involved in 'swimming with' marine mammals. To achieve these ends, the chapter adopts an ethical framework set out by Hughes (2001), which groups ethical concerns about the nature of tourist–wildlife interactions into three theoretical perspectives: environmental ethics, animal welfare and animal rights. Some conclusions are then drawn regarding the importance of ethical considerations in the management and regulation of marine wildlife tourism.

Ethical Perspectives on Human–Wildlife Interactions in Marine Tourism

In discussing the ethical dimension of the relationship between tourism and animals, Hughes (2001) notes that the extension of moral considerability to non-human subjects has been a continually evolving process. In general, however, three broad themes or positions can be identified. Each views the relationship between tourism and wildlife from a different ethical perspective and therefore draws different conclusions regarding the morality of specific wildlife tourism activities. This, in turn, leads to different conclusions being drawn about how such activities should best be managed and regulated.

The first position is based on a conservationist approach to the ethical treatment of animals and this has a strong resonance with one of the foundational propositions in environmental ethics known as Leopold's land ethic (Leopold, 1989). This ethical position argues that any action is justifiable so long as it does not endanger the ecological integrity of the ecosystem as a whole. The focus is, therefore, not on individual animals: rather it is the ecosystem as a whole that is accorded moral considerability. Those holding such a position would thus countenance the killing of individual animals (e.g. fish killed during a game angling trip), provided that this does not compromise the effective functioning of the ecosystem, for example, by threatening the survival of a species or by disrupting predator–prey relationships. This line of ethical reasoning is therefore interested in human well-being, although only to the extent that this is likely to be affected by damage to the ecosystems in which humans reside.

The second position, by way of contrast, focuses not on the well-being of the ecosystem as a whole but on the welfare of individual animals. Proponents of this position typically base their arguments on the 'five freedoms' of animal welfare (Wilkins *et al.*, 2005), which were first proposed by the Brambell Committee of the UK parliament in 1965. The five freedoms can be summarized briefly as.

- Freedom from hunger, thirst and malnutrition;
- Freedom from discomfort;
- Freedom from pain, injury or disease;
- Freedom to express normal behaviour;
- Freedom from fear and distress.

This mode of ethical reasoning is not necessarily incompatible with the position based on environmental ethics, in that the welfare of individual animals clearly depends on them being part of a viable population and supported by a well-functioning ecosystem. An animal's welfare may, however, be directly impacted by marine wildlife tourism activities that have the potential to inflict pain or suffering, and this is the major concern of this particular ethical position. Individual animals are therefore afforded some moral considerability, although this need not imply that they are accorded greater or even equal status with humans (see Orams, 2002). Indeed, it would be possible to argue from this position that the suffering of individual animals is morally acceptable provided it is outweighed by some benefit that arises to humans as a result. The nature of such benefits is also important under this mode of ethical reasoning. Thus, for example, causing suffering to a dolphin by keeping it in captivity for the purpose of providing 'therapy' to humans with medical problems might be seen as quite different to keeping it in captivity simply so that it can entertain us by jumping through hoops.

The third ethical position, meanwhile, is based solely on the rights of animals and is therefore at least partly inconsistent with the previous two positions (both of which attempt to *balance* the interests of animals and humans to some extent). According to this position, animals are granted moral considerability because of their sentience and, because this means they can feel both physical pain and psychological stress, any action that harms them is deemed unacceptable. As such, advocates of the animal rights position will often find themselves at odds with the marine wildlife tourism industry, which by facilitating human–wildlife interactions can cause harm to animals either directly or indirectly.

Hughes (2001) examined the implications of these three contrasting positions with respect to the specific case of dolphin-based tourism attractions. The purpose of this chapter is to extend this analysis to look at marine wildlife tourism more generally. The discussion will focus on three particular ethical issues: the killing of marine wildlife, either intentionally or unintentionally; the keeping of marine animals in captivity; and the feeding of wildlife as part of the tourism product offering.

Consumptive Versus Non-consumptive Marine Wildlife Tourism

A good example of the complex ethical nature of marine wildlife tourism is the debate about whether the concept of 'ecotourism' should be defined widely enough to embrace 'consumptive' use of wildlife species. There is a strong affinity between marine wildlife tourism and ecotourism: indeed much of the marine wildlife tourism that goes on around the world would readily be described as ecotourism in the sense that it shares many of its defining characteristics, being essentially nature-based, managed according to the principles of sustainable development, benefiting local communities, providing resources for conservation and so on (Blamey, 1997; Diamantis, 1999; Fennell, 2001; Weaver, 2005a; Donohoe and Needham, 2006). It can be argued, however, that specific marine wildlife tourism product offerings need not adopt any of the components of this special orientation in order to operate successfully and survive in the market place. Indeed, the many reports of poorly managed wildlife tourism taking place in various places around the world confirm this assertion (see, e.g. Shackley, 1996; Green and Giese, 2004; Newsome *et al.*, 2005). It would therefore probably be best to view ecotourism as a subset of wildlife tourism (and marine ecotourism as a subset of marine wildlife tourism, see Garrod, 2003): one that takes a particular ethical view with regard to the relationship between tourism and the natural environment in which it operates.

The crux of the problem in defining ecotourism in ethical terms is how to determine its proper conceptual boundaries. Typifying this debate is a series of papers appearing in the *Journal of Sustainable Tourism* focusing on the particular case of billfish angling (Holland *et al.*, 1998, 2000; Fennell, 2000). The debate centres on the claim made in a paper by Holland *et al.* (1998, p. 111) that 'to the extent that anglers act responsibly to minimise their impacts and billfish angling remains sustainable as a result of their efforts, the ecotourism label seems appropriate for billfish angling'. Fennell's (2000) commentary critiques this view, arguing that while it is right to suggest that other forms of ill-planned wildlife tourism may be just as disruptive to the resource base, billfish angling is quite distinct in terms of the purposeful intent implied (the primary aim of the activity is to catch fish), its capacity to cause pain to individual fish and the essentially 'consumptive' nature of the activity (meaning that animals are harmed and often killed). Fennell believes that these ethical considerations set billfish angling apart from other forms of wildlife tourism, which he considers to be more appropriate candidates for the label of ecotourism. Meanwhile, the rejoinder of Holland *et al.* (2000) stresses the growth in popularity of catch-and-release fishing methods, as well as the need in assessing sustainability considerations to focus on the population as a whole, rather than as individual animals.

This exchange of views demonstrates how the application of different ethical positions to a given problem can result in mutually opposing recommendations. In drawing attention away from individual animals and focusing instead on the sustainability of the population as a whole, Holland *et al.* would seem to be marshalling the arguments espoused in the environmental ethics approach. This position detracts from the potential negative welfare impacts on individual animals and focuses on the wider impacts this individual's death might have on the wider population, the perpetuation of its species or the ecological status of its habitat. The viewpoint expressed by Fennell (2000, p. 345), in contrast, tends to emphasize an animal welfare perspective, arguing that regardless of the ecosystem impacts billfish fishing may or may not have, the activity is simply 'wrong as an ecotourism activity' because of its inherent potential to inflict unacceptable welfare impacts on individual animals.

The issue of whether particular tourism activities should be considered essentially 'consumptive' has long been debated. Tremblay (2001) argues that the tourism industry has traditionally drawn a dividing line between wildlife viewing and hunting–fishing. He goes on to argue, however, that such a distinction is difficult to justify on ethical grounds. A particular problem is that the term 'consumptive' is used both widely and inconsistently. Many people identify 'consumptive' marine wildlife tourism activities as being those involving the killing of animals (see, e.g. Duffus and Dearden, 1990). However, it is not always clear whether it matters that such killing is intentional (as Fennell argues), or whether it is wrong to kill animals for the purpose of sport or recreation, even unintentionally. In the debate above, Holland *et al.* argue that since many billfish anglers seek to return their live catch to the water, they do not actually intend to kill them (even if this may be the result for those that are mishandled, injured by hooks or kept out of the water too long).

Similarly, while for some commentators an important issue is whether the killing of the animal involved causing it physical pain or psychological stress (as Fennell argues in the foregoing debate), others take the view that what happens to an individual animal is of less importance than what happens to the population as a whole (Buckley, 2005). The argument put forward by Holland *et al.* seems to rely upon this reasoning. Bauer and Herr (2004, p. 59), meanwhile, suggest that 'hunting and fishing both use wildlife, both can be humane and professional, or cruel and destructive...[However] both can only be justi-fied...if they are sustainable'. This argument draws out the stark contrasts in reasoning that result from adopting each of the three ethical positions noted in the section above. While those adopting an environmental ethics position would tend to argue that if the activity has a neutral or beneficial impact on the ecosystem concerned (is 'sustainable') then all is well. Proponents of the animal welfare perspective, on the other hand, focus very much on whether the activity is 'humane and professional' or 'cruel and destructive', and would only con-sider the activity acceptable if there were significant mitigating benefits to humans if a degree of cruelty or destruction was unavoidable. Those adopting the animal rights perspective, meanwhile, tend to view any activity that could potentially harm an animal as being morally unacceptable.

The application of ethical reasoning to marine wildlife tourism is seldom a straightforward affair. Indeed, Fennell (2000) points out that to associate only hunting–fishing activities with negative impacts on the wildlife population con-cerned is actually quite misleading. Wildlife-watching activities can also have serious negative consequences for wildlife populations, particularly if such activities are poorly managed or not managed at all. This may not involve kill-ing individual animals and removing them from the population, but the impacts (e.g. subjecting the animals to psychological stress or degrading their habitat) may be just as severe. Orams (2000), meanwhile, argues that even though whale watching has often been presented as a more ethical use of whale popu-lations than the harvesting of whales for commercial products, there is also widespread concern about the impacts that whale watching itself may have on endangered whale populations.

Similarly, Tremblay (2001) points out that the term 'consumptive' has often been linked, wrongly in his view, to the commercialization of the activity. Thus, while commercial fishing may be classified as consumptive because it is undertaken for a profit (thereby reflecting a basically utilitarian approach to the use of the resource), marine angling should be viewed as 'non-consumptive' because those who undertake the activity are interested primarily in the experi-ential dimension – the battle of wits between hunter and prey, the thrill of the chase – allegedly more 'noble' motives. Tremblay argues that such distinctions are wrong. Both wildlife viewing and hunting–fishing activities can involve vary-ing degrees of experiential intensity, both high and low. Moreover, there is no empirical evidence of any greater understanding, education or respect associ-ated with either of these two categories of activity to suggest that one can claim moral superiority over the other.

Although the ethical considerations discussed in this section of the chap-ter have been set out in reference to the concept of ecotourism, they are also readily applicable in the context of wildlife tourism. Indeed, Newsome *et al.* (2005) explicitly exclude trophy hunting and fishing from the remit of their book on the grounds that these are essentially consumptive activities. Their argument is based on the animal welfare position, which rejects the killing of wildlife for recreational purposes. Such activities are therefore con-sidered to lie 'outside of both our interest and writing' (Newsome *et al.*, 2005, p. 9).

Marine Wildlife Tourism in Captivity

Tourism experts, both academics and practitioners, also seem to disagree on whether the term 'ecotourism' should be reserved for activities that occur in natural areas or could in fact embrace activities taking place in highly modified environments, perhaps even in captivity (Higham and Lück, 2002; Garrod, 2003; Weaver, 2005b). Such debates are also apparent in the broader field of wildlife tourism. Thus, for example, Roe *et al.* (1997) restrict their definition of wildlife tourism to activities that take place in natural areas (see Box 14.1). Newsome *et al.* (2005, p. 9), in contrast, concede that wildlife tourism can take place in captive and semi-captive contexts (see Box 14.1), although they go on to argue that it is 'best carried out in the wild'.

Higginbottom (2004), meanwhile, argues that wildlife tourism is equally valid as a concept in wild and in captive settings (see Box 14.1), and in consequence her book includes a chapter on zoo tourism by Tribe (2004). Higginbottom (2004, p. 3) justifies the inclusion of zoo tourism, which she defines as 'viewing animals in man-made confinement, principally zoos, wildlife parks, animal sanctuaries and aquaria', on pragmatic grounds. Wildlife tourism product offerings are dynamic and tend to follow market trends, and there has arguably been a blurring of the distinction between wild, semi-wild and captive settings in supplying them. Wildlife management practices have also changed considerably in recent times. Thus, for example, a large fenced nature reserve established to conserve a particular habitat would traditionally be considered to contain free-ranging animals, while an open-range zoo, containing the same number and species of animals, would not. Moreover, zoos are increasingly using vegetation types in their compounds to closely resemble the animals' natural habitats (Melfi *et al.*, 2004). Furthermore, these compounds may actually be larger in size than a nature reserve. Free-ranging animals are also often provisioned with food in order to facilitate close interaction with tourists, and we will consider this issue further in the following section.

Box 14.1. Three definitions of wildlife tourism.

'Tourism that includes, as a principle [*sic.*] aim, the consumptive and non-consumptive use of wild animals in natural areas. It may be high volume mass tourism or low volume/low impact tourism, generate high economic returns or low economic returns, be sustainable or unsustainable, domestic or international, and based on day visits or longer stays' (Roe *et al.*, 1997, p. 3).

'Wildlife tourism is tourism based on encounters with non-domesticated (non-human) animals. These encounters can occur in either the animals' natural environment or in captivity. It includes activities historically classified as 'non-consumptive', including viewing, photography and feeding, as well as those that involve killing or capturing animals, particularly hunting . . . and fishing' (Higginbottom, 2004, p. 2).

'Wildlife tourism is tourism undertaken to view and/or encounter wildlife. It can take place in a range of settings, from captive, semi-captive, to in the wild, and it encompasses a variety of interactions from passive observation to feeding and/or touching the species viewed' (Newsome *et al.*, 2005, pp. 18–19).

Tribe (2004) argues that zoos evoke two distinct ethical positions. The first considers that zoos can never reproduce the complex habitats in which wild animals live. Such habitats have considerable temporal and spatial variation, and could never be reproduced artificially. Consequently, no captive setting could ever ensure the well-being of a wild animal. This would seem to accord with the animals rights perspective set out earlier in this chapter. The second ethical position relates to the conditions under which the animals are kept in zoos. Because of the history of zoos as menageries aimed at human entertainment, the historical tendency to overlook animal welfare and the paucity of funds that have been available for zoos to improve their infrastructures, zoo animals tended to be kept in enclosures that were cramped, highly artificial and provided them with little stimulation. This has left many zoo visitors with the feeling that zoo conditions are cruel and this perception has been remarkably persistent in society (Melfi *et al.*, 2004), especially in the more developed countries of Europe, North America and Australasia where such issues have received increasing media attention (Tribe, 2004). In the last 20 years, therefore, zoos have been forced to face up to mounting public scrutiny of how they treat their animals (Mason, 2000).

Tribe (2004) goes on to argue that while there is in practice little that zoos can do to address the reservations of supporters of animals rights, in recent times zoos have made significant efforts to address animal welfare concerns. Consequently, zoos have invested considerable funds in improving their enclosures, for example, by creating more stimulating physical environments for their animals and in providing refuge areas or screening so that the animals can escape the constant gaze of visitors (Blaney and Wells, 2004).

In the light of the ethical objections noted above, zoos have also been forced to review their fundamental mission. Tribe (2004) notes that zoos no longer see themselves as collections of captive wildlife but as conservation centres, engaging in both *ex situ* conservation work, particularly captive breeding programmes for endangered species, and *in situ* conservation work in the form of animal rescue, habitat protection and restoration, species reintroduction and so on. It is notable that such *in situ* conservation projects make an excellent focus for wildlife tourism product offerings, particularly in the form of volunteer holidays. This wider concern for both species and habitat conservation resonates strongly with the environmental ethics approach to determining the moral status of wildlife tourism.

In the case of marine wildlife tourism, perhaps the most prominent example is the interaction between humans and cetaceans. Hughes (2001) notes that in the case of dolphin-based tourism this has taken two forms, the first involving the captive display of dolphins in marine aquaria, 'dolphinaria' and marine theme parks, and the second exploiting the tendency of wild dolphins to seek out and apparently enjoy contact with humans. He goes on to indicate that the recent growth in the dolphin tourism industry has centred mainly on the second of these formats because this is deemed more compatible with the growth of the animal rights movement worldwide. By way of illustration, Hughes uses the example of how in the early 1990s, animal rights protestors effectively forced the closure of Morecambe Marineland, a dolphin-based tourist attraction in the

UK. This quickly led to the closure of all of the remaining dolphinaria across the country. At the same time, the market for watching dolphins (and other cetacean species) has expanded significantly, not only in the UK but worldwide (Hoyt, 2000), and Hughes notes that while this raises different ethical considerations, the availability of an alternative means by which humans can interact with dolphins has undoubtedly hastened the transition of the industry from being one based mainly on captive individuals to one based on experiencing dolphins in the wild. Cloke and Perkins (2005) also note a similar tendency from the perspective of human–cetacean interactions as a form of 'spectacle', the 'performance space' no longer being the dolphinarium but the open sea.

It is important to note, however, that the great majority of tourist–dolphin interactions still take place in captivity (Hughes, 2001; Curtin, 2006). There has been a great deal of research into the impacts on the welfare of dolphins kept in captivity conditions. Hughes (2001), for example, notes four main areas of concern:

- The disruption of family groups and wider social structures during transport;
- Death during transport;
- Lack of space, sunburn, ingestion of toxic paint and other adverse health impacts while in captivity;
- Encouragement of unnatural behaviour as part of the training of dolphins to perform displays for the entertainment of tourists.

Curtin and Wilkes (2007), meanwhile, focus on the issue of stress among captive dolphins, linking increased levels of stress both to the premature death of individuals and the increased display of aggressive behaviour. Wild dolphins are likely to travel long distances daily in search of food and the limited space available within any aquarium pool or sea cage may serve as a physical stressor (Frohoff, 2004). This having been said, the major contributor to stress in captive dolphins is likely to be psychological in origin (Frohoff, 2004). The removal of individuals from the wild takes them away from the complex social group in which they flourish. To that extent, Curtin and Wilkes (2007) argue that no holding pool could ever recreate the natural habitat of a dolphin and this would seem to reflect an animal rights-based approach to the issue.

To proponents of the animal welfare position, such impacts might perhaps be considered acceptable provided there are mitigating benefits to humans. It might be argued that dolphins are kept in captivity for one or more of three basic reasons: to entertain us, to educate us or to provide us with therapy. The last of these possible justifications is perhaps the most interesting in ethical terms because the potential benefits to humans, if they can be substantiated, could be considerable. Researchers have found that swimming with dolphins can help to alleviate mild to moderate depression in adults (e.g. Antonioli and Reveley, 2005). Supporters of dolphin-assisted therapy also claim that it can help children with learning disabilities such as autism, cerebral palsy and Down's syndrome, as well as those with physical disabilities (Carwardine *et al.*, 1998). Other authorities, however, have challenged these claims, either on the basis of the poor elaboration of cause-and-effect in explaining the claimed health benefits or on the basis of apparent flaws in the research methodologies that have been used (e.g. Basil and Mathews, 2005). Even so, there may be ethical arguments to be made from an animal welfare perspective to allow some dolphins to be kept in captivity for such reasons, provided that every effort is made to maximize the well-being of those few animals.

While the above arguments serve to make a special ethical case of keeping dolphins in captivity in order to provide dolphin-assisted therapies, the WDCS (2007), a prominent non-governmental organization concerned with whale and dolphin conservation, points out that there is no strong evidence to suggest that the therapeutic benefits associated with swimming with dolphins are any greater than those available through interaction with pets and domesticated animals. It might also be argued that swimming with dolphins can also be undertaken in the wild, as indeed can dolphin watching and dolphin-oriented educational tourism. Swimming with wild or semi-wild dolphins is in fact now a widespread global phenomenon, usually involving the placement of swimmers in the water with the dolphins under somewhat regulated conditions (Orams, 2004; Scarpaci et al., 2004). In some countries, however, swimming with dolphins in the wild is completely restricted by law. Similarly, while scuba diving with dolphins and other cetaceans is permitted in some countries, it is illegal in others (Curtin and Garrod, 2007). The WDCS (2007) argues that it is so difficult to ensure that swimming with wild dolphins is not intrusive and stressful for the animals involved that they are unable to recommend their public support to commercial 'swim-with' dolphin tours.

Where swimming with dolphins is undertaken by tour operators on a commercial basis, food is often used as a bait to encourage dolphins to remain in the local area and to accept the presence of tourists in the water with them. Technically, therefore, such dolphins are neither wild nor in captivity, but semi-wild. This raises a number of ethical issues, to which this chapter now turns.

Feeding of Wildlife in Marine Wildlife Tourism

It is interesting to note that in spite of Newsome et al. (2005) rejecting the validity of wildlife tourism in captive settings, their definition clearly allows for the feeding of wild animals (see Box 14.1). This is perhaps surprising given their complete rejection of consumptive and captive forms of wildlife tourism on ethical grounds. Indeed, many commentators argue that feeding wild animals is morally unacceptable. For example, the WDCS (2007) strongly advises against feeding cetaceans as part of a tourism experience. Many whale-watching and dolphin-watching codes of ethics also caution against the practice (Garrod and Fennell, 2004).

Orams (2002) notes that feeding may be part of a marine wildlife tourism experience for two rather different reasons. First, food might be provisioned in order to attract particular species of interest, thereby enabling tourists to watch or interact closely with animals that would otherwise be difficult to observe or unpredictable to encounter. Secondly, feeding might be done either by the tour operators in order to encourage spectacular behaviour or by the tourists themselves as part of the tourism experience. Newsome *et al.* (2005) argue that feeding animals seems to satisfy certain emotional needs in humans, including the need to maintain kinship with the animal world and, in an increasingly urbanized world, to 'reconnect' with nature. Feeding animals may also serve to satisfy the nurturing instinct that is also evident in the human tendency to keep animals as household pets. It may also have secondary benefits for tourism by encouraging the animals to come very close to tourists, close enough perhaps that a tactile encounter may even be possible. Orams (2000) argues that there is an assumption among wildlife tourism operators that tourists desire 'close-up' encounters and this is undoubtedly based on the strong emotional reactions that tourists display when such interactions do occur. Feeding may be important in this sense even if, as Orams' research argues, close-up interactions are not actually required for the tourist to gain a satisfying experience from the tourism product offering in question. The tendency for zoos to emphasize 'feeding times' as 'peak' experiences for their visitors also seems to suggest that people like to watch animals when they are feeding. Indeed, this is often the time when the animals display their most spectacular, interesting and charismatic behaviour. It is also often considered a good time for zoos (and indeed wildlife tourism operators) to impart their educational messages to the tourists.

It is clear that humans have much to gain from feeding marine wildlife. From an ethical perspective, the important issue is whether such benefits may be said to outweigh any negative consequences for the animals concerned. Indeed, studies suggest that there can be a number of significant negative impacts associated with the feeding of wildlife, including the potential to cause behaviour change (e.g. alteration to an animal's daily activity budget, breeding activities or migratory pattern), dependency, habituation and aggression (Orams, 2002; Newsome *et al.*, 2005). Cater (2007) notes growing evidence that the practice of throwing food into the water to attract sharks, known as 'burleying' or 'chumming', is resulting in these animals becoming more aggressive towards humans. Proponents of the animal welfare ethic would tend to argue that these negative impacts on animals' well-being could only be justified if the positive welfare benefits for humans outweighed them. However, it might be argued that these benefits are largely trivial, being associated more with our desire to be entertained than with our need to connect with animals. Moreover, such benefits could arguably be achieved without feeding taking place, albeit not to the same degree. Supplementary feeding clearly enhances marine wildlife tourism experiences, but it is arguably not essential to the marine wildlife tourism product offering.

Another possible justification for the feeding of animals by tourists is when it is part of a wider programme of wildlife conservation, for example, to ensure the survival of a species in its natural habitat or to assist in its reintroduction to the wild. The principal benefit is that instead of the provisioning being done by marine conservation scientists, the same food is provided (and paid for) by marine wildlife tourists. This arrangement could then be compatible with the environmental ethics position, in that the supplementary feeding of wildlife may be deemed necessary to maintain the viability of the ecosystem in which the animals play a vital part. It is nevertheless difficult to find many examples of where the supplementary feeding of particular marine wildlife species might be justified on these grounds, most of the examples cited in the literature tending to be birds of prey, such as the Mauritius kestrel and the Californian condor (Newsome *et al.*, 2005).

As we have seen, the ethical issues involved in marine wildlife tourism are rarely straightforward. Orams (2002) points out that in many cases, animals being fed by tourists are already significantly influenced by human activities. For example, it has been calculated that the major food source of dolphins in Moreton Bay, Australia, is actually by-catch from shrimp trawlers. If these animals are instead fed by tourists, then they are simply substituting one human food source for another. The issue is then no longer one of whether wild animals are fed by humans but of how they are fed. Orams also notes that tourists are not the only ones to feed wildlife. Scientists have long used food provisioning as a means of obtaining reliable behavioural data and for observing species that are difficult to locate in the wild. They have also from time to time deliberately altered the amount of food available to certain animal populations in order to study how they react to varying levels of food abundance (Orams, 2002). This might be justified as being ethically superior to observing the same animals in captivity, but the outcome is that such animals are nevertheless being fed by humans. Supporters of animal rights may consider this outcome morally unacceptable under any circumstances.

Conclusion

This chapter has argued that ethical issues are likely to be particularly important in the marine wildlife tourism context because of the nature of the environment in which the activity takes place. While the range of ethical issues for marine wildlife tourism is really no different to that which applies to terrestrial forms of wildlife tourism, such issues are in many instances more pressing in the marine context. This is because of the often highly embodied nature of marine wildlife tourism, the lack of experience the typical tourist has with marine ecosystems and wildlife species, and the nature of the marine environment.

The chapter then went on to examine three particular ethical issues in marine wildlife tourism: environmental ethics, animal welfare and animal rights. In each case the three different ethical positions (as per Hughes, 2001) were applied and the recommendations of each position compared. This analysis serves to demonstrate the sharply contrasting recommendations made by applying the three different ethical approaches. This reveals just how deep-rooted, complex and dynamic the ethical considerations involved in marine wildlife tourism actually are.

Ethical considerations are nevertheless of vital importance. Ultimately, if tourism is to be made more sustainable, tourists' behaviour must change. This condition applies equally to marine wildlife tourism, terrestrial wildlife tourism, ecotourism, urban tourism or, indeed, conventional mass tourism. Fennell (2003) makes the point, however, that behavioural change is achieved not through changing people's attitudes – how they respond when questioned about something – but by confronting their deep-seated values and beliefs. Orams (1995) also argues that interpretation plays an important role in managing tourism through creating 'dissonances' in the conscience of the tourist – ethical dilemmas that the tourist will want to resolve. Such values and beliefs clearly have an ethical basis, representing the fundamental moral sounding board people use to inform their decision making. Only if these core ethical beliefs can be accessed, reshaped and reapplied can the tourism industry ever hope to achieve the ambitious sustainability goals that have been set for it. Ethical considerations are therefore important because they represent both the ultimate goal for sustainable tourism and the fundamental means by which we are expected to get there.

References

Antonioli, C. and Reveley, M.A. (2005) Randomised controlled trial of animal facilitated therapy with dolphins in the treatment of depression. *British Medical Journal* 331, 1231–1234.

Basil, B. and Mathews, M. (2005) Human and animal health: strengthening the link – methodological concerns about animal facilitated therapy with dolphins. *British Medical Journal* 331, 1407.

Bauer, J. and Herr, A. (2004) Hunting and fishing tourism. In: Higginbottom, K. (ed.) *Wildlife Tourism: Impacts, Management and Planning*, Common Ground/CRC, Altona, Australia, pp 57–77.

Blamey, R.K. (1997) Ecotourism: the search for an operational definition. *Journal of Sustainable Tourism* 5(2), 109–130.

Blaney, E.C. and Wells, D.L. (2004) The influence of a camouflage net barrier on the behaviour, welfare and public perceptions of zoo-housed gorillas. *Animal Welfare* 13(2), 111–118.

Buckley, R. (2005) In search of the narwhal: ethical dilemmas in ecotourism. *Journal of Ecotourism* 4(2), 129–134.

Butcher, J. (2003) *The Moralisation of Tourism*. Routledge, London/New York.

Carwardine, M., Hoyt, E., Fordyce, R.E. and Gill, P. (1998) *Whales & Dolphins: The Ultimate Guide to Marine Mammals*. HarperCollins, London.

Cater, C. (2007) Perceptions of and interactions with marine environments: diving attractions from great whites to pygmy seahorses. In: Garrod, B. and Gössling, S. (eds) *New Frontiers in Marine Tourism: Diving Experiences, Sustainability, Management*. Elsevier, Oxford (forthcoming) pp. 49–64.

Cater, E. and Cater, C. (2001) Marine environments. In: Weaver, D.B. (ed.) *The Encyclopedia of Ecotourism*. CAB International, Wallingford, UK, pp. 265–285.

Cloke, P. and Perkins, H.C. (2005) Cetacean performance and tourism in Kaikoura, New Zealand. *Environment and Planning D: Society and Space* 23(6), 803–924.

Curtin, S. (2006) Swimming with dolphins: a phenomenological exploration of tourist recollections. *International Journal of Tourism Research* 8(4), 301–315.

Curtin, S. and Garrod, B. (2007) Vulnerability of marine mammals to diving tourism activities. In: B. Garrod and S. Gössling (eds) *New Frontiers in Marine Tourism: Diving Experiences, Sustainability, Management*. Elsevier, Oxford (forthcoming), pp. 93–113.

Curtin, S. and Wilkes, K. (2005) British wildlife tourism operators; Current issues and typologies. *Current Issues in Tourism* 8(6), 455–478.

Curtin, S. and Wilkes, K. (2007) Swimming with captive dolphins: current debates and post-experience dissonance. *International Journal of Tourism Research* 9(2), 131–146.

Diamantis, D. (1999) The concept of ecotourism: evolution and trends. *Current Issues in Tourism* 2(2 & 3), 93–122.

Donohoe, H.M. and Needham, R.D. (2006) Ecotourism: the evolving contemporary definition. *Journal of Ecotourism* 5(3), 192–210.

Duffus, D.A. and Dearden, P. (1990) Non-consumptive oriented recreation: a conceptual framework. *Biological Conservation* 53(3), 213–231.

Fennell, D.A. (2000) Ecotourism on trial: the case of billfish angling as ecotourism. *Journal of Sustainable Tourism* 8(4), 341–345.

Fennell, D.A. (2001) A content analysis of ecotourism definitions. *Current Issues in Tourism* 4(5), 403–421.

Fennell, D.A. (2003) *Ecotourism*, 2nd edn. Routledge, Milton Park, UK/New York.

Fennell, D.A. (2006) *Tourism Ethics*. Channel View, Clevedon, UK.

Fennell, D.A. and Malloy, D.C. (1999) Measuring the ethical nature of tourism operators. *Annals of Tourism Research* 26(4), 928–943.

Frohoff, T.G. (2004) Stress in dolphins. In: Beckoff, M. (ed.) *Encyclopedia of Animal Behaviour*. Greenwood Press, Westport, Connecticut, pp. 1158–1164.

Garrod, B. (2003) Defining marine ecotourism: a Delphi study. In: Garrod, B. and Wilson, J.C. (eds) *Marine Ecotourism: Issues and Experiences*. Channel View, Clevedon, UK, pp. 17–36.

Garrod, B. and Fennell, D.A. (2004) An analysis of whalewatching codes of conduct. *Annals of Tourism Research* 31(2), 334–352.

Garrod, B. and Wilson, J.C. (2004) Nature on the edge? Marine ecotourism in peripheral coastal areas. *Journal of Sustainable Tourism* 12(2), 95–120.

Green, R. and Giese, M. (2004) Negative effects of wildlife tourism on wildlife. In: Higginbottom, K. (ed.) *Wildlife Tourism: Impacts, Management and Planning*. Common Ground/CRC, Altona, Australia, pp. 81–97.

Higginbottom, K. (2004) Wildlife tourism: an introduction. In: Higginbottom, K. (ed.) *Wildlife Tourism: Impacts, Management and Planning*. Common Ground/CRC, Altona, Australia, pp. 1–14.

Higham, J. and Lück, M. (2002) Urban ecotourism: a contradiction in terms? *Journal of Ecotourism* 1(1), 36–51.

Holden, A. (2003) In need of a new environmental ethic for tourism? *Annals of Tourism Research* 20(1), 94–108.

Holland, S., Ditton, R. and Graefe, A. (1998) An ecotourism perspective on billfish fisheries *Journal of Sustainable Tourism* 6(2), 97–116.

Holland, S., Ditton, R. and Graefe, A. (2000) Response to ecotourism on trial: the case of billfish angling as ecotourism. *Journal of Sustainable Tourism* 8(4), 346–351.

Honey, M. (1999) *Ecotourism and Sustainable Development: Who Owns Paradise?* Island Press, Washington, DC.

Hoyt, E. (2000) *Whalewatching 2000: Worldwide Tourism Numbers, Expenditures and Expanding Socioeconomic Benefits*. International Fund for Animal Welfare, Yarmouth, USA.

Hughes, P. (2001) Animals, values and tourism – structural shifts in UK dolphin tourism provision. *Tourism Management* 22(4), 321–329.

Hultsman, J. (1995) Just tourism: an ethical framework. *Annals of Tourism Research* 22(3), 553–567.

Larsen, F. (1999) Global economic and financial development in the 1990s and implications for monetary policy. International Monetary Fund, 27th Economics Conference: Possibilities and Limitations of Monetary Policy, Vienna, July.

Leopold, A. (1989) *A Sand County Almanac and Sketches Here and There*. Oxford University Press, Oxford/New York.

Mason, P. (2000) Zoo tourism: the need for more research. *Journal of Sustainable Tourism* 8(4), 333–339.

Melfi, V.A., McCormick, W. and Gobbs, A. (2004) A preliminary assessment of how zoo visitors evaluate animal welfare according to enclosure style and the expression of behavior. *Anthrozoos* 17(2), 98–108

Newsome, D., Dowling, R. and Moore, S.A. (2005) *Wildlife Tourism*. Channel View, Clevedon, UK.

Orams, M.B. (1995) Towards a more desirable form of ecotourism. *Tourism Management* 16(1), 3–8.

Orams, M.B. (2000) Tourists getting close to whales, is that what whale-watching is all about? *Tourism Management* 21(5), 561–569.

Orams, M.B. (2002) Feeding wildlife as a tourism attraction: a review of the issues and impacts. *Tourism Management* 23(3), 281–293.

Orams, M.B. (2004) Why dolphins may get ulcers: considering the impacts of cetacean-based tourism in New Zealand. *Tourism in Marine Environments* 1(1), 17–28.

Payne, D. and Dimanche, F. (1996) Toward a code of conduct for the tourism industry: an ethics model. *Journal of Business Ethics* 15(9), 997–1007.

Reynolds, P.C. and Braithwaite, D. (2001) Towards a conceptual model for wildlife tourism. *Tourism Management* 22(1), 31–42.

Roe, D., Leader-Williams, N. and Dalal-Clayton, B. (1997) *Take Only Photographs: Leave Only Footprints*. IIED Wildlife and Development Series No.10, IIED, UK.

Scarpaci, C., Hugegoda, D. and Cokeron, P.J. (2004) No detectable improvement in compliance to regulations by 'swim-with-dolphin' operators in Port Phillip Bay, Victoria, Australia. *Tourism in Marine Environments* 1(1), 41–48.

Shackley, M. (1996) *Wildlife Tourism*. International Thomson Business Press, London.

Sırakaya, E. (1997) Attitudinal compliance with ecotourism guidelines. *Annals of Tourism Research* 24(2), 919–950.

Tremblay, P. (2001) Wildlife tourism consumption: consumptive or non-consumptive? *International Journal of Tourism Research* 3(1), 81–86.

Tribe, A. (2004) Zoo tourism. In: Higginbottom, K. (ed.) *Wildlife Tourism: Impacts, Management and Planning*. Common Ground/CRC, Altona, Australia, pp.35–56.

Weaver, D.B. (2005a) Comprehensive and minimalist dimensions of ecotourism. *Annals of Tourism Research* 32(2), 439–455.

Weaver, D.B. (2005b) *Sustainable Tourism: Theory and Practice*. Elsevier Butterworth Heinemann, Oxford/Burlington, Vermont.

WDCS (2007) WDCS policy on swimming with whales and dolphins. Whale and Dolphin Conservation Society. Available at: http://www.wdcs.org/dan/publishing.nsf/allweb/60D236DE0E661D0180256FD90033D2C7.

Wilkins, D.B., Houseman, C., Allan, R. and Appleby, D. (2005) Animal welfare: the role of non-government organisations. *Revue Scientifique et Techinque de L'Office International des Epizooties* 24(2), 625–638.

WTO (2001) Sustainable development of tourism. Report of the Secretary-General to the Organizational Session of the World Summit on Sustainable Development. World Tourism Organization: Madrid, Spain.

15 Protecting the Ocean by Regulating Whale Watching: The Sound of One Hand Clapping

P.H. FORESTELL

Il faut aller voir. ('We must go and see.')

Jacques Cousteau, motto of the Calypso

[A]ll conservation of wildness is self-defeating, for to cherish we must see and fondle, and when enough have seen and fondled, there is no wildness left to cherish.

Aldo Leopold, *A Sand County Almanac*

Introduction

The history and current status of interactions between humans and marine mammals provide an important focus in the discussion of wildlife tourism and conservation. The relationship between whale watching, marine tourism and wildlife based tourism (WBT) in general has been described by Curtin (2003), as illustrated in Fig. 15.1. Whether we consume whales (Worm *et al.*, 2007) or merely watch them (Orams, 1999), the impact of humans on marine species and ecosystems can be dramatic and detrimental. Yet, in contrast to impact assessment of human activities on land, our ability to monitor degradation of the marine environment is severely constrained by: the relative remoteness and apparent impenetrability of the ocean; our general lack of knowledge about range patterns and habitat needs of pelagic animals; and the tremendous cost and labour required to carry out the necessary science (Norse *et al.*, 2005).

Recreational excursions to view marine mammals have developed as among the most visible, accessible and frequent of human activities in the marine environment. Such activity takes place primarily on the ocean surface, requires no more specialized equipment than a floating platform, and is one of the world's fastest-growing sectors of marine tourism (Hoyt, 2001, 2005). This chapter will consider whale and dolphin watching from the perspective of whether marine wildlife tourism and marine conservation are antagonistic or complimentary concepts, and the significance of legislation in mediating the relationship between the two.

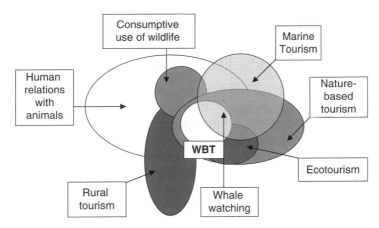

Fig. 15.1. The relationship between whale watching and other forms of human interactions with animals including wildlife based tourism (WBT). (From Curtin, 2003.)

Human Interest in Cetaceans

Our fascination with whales and dolphins (and to a lesser degree pinnipeds, polar bears, manatees, dugongs and sea otters) has a history extending back at least as far as the Stone Age (e.g. Norwegian petroglyphs, drawn 6000–9000 years ago, depict whales and dolphins; Soggnes, 2003). Across millennia, human interaction with cetaceans has changed from localized subsistence hunting, to whaling on a global scale, and eventually to fascination with their biological and aesthetic importance, also on a global scale (Forestell, 2002). The change in the public attitude towards whales and dolphins has been most profound during the last 50 years (Lavigne *et al.*, 1999). On the face of it, the three phases may represent a general transition from a consumptive to a non-consumptive point of view (Lavigne *et al.*, 1999), although others describe them more as a move from direct consumption to indirect consumption (Forestell, 2002; Corkeron, 2004).

Until the end of the Second World War whales were widely viewed as a source of food and by-products that challenged industrialized nations to turn their ingenuity towards maximum extraction of the resource with maximum efficiency (Robertson, 1954). However, as the 1960s dawned, the public perception of whales and dolphins began a dramatic and relatively rapid transformation (Forestell, 2002). Television allowed millions to follow the exciting underwater escapades of Jacques Cousteau and his team of divers. Neuroscientist John Lilly argued that dolphins are highly intelligent and live complex social lives (Lilly, 1975). Popular books by conservationists and scientists like Aldo Leopold (*A Sand County Almanac*, 1949), Rachel Carson (*Silent Spring*, 1962) and Victor Scheffer (*The Year of the Whale*, 1969) raised the collective consciousness about environmental concerns. A growing number of aquariums displayed dolphins and small-toothed whales in tanks and shows for the public (Norris, 1974). Throughout North America, the UK, Europe and Australia/New Zealand during the 1960s there grew a perception of whales and dolphins as intelligent, entertaining and endangered. That general perception provided the backdrop for the development of two dramatic human impacts on the marine environment: the movement to save whales and dolphins from commercial hunting, and the great rush to see them in the wild.

Growth of Whale Watching

Prior to the mid-1960s, whale watching was a localized, relatively low-key activity known only in the USA. In Hawaii, a loosely organized 'whale watchers' club on the island of Oahu made occasional reports of humpback whales during the winter breeding season (Herman, 1979), but sightings were few, interest waned and the group was discontinued. In California, however, grey whales could be seen in fairly large numbers from shore during their annual migration from Alaska to Mexico, and they attracted the attention of Carl Hubbs, a professor of zoology at Scripps Institute near San Diego. Hubbs organized his graduate students to conduct an annual shore-based census of the whales starting in the 1940s (Norris, 1974), and interest in the migration began to grow. In 1950, a popular whale-watching lookout site for the public was established at the Cabrillo National Monument in San Diego (Hoyt, 2002). In January of 1953, the US Department of the Interior's Fish and Wildlife Service issued a press release announcing the whales' 'little-known but spectacular wildlife migration' (US Department of the Interior, 1953). The government's interest in the grey whales was to monitor population size with a view to recommencing shore-based whaling under the terms of the International Whaling Commission (IWC). 'The once great American whaling industry – 100 years ago over 700 whaling ships involving an investment of $40 million were engaged in the business – is temporarily dormant because of a lack of demand for the products. The business may be revived because the flesh of whales is similar to beef in flavour and texture, and is extensively used for food in Japan, as well as in several European countries, and has recently been introduced in the USA' (US Department of the Interior, 1953).

Two years later the first commercial whale-watching trips began operating out of San Diego (Hoyt, 2002). Within 6 years, Raymond Gilmore (the biologist hired by the Fish and Wildlife Service to oversee the issuance of whaling permits and operation of whaling stations in California) became the first onboard naturalist for whale watches operated by the San Diego Natural History Museum (Nickerson, 1977). Despite the government's focus on whale consumption, the public seemed more interested in live whales than dead ones. Gilmore developed a devoted following of avid whale watchers, and continued his popular trips for 25 years until he died of a heart attack as he boarded a whale-watching boat for one more trip at the age of 77 (New York Times, 1984).

Grey whale watching expanded to the north and south from its San Diego base throughout the 1970s, and other species began to attract attention in Canada (fin, minke and beluga whales on the Saint Lawrence River), the north-east USA (humpback whales, fin whales) and Hawaii (humpback whales; Forestell, 1991;

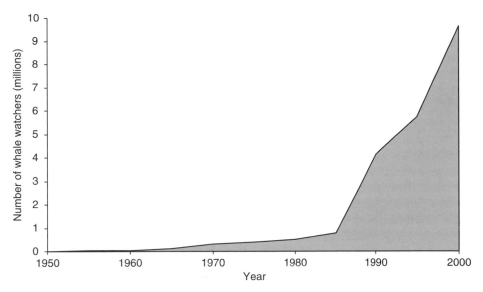

Fig. 15.2. Estimated millions of whale watchers per year, worldwide. (Modified from Hoyt, 2001.)

Hoyt, 2001, 2002). During the following decade, commercial whale-watching operations grew rapidly throughout the world (Hoyt, 2002), and the concept of 'whale watch' grew to include both whales and dolphins. Since 2000, some 10 million people have been spending nearly half a billion US dollars a year to watch whales and dolphins in more than 85 countries (Hoyt, 2001). Whale watching has been the fastest growing wildlife-based activity in the world (Lien, 2001). A general picture of the growth of whale watching between 1950 and 2000 is shown in Fig. 15.2, based on information provided by Hoyt (2001).

Environmental Impacts of Whale Watching

As whale watching has developed in hundreds of locations around the world, increasing attention has been paid to its impact on individual species (Bejder and Samuels, 2003), the local ecosystem (Olesinski, 1994), the host community (Forestell, 2005) and the global environment (IFAW, 1999). Whale watching shares with all other marine tourism activities the potential for degradation of species and their marine environment (Swartz, 1989; Miller, 1993; Orams, 1999). Corkeron (2004) notes that whale watching is unique, however, in that many conservation groups actively promote whale watching as a net benefit overall, despite short-term negative impacts to target species. He cites four justifications used by conservation groups for whale watching:

- Observing whales in their own environment will induce support for marine conservation.
- Commercial whale-watching platforms provide opportunity for scientific research.

- Whale watching will reduce the need for captive display facilities.
- Observing whales in their own environment will reduce support for commercial whale hunting.

Corkeron (2004) argues that encouragement of whale watching has a trivial impact on improving cetacean conservation. Whale watching, he believes, is an outdated manifestation of the iconic value of whales developed during the 1970s, as a reaction to rampant industrialized whaling. The 'whales as icons' perspective is now problematic he suggests, because it interferes with a rational ability to address such current questions as 'how much of the reduced productivity of the oceans and coasts should remain available to whales' (Corkeron, 2004, p. 848). Whale watching is, in his view, an outmoded front for anti-whaling advocates. He raises the question as to whether it is time to 'spread new messages' to whale watchers that 'whale populations will fare better under an internationally controlled regime of sustainable hunting rather than under culls instigated by individual nations' (Corkeron, 2004, p. 848).

Indeed, international non-profit organizations do promote a wide range of marine tourism activities that target marine mammals: World Wildlife Fund (http://worldwildlife.org/travel/), International Fund for Animal Welfare (http://www.ifaw.org), Whale and Dolphin Conservation Society (http://www.wdcs.org/whalewatching), American Cetacean Society (http://www.acsonline.org/) and Pacific Whale Foundation (http://pacificwhale.org/) are all international marine conservation organizations that actively promote responsible whale watching. None of them supports a return to commercial whaling. All of them recognize the rapid growth in whale watching (averaging 12% per year between 1990 and 2000; Hoyt, 2001), and the need to reduce the negative impact of tourism on the target species through the development of responsible whale watching.

The relationship between commercial whale watching and commercial whale hunting has become a more visible issue with the increasing interest by the members of the IWC in whale watching. In 1993, the IWC, an international organization charged with providing for conservation of whale stocks and the orderly development of the whaling industry, undertook initial investigations of whale watching as a sustainable use of cetacean resources. Following 3 years of preliminary collection of information, the IWC adopted a resolution in 1996 to develop guidelines for the management of whale watching to ensure ecological sustainability and satisfy, to the extent possible, the requirements of the industry and expectations of the wider community (available at: http://www.iwcoffice.org/conservation/whalewatching.htm). At a workshop in South Africa in 2004, a recommendation was made by representatives of the member nations to ensure that 'the best science is available for the sustainable management of whale watching' (IWC, 2004).

Established whale-watching venues consist of complex assemblages of tourism support services that develop in a relatively predictable series of phases (Forestell and Kaufman, 1994, 1996). Following some period of limited and informal public interest in the presence of one or more marine mammal species in a particular location (the Discovery phase), commercial entities either

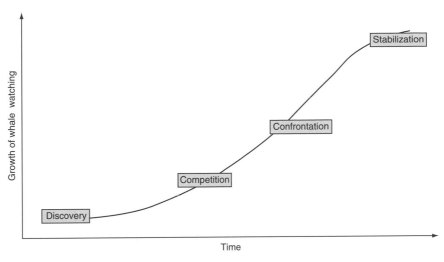

Fig. 15.3. Proposed stages in development of whale watching. (From Forestell and Kaufman, 1994, 1996.)

develop or refocus to exploit the tourists' needs for food, accommodation, transportation, equipment and guide services (the Competition phase). A by-product of the Competition phase is growth in the number of people who participate in the activity (Forestell and Kaufman, 1996). That growth frequently leads to conflicts among a range of stakeholders, including business owners, members of the local community (who may also be business owners), researchers, conservationists, resource managers and tourists (the Confrontation phase). Eventually, the conflicts are brought to some state of resolution and a period of Stabilization ensues (Fig. 15.3). Depending upon a variety of local considerations the Stabilization phase may remain as a steady state or occasionally cycle between Stabilization and Confrontation (IFAW, 1999).

Regulation of Whale Watching

One mechanism for moving stakeholders from the Confrontation to the Stabilization phase has been the enactment of various forms of regulations developed at local, regional, state, federal and international levels. The introduction of regulations may serve as a marker that the Confrontation phase is underway (Forestell and Kaufman, 1996). Many efforts to regulate the impact of whale watching on the community, the target species, or the environment develop retroactively. The passage of the legislation post-dates the realization that there might be a problem. During the early years of legislative control of whale watching (1975–1985), uncertainty about the impacts of whale watching and the focus of legislative mandates on balancing conservation with development made decision making difficult: 'The primary objectives of guidelines and legislation regarding whale watching are to protect whales from harmful effects and minimize disturbance of whales during this activity. Complementary objectives include the sustainable development of the whale-watching industry, educating the public and ensuring the safety of operators and the public involved' (IFAW, 2000, p. 18). In a context of competing demands and limited knowledge, legislation is likely to serve more as a retroactive tool for negotiating resolution of disparate constituent demands than a proactive mechanism for anticipating and preventing potential negative impacts (Thorne-Miller and Catena, 1991; IFAW, 2000; Garrod and Fennell, 2004; Reynolds, 2005).

Regulations are a type of intervention, and can range in degree of formality from government legislation to suggested guidelines, or voluntary codes of conduct (Garrod and Fennell, 2004). During the last five decades, a wide range of guidelines and regulations directed at whale watching have developed around the world (Carlson, 2004). At least 32 countries (including territories and dependencies) have enacted hundreds of regulatory actions specifically directed at whale (i.e. cetacean) watching. A review of the regulations, guidelines and codes in Carlson (2004) shows that current regulatory actions aim to control disturbance to specific marine mammal species from human behaviour and platform characteristics during whale-watching activities, taking into consideration a range of species-specific temporal, geographical and biological factors, while allowing for some level of tourism development.

Despite the developing proliferation of controls (voluntary or otherwise) on whale watching and other anthropogenic marine activities since the mid-1990s or so, there remains a concern that such controls have either trivial (Corkeron, 2004) or insufficient (Meffe et al., 1999) effects on the reduction of harmful impacts on marine mammals. Two possible reasons for the failure of whale-watching regulations to significantly enhance environmental recovery are that: (i) the regulations are not being effectively followed; or (ii) the regulations do not correctly address the impact of concern.

Managers and regulators must negotiate compromises between the needs of naturally evolving ecosystems and the needs of increasingly exploitative humans (Thorne-Miller and Catena, 1991). As Aldo Leopold (1949) pointed out, it is not ecosystems that can be managed, but the humans who exploit them. Although whale-watching regulations are for the recovery and/or protection of cetaceans, they are directed at modifying behaviours of humans. To be effective, regulations must be coincident with the psychological make-up of that group expected to comprehend the regulation and comply with it (Gardner and Stern, 1996). Unfortunately, very little work has been done to develop a systematic understanding of the relationship between regulations and the psychological or physical predispositions of whale watchers (Forestell, 1995). Garrod and Fennell (2003) undertook a comparative content analysis of 58 whale-watching codes of conduct compiled by Carlson (2001) in an effort to assess their overall consistency. The analysis was carried out on the basis of three themes (controls on approaching cetaceans, controls on interacting with cetaceans and the overall management orientation of guidelines). The authors found considerable variability among the codes, and concluded that the lack of systematic development of codes on a global basis threatened both the sustainability of whale watching and the conservation of cetaceans.

Of particular interest in Garrod and Fennell's (2004) analysis was the finding that more than 90% of the codes they reviewed were based on a deontological perspective (correct action is mandated by authority), rather than a teleological one (correct action derives from a pursuit of best consequences). Piaget (1977), Kohlberg (1981) and others (e.g. Gardner on Stern, 1996) have emphasized the dependence of an individual's behaviour on the level of cognitive development. Their work suggests that deontological and teleological motivations to behave will influence different segments of the human population, based on factors such as life-span development, general cognitive capability and education. Garrod and Fennell's (2004) findings, however, show that the ethical orientation of the codes they reviewed was not based on careful consideration of the human population meant to follow the codes, but was a by-product of the fact that most of the codes were developed by government agencies, which are generally in the business of telling people how to behave, rather than why certain behaviours may be desirable. The result is that codes of behaviour most often result in a 'lack of ownership' (Garrod and Fennell, 2004) or a level of outright antagonism (Marion and Reid, 2007) that limits compliance.

Development of 'values-based' regulation by government agencies is restricted by the fact that enforcement must generally be based on the observable behaviours of the human rather than the internal states of either the human or the whale, or the potential long-term effects of a given behaviour (Forestell, 1995). Deontological, action-based codes are more likely to be proposed by regulatory agencies responsible for ensuring compliance. Unfortunately, this can create further problems if deontological regulations are implemented that are out of step with the competencies of the group that is expected to comply with them. Recreational boaters not experienced at differentiating between species will have difficulty following species-specific regulations. Distance limitations on approaching pods with calves are of little utility for those who cannot determine whether a calf is present until after the limitation is exceeded. The most frequently invoked regulation in the codes reviewed by Garrod and Fennell (2004) concerned approach distance (found in 88% of the codes). A restriction on approach distance was first employed in 1976 in a public notification as to how National Marine Fisheries Service would define 'harassment' in its efforts to protect humpback whales from human activity in Hawaii. Since then, the control of approach distance has become an integral part of regulations throughout the world. The prevalence of distance limitations is somewhat surprising, given that humans are relatively poor at estimating the distance between a boat they are on and pods of whales they are observing (Baird and Burkhart, 2000).

Formally established codes of conduct (whether legislated or voluntarily accepted) frequently suffer from inflexibility and intractability. The ability to track changes following implementation of regulations and the ability to modify regulations on the basis of change in a timely and effective manner are critical components of successful management and stakeholder acceptance (Hilborn, 2005). In 1991, the state of Hawaii passed a law banning commercial jet ski and parasail operations in certain near-shore waters of Maui during the winter time while humpback whales were present. It was argued that these operations might be responsible for displacing mothers and calves from preferred near-shore areas. No systematic surveys had ever been undertaken to establish that mothers and calves had, in fact, been displaced from shore (although some anecdotal observations supported that conclusion, e.g. Glockner-Ferrari and Ferrari, 1990); nor were data available regarding the effect of jet ski and paras-ail operations on whales in Maui waters, relative to other activities such as whale watching, near-shore development, military activity on a near-by island and so on. None the less, it was generally felt that, given the endangered status of humpback whales and the proliferation of human activities in the near-shore waters of Maui, it was best to 'err on the side of caution' and ban the activities in question.

Following the implementation of the ban, no studies were carried out to determine whether any observable change in the distribution or behaviour of humpback whale mothers and calves resulted. Consequently, little or nothing was learned that might be of value to managers in other areas with respect to how to deal with potential problems associated with similar activities. No evidence has been established to verify either that such activities have a measurable impact, or that any effects of such activities have been mitigated following imposition of regulatory control. Any manager operating elsewhere must fight the same battle from scratch, unaided by any precedents that might have emerged from the Hawaii experience. In 2004, a US District court ruled the ban unconstitutional (Kubota, 2004), and although further legislative efforts have been taken to reinstate the ban, the matter continues to be debated without the benefit of scientific evidence.

A further limitation on the ability of whale-watching regulations to improve the conservation status of cetaceans is lack of enforcement (Beaubrun, 2002). As the economic significance of the whale-watching industry grows, the lobbying capabilities and overall political influence of operators become a major force in the determination of how, or whether, regulations are enforced. Especially during the Competition phase (Forestell and Kaufman, 1996) the stakes increase, and so may the degree of self-interest and perceived need to protect the status quo. Establishing a pattern of equitable and effective enforcement as quickly as possible may decrease the likelihood that regulations will become neutered or abandoned in the interests of economic success. Rules should be enforced. While this may appear obvious, it is frequently the case that regulations are developed even when there may be little hope of enforcing them. Often this is because the resources are not available to permit monitoring, enforcement or prosecution. In some cases the activity may take place in relatively inaccessible areas, in other cases the cost of enforcement is prohibitive.

The author's personal experience at various levels in the development and growth of whale watching in Hawaii, Japan, Australia, Costa Rica and Ecuador since the early 1980s has provided a number of disparate views of enforcement. Perhaps the most well-funded and stringent enforcement occurs in the USA (viz. Hunter, 2007). Enforcement is also well funded in Australia, but in many cases, the view that operators know best prevails (May, 1994), and the development and oversight of regulation are heavily influenced by the industry.

In Japan, an age-old network of local fishermen's cooperatives is used to manage inshore resources equitably among the members (Acheson, 2005), and the system has been adapted to allow operators to manage whale watching in some areas (IFAW, 1997). In Costa Rica, enforcement is not funded at the local level, although federal regulations prohibit swimming with cetaceans (L. May-Collado, San Jose, Costa Rica, 2005, personal communication). Ecuador does not yet have regulations controlling whale watching. While there is some degree of control over commercial operations, it is aimed primarily at passenger safety (C. Castro, Puerto Lopez, Ecuador, 2007, personal communication). Fewer than half the countries reported by Hoyt (2001) to promote whale watching are included in the list of those that have regulations in place to control the activity (Carlson, 2004), and even where regulations are in place there is ample evidence that regulation without enforcement is an exercise in futility.

What Does Regulation of Whale Watching Accomplish?

Regulation of whale watching has been driven by concerns that particular species may be irreversibly damaged by the presence and behaviour of humans on boats or in the water (IFAW, 1995). An increasing number of studies have shown that short-term behaviour (e.g. vocalization patterns, respiration rate, diving pattern, direction and speed of movement, activity state) of many cetacean species is significantly altered by the presence and activity of whale watching boats and/or humans entering the water to engage in a variety of 'swim-with' programmes (Constantine, 1999; Bejder and Samuels, 2003; Samuels *et al.*, 2003). The short-term damage cetaceans can experience from interactions with 'curious' humans has been demonstrated many times in all parts of the world. The last intentional killing of a right whale in the USA occurred off the coast of Florida in 1935, when a group of recreational deep-sea fishermen repeatedly shot and harpooned a mother and calf over a 6 h period, until the calf finally died (Kraus and Rolland, 2007a). The calf was then tied to the boat and dragged back to shore for public display. Doak (1988) reviews a number of incidents, extending over a 30-year period, in which lone 'sociable' dolphins were injured or killed following interactions with humans. In Brazil, a lone bottle-nose dolphin approached and swam with humans over a 15-month period beginning in March 1994 (Santos, 1997). Swimmers repeatedly attempted to touch, grab and climb on to the dolphin's back. Attempts were made to stick objects in the dolphin's blow-hole. The dolphin injured a number of swimmers in response to harassment, finally killing a 30-year-old man with its tail flukes. In Australia, unruly behaviour of tourists attempting to feed wild dolphins at two locations (Shark Bay in Western Australia and Tin Can Bay in Queensland) has resulted in government regulations to control (but not prevent) such activity (Constantine, 1999; Mann and Kemp, 2003; Samuels *et al.*, 2003).

The evidence is overwhelming (both quantitatively and qualitatively) that unregulated interactions between humans and dolphins are almost always more dangerous for the dolphin (Doak, 1988). For that reason it is not only appropriate, but critical, that limits on feeding and touching wild cetaceans be imposed to prevent the most egregious forms of human behaviour. Samuels *et al.* (2003) provide evidence that such limits can have positive outcomes. However, even with regulations (whether legislated or voluntary) in place there remains a need to engage in proactive efforts to inform the public about appropriate behaviour, and to maintain ongoing monitoring and enforcement (Orams, 1995b; Santos, 1997).

Although it seems clear that regulating human behaviour in the vicinity of whales and dolphins can reduce direct, short-term impacts, the greater concern that has emerged since the mid-1990s is the need to address the long-term effect of anthropogenic impacts (IFAW, 1995). As noted by Meyers and Ottensmeyer (2005, p. 59): '[I]t is the process of extinction that is important, not the recording of the last individual.' Whale researchers are rising to the challenge of those who call for rigorous, quantitative and systematic evidence of long-term effects of whale watching. A comprehensive review by Bejder and Samuels (2003) provides a much-needed framework for not only conducting future studies, but also putting current and past efforts into a more unified context. The use of methodologies and analyses such as those described by Bejder and Samuels (2003) and Lusseau (2003) has led to a more rigorous and nuanced understanding of the effects of tourism on marine mammal behaviour. Lemon *et al.* (2006) have shown that the presence of boats can change the behaviour and movement direction of resident bottlenose dolphins without changing the rate of whistling or duration of echolocation bouts. The effect of boats was found even beyond the 30 m distance limit imposed on boats by federal regulations in place in that area (Lemon *et al.*, 2006). Bejder *et al.* (2006) found that some, but not all, bottlenose dolphins left an area when boat approaches were newly introduced, although no change in residence was documented in an area where boat approaches had been in place for some time. The finding is important because it bears directly on claims that dolphins (or other marine mammals) habituate to the presence of boats over time (Watkins, 1986). In reality, it may simply be the case that the more disturbed animals leave an area permanently, and only those who may not be able to change their habitat-use patterns remain (Bejder *et al.*, 2006). The result is a kind of 'double whammy' on resident populations, since sensitized animals are driven into less preferred habitat, while the remaining animals are left to bear the brunt of boat disturbance. These studies have significant importance for their ability to inform a rational, science-based development and use of regulations to control the impact of whale watching on cetacean populations.

Whale watching has been identified as a potential threat to local populations of marine mammals (Constantine, 1999; Marsh *et al.*, 2003; Samuels *et al.*, 2003), particularly in the case of endangered species. Physical damage or behavioural disturbance by boats (Marsh *et al.*, 2003) and harassment by human swimmers (Samuels *et al.*, 2003) are of particular concern. The stark reality, however, is that even if we were to win the battle of protecting specific marine mammal groups in particular areas from the impacts of marine tourism by successfully controlling the behaviour of whale watchers through legislative initiatives, it is unlikely that we would have done much to win the war of protecting endangered marine mammal species or the ocean in which they live. Simply put: 'In the big picture of conservation concerns for cetaceans . . . the effects of whale watching are pretty trivial' (Corkeron, 2004, p. 848). Even when considering one of the most endangered cetacean species in the world, the North Atlantic right whale, Kraus and Rolland (2007a) write: 'Although whale watching could potentially have some effect on the whales by distracting or stressing them, it is difficult to imagine this is a significant problem compared to the fatal threats posed by large ships and fixed fishing gear' (p. 25). Recent compendia of detailed scientific analyses of issues considered most detrimental to marine biodiversity in general (Norse and Crowder, 2005a) and marine mammals in particular (Twiss and Reeves, 1999; Gales *et al.*, 2003; Reynolds *et al.*, 2005; Estes *et al.*, 2006; Kraus and Rolland, 2007b) are, with one exception (Gales *et al.*, 2003), devoid of any mention that whale watching presents a substantial threat to the overall protection of marine mammals or the global marine environment. Commercial hunting, fisheries by-catch and entanglement, ship strikes, toxic run-off from land, marine debris, noise pollution, habitat degradation and global warming all present far greater threats to cetaceans than whale watching (see Twiss and Reeves, 1999; Reynolds *et al.*, 2005 for detailed treatment of these effects). Kraus and Rolland (2007c) argue that many species of cetaceans, including right whales, killer whales, beluga whales and Indo-Pacific humpbacked dolphins suffer from what they term 'urban whale syndrome' – increased mortality, decreased reproduction, compromised health and habitat loss as a result of exposure to anthropogenic factors. While whale watching might add to the problems, it is not considered a major contributor to the syndrome.

As whale watching continues to grow and thrive worldwide, global perspectives are becoming more and more important in shaping conservation agendas. The conversation has shifted from protection of species to protection of ecosystems (Meffe *et al.*, 1999). There is a continuing debate about the relative importance of observing individual organisms acting locally in the short term (Greene, 2005; Parrish, 2005) versus clusters of species interacting in complex benthic webs across vast temporal and spatial scales (Paine, 2006). Despite the challenges to understanding how large-scale marine ecosystems have been impacted by current and historical anthropogenic impact patterns (Jackson, 2007), the emerging view of appropriate strategies for conservation of marine species is tipping towards 'whole of ocean' strategies (Soulé, 2005). The focus of legislation and management (and the science to support them) based on a 'whole ocean' approach is to maintain or restore the natural structure and function of ecosystems, including their biodiversity and productivity (Currie, 2006). Ecosystem Based Management (EBM) faces the challenge of protecting resources across a wide range of political and cultural agendas protected by the 1982 Convention of the Law of the Sea out to the limits of each coastal country's Exclusive Economic Zone (generally 200 nautical miles from shore), and the 'frontier exploitation' (Norse, 2005) that has characterized human behaviour on the open ocean.

Perhaps one of the most widely recognized tools available to assist in the promotion of EBM is the use of 'place-based management' strategies (Norse *et al.*, 2005). The first images taken of Earth from space brought realization to many that we live on the 'Ocean Planet' (Benchley, 1994). Those images may have also reinforced the notion of a vast, featureless realm, like the big blue spaces found on maps of the world. The real ocean, of course, is a complex web of horizontal and vertical patterns of movement of great portions of sea-water driven by wind, tide and temperature, but shaped by interactions with diverse geological and biological formations on the ocean floor, stretching from coastal reef structures all the way out to deep ocean sea mounts and trenches (Thorne-Miller and Catena, 1991). The biological diversity of the ocean is, in turn, largely determined and distributed by the presence of 'hot spots' generated by a range of these heterogeneous sub-surface features that are largely hidden from view (Norse *et al.*, 2005). Marine place-based management aims to identify and conserve the resources associated with specific 'hot spots'. In doing so, such an approach may be viewed as consistent with the perceived need to protect habitats and ecosystems of biological significance (Ragen, 2005).

Place-based management efforts have led to a wide range of decisions regarding the location and size of places to be managed, and the activities to be regulated within them. Ideally, management schemes should recognize the interconnectedness and interdependence of physical and biological systems throughout the marine environment: 'There is only one world ocean system and all the water circulates throughout it' (Hoyt, 2005, p. 69). In reality, however, only one half of 1% of the entire ocean outside the limits of the Exclusive Economic Zones is subject to some form of fisheries-related protective regulation (Roberts, 2005). Hoyt (2005) has provided a very detailed and informative summary of more than 500 marine protected areas (MPAs) proposed or in place, in all oceans of the world, that are in whole, or in part, aimed at providing some form of protection to marine mammals. Hoyt (2005) also reviews a number of legally and functionally different definitions, listing 68 different designations that may broadly be considered MPAs. In fact, many of these designations provide little substantive protection because they are too small in size, too limited in biological focus or inadequately managed (Roberts, 2005). Hoyt recognizes these challenges, but argues that MPAs for cetaceans are an important beginning, and provide a necessary framework for future improvements in marine conservation. He also emphasizes repeatedly that aggressive pursuit of well-defined and managed MPAs for cetaceans is critical because 'around the world, cetacean habitat, inside and outside protected areas and international sanctuaries, is little recognized, largely undescribed, marginally protected at best and being degraded every day' (Hoyt, 2005, p. 11). Marsh *et al.* (2003) argue that with the appropriate focus on protecting the full range of habitat requirements, incorporation of core areas and 'no-take' zones to protect prey abundance, and embedded within broader networks of protected areas to ensure ecological sustainability, MPAs can serve as critical mechanisms for recovery and protection of marine mammal species. Regional MPAs aimed at controlling direct impacts of wildlife tourism on local groups of animals do not come close to meeting that description (Hoyt, 2005).

The Sound of Two Hands Clapping

Regulation of whale watching as a means of affording global protection to marine mammals and their habitat is akin to putting a band-aid on a gaping, haemorrhaging wound. Regulation might reduce local disturbance to some resident populations targeted by tourism operators, or it might prevent added stress to populations suffering debilitation from other impacts. It will not prevent or reverse the range of more serious human-induced impacts that continue to threaten a number of marine mammal species that, for the most part, are not even targeted by marine tourism. These include the North Atlantic right whale, Florida manatee, Hawaiian and Mediterranean monk seals, Stellar sea lions, southern sea lion, spotted and spinner dolphins in the Eastern Tropical Pacific, beluga whales, killer whales, Asian river dolphins, Western North Pacific grey whales, vaquita, Indo-Pacific humpbacked dolphin, striped dolphin, baiji, boutu, Okinawan dugong, sea otter and polar bear to name a few (Perrin, 1999; Ragen *et al.*, 2005). In the grand scheme of marine mammal protection and conservation of marine biodiversity, whale watching is not the problem.

Sadly, however, there is little evidence to date that whale watching is part of the solution.

It need not be that way. With 10 million people a year participating in commercial whale-watching operations (Hoyt, 2001) there is a continuing opportunity to 'turn tourists into Greenies' (Orams, 1997; Johnson, 2003). Forestell (1991, 1993) first made the case that commercial whale-watching trips were an important, generally untapped, venue for promoting environmental awareness, enhancing appreciation for endangered species, and motivating participants to undertake behaviours that would reduce land-based activity that could threaten the oceans. Orams (1995a,b; 1997) further developed the use of interpretation as a mechanism for changing attitudes and behaviours of participants in a dolphin-feeding programme at Tangalooma Resort on Australia's east coast. He showed promising evidence that as a result of educational intervention behaviour change could occur (Orams, 1995b). Despite encouraging signs that interpretation programmes would become an integral part of commercial whale-watching operations around the world (IFAW, 1997), there has been little evidence that such programmes have ever been widely developed. Although onboard interpretation is mandated as a condition of obtaining commercial whale-watching permits in Australia and New Zealand, there is little effort made to ensure the quality of such programmes (IFAW, 1997). Where programmes are in place, their focus is often limited to the local species and environment, and is not based on a structured approach, and fails to connect the whale watcher with global environmental concerns (Curtin, 2003; Lück, 2003). Exceptions have been Orams' (1997) work with tourists at Tangalooma, and programmes offered by Pacific Whale Foundation in Hawaii (Forestell, 2005). In both cases, careful consideration has been given to the structured content of the programmes, the training of the interpretation staff and the need to connect the participant to broader environmental concerns.

The historical focus on whale watching as a place-based, species-specific, locally controlled activity has resulted in a disconnected network of hundreds of whale-watching operations around the world (Hoyt, 2001) characterized by a wide range of inconsistently developed and enforced codes of conduct (Carlson, 2004) carried out in a vast patchwork quilt of MPAs (Hoyt, 2005). In many cases, the designation of protected areas is little more than a 'national or international statement of good intention' (Hoyt, 2005, p. 23). While there are clearly needs for at least some of the management regimes that have emerged from such a focus (Marsh *et al.*, 2003), one cannot help but notice that in spite of the efforts to keep whale watchers from directly harming whales (and other marine mammals) while watching them, the rate continues to escalate at which the oceans are being emptied of marine life and filled with noise, debris and poison (Norse and Crowder, 2005b). The 10 million people going whale watching every year are part of the reason why that is happening, but with few exceptions, nobody is effectively delivering that message.

It is frequently noted by scientists, resource managers and educators that any effort to reduce the harmful impacts of human use of the ocean and conserve marine biodiversity must incorporate an understanding of human behaviour and include well thought out educational programmes (IFAW, 1999; Meffe *et al.*, 1999; Marsh *et al.*, 2003; Hoyt, 2005). There is excellent evidence that wildlife tourists want to be educated about global issues (Lück, 2003, Fig. 15.4) and that well-designed education programmes can affect their behaviour (Orams, 1997, Marion and Reid, 2007). Despite the best intentions, however, high-quality education programmes are either absent or are overshadowed by 'animal protectionists (who) mobilize concerned, yet uninformed citizens to

Fig. 15.4. There is excellent evidence that whale watchers want to be educated about marine conservation issues (Lück, 2003). (Photograph Pacific Whale Foundation.)

clamour for an end to any animal use' (Marsh *et al.*, 2003, p. 5). As Corkeron (2004) has suggested, we need to 'refashion' the iconic value of whales and use them to 'help spread new messages about marine conservation' (p. 848).

There are a number of land-based models of effective interpretation and education programmes that can inform similar efforts in marine settings (Marion and Reid, 2007). However, whale-watching tours are often relatively brief (2–3 h), and are frequently conducted aboard small boats with little stability, noisy engines and limited space and freedom to move about. These factors provide unique challenges to effective delivery and follow-up (Orams, 1999), particularly if half the audience is fighting nausea and/or fear that the boat may sink or be hit by a whale (Forestell, 1992). One of the things that we have learned about whale watchers is that they tend to travel widely and engage in whale watching in a number of venues (Forestell and Kaufman, 1994). Rather than worry about designing 'one-size-fits-all' education programmes, it may well be better to focus more on matching message to platform, with the long-term view of 'modularizing' the information concerning global issues and marine environmental concerns. Another way to extend the opportunity to educate whale watchers is to redefine the whale-watching experience to include time spent planning and preparing for the experience and time spent returning home following the experience (IFAW, 1999). Nature shows on television, air-line flights, passage aboard cruise ships, interactions with travel agencies, rental car agencies, booking agencies, accommodations, dining and auto service, all provide novel and potentially important venues to amplify and reinforce an understanding of anthropogenic threats to the marine environment and steps that can be taken to mitigate them.

Not only is it important to broaden and re-imagine the venues available for education and interpretation as part of the whale-watching experience, it is also important that the venues involved model desired behaviours and provide participants the opportunity to engage in well-defined and specific activities that help mitigate environmental concerns. For example, Pacific Whale Foundation in Hawaii, which takes nearly 200,000 people a year on a variety of marine tourism activities (G. Kaufman, Maui, Hawaii, 2007, personal communication), has introduced a number of environmentally sound practices: they use purpose-built boats with special noise abatement features to reduce noise, increase fuel efficiency and minimize disturbance to marine mammals; they burn bio-fuels; they develop pump-out facilities to avoid ocean discharge; they use recycled food receptacles; they choose menu items that promote sustainable practices; and their staff are trained naturalists with the ability to educate passengers about the local ecology and its linkage with global concerns (Forestell, 2005). There is encouraging news that many sectors of the travel industry are introducing new initiatives to not only reduce their own carbon emissions, but also to provide customers the opportunity to calculate and offset the 'carbon footprint' generated by their own travel (Boehmer, 2006).

One major issue that has yet to be effectively recognized and incorporated within whale-watching interpretation/education programmes is the need to better understand the attitudes, motivations and cultural values of the target audience (Ham and Krumpe, 1996). All too often, tourists are viewed as the uninformed 'dupes' of animal rights advocates who need to be told the facts by more knowledgeable scientists (Marsh et al., 2003; Corkeron, 2004). This perspective drives a continuing tension between science-dominated environmental research and socio-cultural methodologies aimed at changing behaviour through teleological rather than deontological approaches (Head et al., 2005). Much more work remains to be done to build effective interpretation/education programmes that derive their techniques from a rigorous understanding of the 'zeitgeist' of the whale watcher (Higham, 1998).

Summary and Conclusion

Interest in watching whales, dolphins and other marine mammals has grown exponentially over the last five decades, and shows little evidence of abating. The global reach of marine wildlife tourism ventures focusing on marine mammals raises valid concerns about the potential impact on endangered species and the sustainability of ever-expanding commercial operations. In response, a number of local, national and international efforts have been undertaken to protect marine mammals from short-term and cumulative impacts of whale watching. The long-term effects have yet to be established, but there is growing encouragement for the use of the 'precautionary principle' in developing strategies for protection (IFAW, 1995). Whale watching takes place in more than 80 countries around the world (Hoyt, 2001). Less than half of those have formal regulations or voluntary guidelines controlling the actions of whale-watching platforms or swimmers in the vicinity of marine mammals (Carlson, 2004). More than 500 MPAs have been, or are, proposed to be put in place around the world (Hoyt, 2005). There is good evidence that some of the more obvious and flagrant forms of local, short-term disturbance to marine mammals may be prevented or mitigated by the regulation of whale watching. There are two general problems associated with depending primarily on regulation for the conservation of marine mammals and their ocean environment. First, the marine mammals most threatened by anthropogenic factors are faced with problems generally believed to be far more serious than whale watching. Second, even if we are successful in preventing all disturbance to marine mammals associated with whale watching, we will have contributed little or nothing to the overall protection of marine biodiversity.

Although whale watching does need to be regulated to some degree to ensure it does not add to the greater damage of other anthropogenic effects, such regulation is of little long-term utility on its own. Coupled with appropriate and well thought out education programmes, however, it may be possible to make greater headway in addressing the need to change a range of human behaviours associated with damaging the ocean. Appropriate educational efforts need to go beyond the local context of the particular whale-watching location, species and concerns and extend to the 'big picture' challenges of overpopulation, overfishing, pollution from debris, toxins and noise, habitat destruction and global warming. A two-handed approach of regulation coupled with education could help to significantly increase the number of people who demonstrate their understanding of the degree to which degradation of the ocean threatens us by changing behaviours that contribute to the problem. Success in such a venture needs a broadened framework that will include not just the stakeholders in the target whale-watching experience, but all those people and agencies involved in getting whale watchers to and from their destinations.

References

Acheson, J.M. (2005) Developing rules to manage fisheries: A cross-cultural perspective. In: Norse, E.A. and Crowder, L.B. (eds) *Marine Conservation Biology: The Science of Maintaining the Sea's Biodiversity*. Island Press, Washington, DC, pp. 351–361.

Baird, R.W. and Burkhart, S.M. (2000) Bias and variability in distance estimation on the water: Implications for the management of whale-watching. International Whaling Commission Meeting Document SC/52/WW1, Adelaide, Australia, July 3–6.

Beaubrun P.C. (2002) Disturbance to Mediterranean cetaceans caused by whale watching. In: Notarbartolo di Sciara, G. (ed.) *Cetaceans of the Mediterranean and Black Seas: State of Knowledge and Conservation Strategies*. A report to the ACCOBAMS Secretariat, Monaco. Section 12.

Benchley, P. (1994) *Ocean Planet*. H.N. Abrams, New York.

Boehmer, J. (2006) Virgin leading effort to reduce CO_2 emissions. *Business Travel News Online*, October 9. Available at: http://www.btnmag.com/businesstravelnews/headlines/frontpage_display.jsp?vnu_content_id=1003221668

Bejder, L. and Samuels, A. (2003) Evaluating the effect of nature-based tourism on cetaceans. In: Gales, N., Hindell, M. and Kirkwood, R. (eds) *Marine Mammals. Fisheries, Tourism and Management Issues*. CSIRO, Collingwood, Victoria, Australia, pp. 229–256.

Bejder, L., Samuels, A., Whitehead, H. and Gales, N. (2006) Interpreting short-term behavioural responses to disturbance within a longitudinal perspective. *Animal Behaviour* 72, 1149–1158.

Carlson, C. (2001) *A Review of Whale Watching Guidelines and Regulations Around the World*. Report for the International Fund for Animal Welfare. Yarmouth Port, Massachusetts.

Carlson, C. (2004) *A Review of Whale Watching Guidelines and Regulations Around the World*. Report for the International Fund for Animal Welfare, Yarmouth Port, Massachusetts.

Carson, R. (1962) *Silent Spring*. Houghton Mifflin, Boston, Massachusetts.

Constantine, R. (1999) *Effects of Tourism on Marine Mammals in New Zealand*. Department of Conservation, Wellington, New Zealand.

Currie, D.E.J. (2006) *Ecosystem-based Management in Multilateral Environmental Agreements: Progress Towards Adopting the Ecosystem Approach in the International Management of Living Resources*. World Wildlife Fund International Global Species Programme, Rome, Italy.

Curtin, S. (2003) Whale watching in Kaikoura: Sustainable development? *Journal of Ecotourism* 3, 173–195.

Corkeron, P.J. (2004) Whale watching, iconography, and marine conservation. *Conservation Biology* 18, 847–849.

Doak, W. (1988) *Encounters with Dolphins*. Hodder & Stoughton, Auckland, New Zealand.

Estes, J.A., Demaster, D.P., Doak, D.F., Williams, T.M. and Brownell, R.L. (2006) *Whales, Whaling, and Ocean Ecosystems*. University of California Press, Los Angeles, California.

Forestell, P.H. (1991) Marine education and ocean tourism: Replacing parasitism with symbiosis. In: Miller, M.L. and Auyong, J. (eds) *Proceedings of the 1990 Congress on Coastal and Marine Tourism, Volume I*. National Coastal Resources Research and Development Institute, Newport, Oregon, NCRI-T-91-010, pp. 35–39.

Forestell, P.H. (1992) The anatomy of a whalewatch: Marine tourism and environmental education. *Current: The Journal of Marine Education* 11, 10–15.

Forestell, P.H. (1993) If Leviathan has a face, does Gaia have a soul? Incorporating environmental education in marine eco-tourism programs. *Ocean and Coastal Management* 20, 267–282.

Forestell, P.H. (1995) Ensuring science based proposals for whale watching regulations are practicable. *MWW/95/46* presented to the Workshop on the Scientific Aspects of Managing Whale Watching. Montecastello de Vibio, Italy, March 30–April 4.

Forestell, P.H. (2002) Popular culture and literature. In: Perrin, W., Würsig, B. and Thewisson, H.G.M. (eds) *Encyclopedia of Marine Mammals*. Academic Press, San Diego, California, pp. 957–974.

Forestell, P.H. (2005) The times, they are a-changing: From Biophilia to nature deprivation disorder. Invited Address, National Marine Educators Association, Maui, Hawaii, July 15.

Forestell, P.H. and Kaufman, G.D. (1994) Resource managers and field researchers: Allies or adversaries? In: Postle, D. and Simmons, M. (eds) *Encounters with Whales '93*. Great Barrier Reef Marine Park Authority, Townsville, Queensland, Australia, pp. 17–26.

Forestell, P.H. and Kaufman, G.D. (1996) The development of whalewatching in Hawaii and its application as a model for growth and development of the industry elsewhere. In: Colgan, K. (ed.) *Encounters with Whales '95*. Australian Nature Conservation Agency, Canberra, Australia, pp. 53–65.

Gales, N., Hindell, M. and Kirkwood, R. (2003) *Marine Mammals: Fisheries, Tourism and Management Issues*. CSIRO, Collingwood, Victoria, Australia.

Gardner, G.T. and Stern, P.C. (1996) *Environmental Problems and Human Behavior*. Allyn & Bacon, Needham Heights, Massachusetts.

Garrod, B. and Fennell, D.A. (2004) An analysis of whalewatching codes of conduct. *Annals of Tourism Research* 31, 334–352.

Glockner-Ferrari, D.A. and Ferrari, M. (1990) Reproduction in the humpback whale (*Megaptera novaeangliae*) in Hawaiian waters, 1975–1988: The life history reproductive rates and behavior of known individuals identified through surface and underwater photography. *Reports of the International Whaling Commission* (Special Issue 12), 161–169.

Greene, H.W. (2005) Organisms in nature as a central focus for biology. *Trends in Ecology and Evolution* 20, 23–27.

Ham, S.H. and Krumpe, E.E. (1996) Identifying audiences and messages for nonformal environmental education – a theoretical framework for interpreters. *Journal of Interpretation Research* 1, 11–23.

Head, L., Trigger, D. and Mulcock, J. (2005) Culture as concept and influence in environmental research and management. *Conservation and Society* 3, 251–264.

Herman, L.M. (1979) Humpback whales in Hawaiian waters: A study in historical ecology. *Pacific Science* 33, 1–15.

Hilborn, R. (2005) Are sustainable fisheries achievable? In: Norse, E.A. and Crowder, L.B. (eds) *Marine Conservation Biology: The Science of Maintaining the Sea's Biodiversity*. Island Press, Washington, DC, pp. 247–259.

Higham, J.E.S. (1998) Sustaining the physical and social dimensions of wilderness tourism: The perceptual approach to wilderness management in New Zealand. *Journal of Sustainable Tourism* 6, 26–51.

Hoyt, E. (2001) *Whale Watching 2001: Worldwide Tourism Numbers, Expenditures, and Expanding Socioeconomic Benefits*. International Fund for Animal Welfare, Yarmouth Port, Massachusetts.

Hoyt, E. (2002) Whale watching. In: Perrin, W., Würsig, B. and Thewisson, H.G.M. (eds) *Encyclopedia of Marine Mammals*. Academic Press, San Diego, California, pp. 1305–1310.

Hoyt, E. (2005) *Marine Protected Areas for Whales, Dolphins and Porpoises: A World Handbook for Cetacean Habitat Conservation*. Earthscan, Sterling, Virginia.

Hunter, D. (2007) Cruise line agrees to fines over dead whale near park. *Anchorage Daily News* January 24, p B1. Available at: http://www.adn.com/news/alaska/wildlife/story/8586518p-8479603c.html

International Fund for Animal Welfare (IFAW) (1995) *Report of the International Workshop on the Scientific Aspects of Whale Watching*. Montecastello de Vibrio, Italy.

International Fund for Animal Welfare (IFAW) (1997) *Report of the International Workshop on the Educational Values of Whale Watching*. Provincetown, Massachusetts.

International Fund for Animal Welfare (IFAW) (1999) *Report of the International Workshop on the Socioeconmic Aspects of Whale Watching*. Kaikoura, New Zealand.

International Fund for Animal Welfare (IFAW) (2000) *Report of the Closing Workshop to Review Various Aspects of Whale Watching*. Tuscany, Italy.

International Whaling Commission (IWC) (2004) *Report of the Workshop on the Science for Sustainable Whalewatching*. Available at: http://www.iwcoffice.org/conservation/whalewatching.htm#workshops

Jackson, J.B.C. (2007) When ecological pyramids were upside down. In: Estes, J.A., Demaster, D.P., Doak, D.F., Williams, T.M. and Brownell, R.L. (eds) *Whales, Whaling, and Ocean Ecosystems*. University of California Press, Berkeley/Los Angeles, California, pp. 27–37.

Johnson, D. (2002) Environmentally sustainable cruise tourism: A reality check. *Marine Policy* 26, 261–270.

Kohlberg, L. (1981) *Essays on Moral Development, I: The Philosophy of Moral Development: Moral Stages and the Idea of Justice*. Harper & Row, San Francisco, California.

Kraus, S.D. and Rolland, R.M. (2007a) Right whales in the urban ocean. In: Kraus, S.D. and Rolland, R.M. (eds) *The Urban Whale*. Harvard University Press, Cambridge, Massachusetts, pp. 1–38.

Kraus, S.D. and Rolland, R.M. (2007b) *The Urban Whale*. Harvard University Press, Cambridge, Massachusetts.

Kraus, S.D. and Rolland, R.M. (2007c) The urban whale syndrome. In: Kraus, S.D. and Rolland, R.M. (eds) *The Urban Whale*. Harvard University Press, Cambridge, Massachusetts, pp. 488–513.

Kubota, R. (2004) Debate goes on over thrillcraft effect on whales. *Honolulu Star Bulletin*, October 31. Available at: http://starbulletin.com/2004/10/31/news/story7.html

Lavigne, D.M., Scheffer, V.B. and Kellert, S.R. (1999) The evolution of North American attitudes toward marine mammals. In: Twiss, J.R. and Reeves, R.R. (eds) *Conservation and Management of Marine Mammals*. Smithsonian Institution Press, Washington, DC, pp. 10–47.

Lemon, M., Lynch, T.P., Cato, D.H. and Harcourt, R.G. (2006) Response of travelling bottlenose dolphins (*Tursiops aduncus*) to experimental approaches by a powerboat in Jervis Bay, New South Wales, Australia. *Biological Conservation* 127, 363–372.

Leopold, A. (1949) *A Sand County Almanac and Sketches Here and There*. Oxford University Press, New York.

Lien, J. (2001) *The Conservation Basis for the Regulations of Whale Watching in Canada by the Department of Fisheries and Oceans: A Precautionary Approach*. Report to the Department of Fisheries and Oceans, Winnipeg, Manitoba, Canada.

Lilly, J.C. (1975) *Lilly on Dolphins*. Anchor Press/Doubleday, New York.

Lück, M. (2003) Education on marine mammal tours as agent for conservation – but do tourists want to be educated? *Ocean and Coastal Management* 46, 943–956.

Lusseau, D. (2003) Effects of tour boats on the behavior of bottlenose dolphins: Using Markov chains to model anthropogenic impacts. *Conservation Biology* 17, 1785–1793.

Mann, J. and Kemp, C. (2003) The effects of provisioning on maternal care in wild bottlenose dolphins, Shark Bay, Australia. In: Gales, N., Hindell, M. and Kirkwood, R. (eds) *Marine Mammals: Fisheries, Tourism and Management Issues*. CSIRO, Collingwood, Victoria, Australia, pp. 229–256.

Marion, J.L. and Reid, S.E. (2007) Minimising visitor impacts to protected areas: The efficacy of low impact education programmes. *Journal of Sustainable Tourism* 15, 5–27.

Marsh, H., Arnold, P., Freeman, M., Haynes, D., Laist, D., Read, A., Reynolds, J. and Kasuya, T. (2003) Strategies for conserving marine mammals. In: Gales, N., Hindell, M. and Kirkwood, R. (eds) *Marine Mammals: Fisheries, Tourism and Management Issues*. CSIRO, Collingwood, Victoria, Australia, pp. 1–18.

May, E. (1994) *We Come in Peace*, Eddie May, Urangan, Queensland, Australia.

Meffe, G.K., Perrin, W.F. and Dayton, P.K. (1999) Marine mammal conservation: Guiding principles and their implementation. In: Twiss, J.R. and Reeves, R.R. (eds) *Conservation and Management of Marine Mammals*. Smithsonian Institution Press, Washington, DC, pp. 437–454.

Meyers, R.A. and Ottensmeyer, C.A. (2005) Extinction risk in marine species. In: Norse, E.A. and Crowder, L.B. (eds) *Marine Conservation Biology: The Science of Maintaining the Sea's Biodiversity*. Island Press, Washington, DC, pp. 58–79.

Miller, M.L. (1993) The rise of coastal and marine tourism, *Ocean and Coastal Management* 20, 181–199.

New York Times (1984) R.M. Gilmore, whale expert, dies on brink of expedition. Associated Press, January 4. Available at: http://query.nytimes.com/gst/fullpage.html?res= 9901E4DD1338F937A35752C0A962948260

Nickerson, R. (1977) *Brother Whale*. Chronicle Books, San Francisco, California.

Norris, K.S. (1974) *The Porpoise Watcher*. W.W. Norton, New York.

Norse, E.A. (2005) Ending the range wars on the last frontier: Zoning the sea. In: Norse, E.A. and Crowder, L.B. (eds) *Marine Conservation Biology: The Science of Maintaining the Sea's Biodiversity*. Island Press, Washington, DC, pp. 422–443.

Norse, E.A. and Crowder, L.B. (2005a) *Marine Conservation Biology: The Science of Maintaining the Sea's Biodiversity*. Island Press, Washington, DC.

Norse, E.A. and Crowder, L.B. (2005b) Why marine conservation biology? In: Norse, E.A. and Crowder, L.B. (eds) *Marine Conservation Biology: The Science of Maintaining the Sea's Biodiversity*. Island Press, Washington, DC, pp. 1–18.

Norse, E.A., Crowder, L.B., Gjerde, K., Hyrenbach, D., Roberts, C.M., Safina, C. and Soule, M.E. (2005) Place-based ecosystem management in the open ocean. In: Norse, E.A. and Crowder, L.B. (eds) *Marine Conservation Biology: The Science of Maintaining the Sea's Biodiversity*. Island Press, Washington, DC, pp. 302–327.

Olesinski, R. (1994) Development of an environmental code and community awareness campaign for whale watchers. In: Postle, D. and Simmons, M. (eds) *Encounters with Whales '93*. Great Barrier Reef Marine Park Authority, Townsville, Queensland, Australia, pp. 75–81.

Orams, M.B. (1995a) Using interpretation to manage nature-based tourism. *Journal of Sustainable Tourism* 4, 81–94.

Orams, M.B. (1995b) Development and management of a wild dolphin feeding program at Tangalooma, Australia. *Aquatic Mammals* 21, 39–51.

Orams, M.B. (1997) The effectiveness of environmental education: Can we turn tourists into Greenies? *Progress in Tourism and Hospitality Research* 3, 295–306.

Orams, M. (1999) *Marine Tourism: Development, Impacts, and Management*. Routledge, London.

Paine, R.T. (2006) Whales, interaction webs, and zero-sum ecology. In: Estes, J.A., Demaster, D.P., Doak, D.F., Williams, T.M. and Brownell, R.L. (eds) *Whales, Whaling, and Ocean Ecosystems*. University of California Press, Berkeley/Los Angeles, California, pp. 7–13.

Parrish, J.K. (2005) Behavioral approaches to marine conservation. In: Norse, E.A. and Crowder, L.B. (eds) *Marine Conservation Biology: The Science of Maintaining the Sea's Biodiversity*. Island Press, Washington, DC, pp. 80–104.

Perrin, W.F. (1999) Selected examples of small cetaceans at risk. In: Twiss, J.R. and Reeves, R.R. (eds) *Conservation and Management of Marine Mammals*. Smithsonian Institution Press, Washington, DC, pp. 296–310.

Piaget, J. (1977) *The Development of Thought: Equilibration of Cognitive Structures* (translated by A. Rosin). The Viking Press, New York.

Ragen, T.J. (2005) Assessing and managing marine mammal habitat in the United States. In: Reynolds, J.E., Perrin, W.F., Reeves, R.R., Montgomery, S. and Ragen, T.J. (eds) *Marine Mammal Research: Conservation Beyond Crisis*. Johns Hopkins University Press, Baltimore, Maryland, pp. 125–136.

Ragen, T.J., Reeves, R.R., Reynolds, J.E. and Perrin, W.F. (2005) Future directions in marine mammal research. In: Reynolds, J.E., Perrin, W.F., Reeves, R.R., Montgomery, S. and Ragen, T.J. (eds) *Marine Mammal Research: Conservation Beyond Crisis*. Johns Hopkins University Press, Baltimore, Maryland, pp. 179–184.

Reynolds, J.E. (2005) The paradox of marine mammal science and conservation. In: Reynolds, J.E., Perrin, W.F., Reeves, R.R., Montgomery, S. and Ragen, T.J. (eds) *Marine Mammal Research: Conservation Beyond Crisis*. Johns Hopkins University Press, Baltimore, Maryland, pp. 1–4.

Reynolds, J.E., Perrin, W.F., Reeves, R.R., Montgomery, S. and Ragen, T.J. (2005) *Marine Mammal Research: Conservation Beyond Crisis*. Johns Hopkins University Press, Baltimore, Maryland.

Roberts, C.M. (2005) Marine protected areas and biodiversity conservation. In: Norse, E.A. and Crowder, L.B. (eds) *Marine Conservation Biology: The Science of Maintaining the Sea's Biodiversity*. Island Press, Washington, DC, pp. 265–279.

Robertson, R.B. (1954) *Of Men and Whales*. Alfred A. Knopf, New York.

Samuels, A., Bejder, L., Constantine, R., Heinrich, S. (2003) Swimming with wild cetaceans in the southern ocean. In: Gales, N., Hindell, M. and Kirkwood, R. (eds) *Marine Mammals: Fisheries, Tourism and Management Issues*. CSIRO, Collingwood, Victoria, Australia, pp. 265–291.

Santos, M.C. (1997) Lone sociable bottlenose dolphin in Brazil: Human fatality and management. *Marine Mammal Science* 13, 355–356.

Scheffer, V.B. (1969) *The Year of the Whale*. Scribner, New York.

Soggnes, K. (2003) On shoreline dating of rock. *Acta Archaeologica* 74, 189–209.

Soulé, M. (2005) Foreword. In: Norse, E.A. and Crowder, L.B. (eds) *Marine Conservation Biology: The Science of Maintaining the Sea's Biodiversity*. Island Press, Washington, DC, pp. xi–xv.

Swartz, S. (1989) The humpback whale. In: Chandler, W.J., Labate, L. and Wille, C. (eds) *Audubon Wildlife Report 1989/1990*. Academic Press, New York, pp. 387–403.

Thorne-Miller, B. and Catena, J. (1991) *The Living Ocean: Understanding and Protecting Marine Bio-Diversity*. Island Press, Covelo, California.

Twiss, J.R. and Reeves, R.R. (1999) *Conservation and Management of Marine Mammals*. Smithsonian Institution Press, Washington, DC.

US Department of the Interior (1953) Whales moving to wintering ground. Fish and Wildlife Service, January 23. Available at: www.fws.gov/news/historic/1953/19530123.pdf

Watkins, W.A. (1986) Whale reactions to human activities in Cape Cod waters. *Marine Mammal Science* 2, 251–262.

Worm, B., Lotze, H.K. and Myers, R.A. (2007) Ecosystem effects of fishing and whaling in the North Pacific and Atlantic Ocean. In: Estes, J.A., Demaster, D.P., Doak, D.F., Williams, T.M. and Brownell, R.L. (eds) *Whales, Whaling, and Ocean Ecosystems*. University of California Press, Los Angeles, California, pp. 334–341.

16 Wildlife and Tourism in Antarctica: A Unique Resource and Regime for Management

P.T. Maher

Introduction

Antarctica is one of the most beautiful and remote places on the planet. The moniker of being the highest, driest, coldest, iciest, windiest, most remote continent, surrounded by the stormiest ocean is well deserved, yet it also acts as quite a draw. Antarctica is a place of mystery, a place of historic exploration, a place where huge icebergs sweep by populous penguin rookeries and where majestic albatross sweep along on wind curling off the polar plateau. These preconceived notions are perhaps why Antarctic tourism has grown substantially since the mid-1980s, now numbering over 30,000 visitors each year (see IAATO, 2007a). Antarctic wildlife may be plentiful across a vast territory of wilderness, but its wildlife is not diverse. The questions are now whether: (i) tourism and wildlife are compatible in the Antarctic; (ii) tourism can support and conserve Antarctic wildlife; and (iii) Antarctic wildlife can support current or increased tourism. This chapter will attempt to reveal and combine some of the known information, but also examine how the balance can be struck to manage a plentiful, unique resource in a regime that is itself a unique situation.

Wildlife

Antarctica covers 50 million km^2, including the surrounding Southern Ocean. The continent alone is 14 million km^2 (Cessford, 1997), which is roughly the size of the Arctic Ocean, or the USA and Mexico combined. Being 98% covered with ice it is easy to imagine why wildlife and tourists come into a position of conflict – much wildlife needs ice-free areas for breeding, and tourists like to visit such areas precisely to experience wildlife 'in action'. Unique because of its harsh physical climate, Antarctica is also notable for its unusual ecology. Consider these facts:

- From diatom, a one-celled organism, to the largest of all animals, the blue whale, there is only one step in the food chain (Campbell, 1993), which Shirihai (2002) terms a short cut in the system.
- If one leaf of one Amazonian palm was counted for mosses, fungi, lichens, mites and insects, there would be more species on it than are found on the entire Antarctic continent (Campbell, 1993).

What the Antarctic ecosystem lacks in terms of diversity, it makes up for in numbers. Chester (1993), using the Scientific Committee on Antarctic Research data, states there are the following populations in Antarctica:

- 1 million pairs of breeding King penguins;
- 2.5 million pairs of Adélie penguins;
- 7.5 million pairs of Chinstrap penguins;
- 3.7 million pairs of Rockhopper penguins (mainly in the subantarctic);
- 315,000 pairs of Gentoo penguins;
- 12 million pairs of Macaroni penguins;
- 200,000 pairs of Emperor penguins (see Fig. 16.1);
- Between 250,000 and 800,000 Weddell seals;
- 200,000 Ross seals;
- 30–70 million Crabeater seals;
- 400,000 leopard seals;
- 600,000 Southern elephant seals;
- 2 million Antarctic fur seals.

These numbers do not even consider the numerous populations of whales, albatross, petrels and krill. Shirihai (2002) states that 37 species of cetaceans are found in Antarctic waters (almost half of the world's recognized species).

Fig. 16.1. Emperor Penguins on the sea ice at McMurdo Station. (Photograph P. Maher.)

The overwhelming marine focus of Chester's (1993) list is due to the fact that the largest terrestrial Antarctic wildlife is 2 mm long springtails and mites (Waterhouse, 2001).

Wildlife Management

Politically and managerially the continent of Antarctica, and thus its wildlife, is treated uniquely because of the difference to other continents. Antarctica is a neutral territory with no military presence other than that used to support scientific research. Although claims of national sovereignty have been made, these have been held in abeyance for several decades, and Antarctica is currently under the international regime of the Antarctic Treaty System (ATS). The ATS effectively provides legal status to all land and resources of the entire Antarctic continent (Hall and Johnston, 1995).

As a management regime, the ATS allows Antarctica to be recognized as a shared resource for all humankind to promote peaceful and scientific purposes and covers the area south of 60°S. Specifically, the Antarctic Treaty states 'that it is in the interests of all mankind that Antarctica shall continue forever to be used exclusively for peaceful purposes and shall not become the scene or object of international discord' (Antarctic Treaty, 2002). Furthermore, 'the continuance of international harmony in Antarctica will further the purposes and principles embodied in the Charter of the United Nations' (Antarctic Treaty, 2002). Thus, the ATS prohibits military measures and establishes freedom of scientific investigation, cooperation and access to all areas. It also prohibits nuclear explosions and disposal of radioactive waste in all of the area defined as Antarctica.

Although the Treaty does not recognize, dispute or establish territorial claims, insomuch no claims are to be asserted while the ATS is in force (Antarctic Treaty, 2002). Together, the 45 nations in the ATS represent decisions made regarding the continent by two-thirds of the world's population. Prosser (1995) believes that the development of the ATS may have resulted from the simple fact that, during the 1950s and 1960s, nations involved saw little economic potential or otherwise for the continent and thus lacked foresight. However, as described by Davis (1992, p. 39), the Antarctic Treaty is today 'one of the most successful international regimes of our time'.

Under the umbrella of the ATS, there are several specific international agreements, which cover additional avenues of concern for Antarctica. The Antarctic Treaty itself was established by the United Nations in 1959 following the International Geophysical Year (IGY, 1957–1958), but was not ratified until 1961. In 1964, the ATS adopted the first major Antarctic conservation regime, the *Agreed Measures for Conservation of Antarctic Flora and Fauna*. Under this, two types of special conservation areas were considered, Specially Protected Areas (SPAs) and Sites of Special Scientific Interest (SSSIs). SPAs preserve both unique and representative examples of the natural ecological systems of areas, which are of outstanding scientific interest. SSSIs protect any kind of scientific investigation or set aside undisturbed reference areas for the needs of a particular science. SSSIs can only be designated where there is a demonstrable risk of harmful interference. These designations are relatively small in size and number, with little management planning and effective implementation (Lucas, 1995). Thus, successive additional designations and governance of Antarctic wilderness have been and are necessary.

The *Protocol on Environmental Protection to the Antarctic Treaty* (Madrid Protocol) is the 1991 agreement by ATS nations that deals with the specifics of environmental management, and promotes Antarctica as a scientific vessel for global understanding. The Protocol sets regulations regarding activities, duration, impact, protection and adverse effects and changes for a number of areas. Essentially, it enhances environmental standards set out in the ATS.

Annex V of the Madrid Protocol sets out the types of values to be considered when assessing whether an area warrants special protection. It also describes the process for preparing and submitting a draft management plan through the Committee for Environment Protection to the Antarctic Treaty Consultative Meetings. Under Annex V, there is designation as Antarctic Specially Protected Areas (ASPAs) and Antarctic Specially Managed Areas (ASMAs) (see Bastmeijer, 2002; Bastmeijer and Roura, 2004). ASPAs are intended to protect areas to be kept free of human impact for comparative purposes, representative examples of major ecosystems, places with important or unusual animal or plant communities, type localities or only known habitats of species, places of value for scientific research, places with outstanding landform attributes, areas of outstanding aesthetic and wilderness value and places of historic value. SPAs and SSSIs are combined as ASPAs.

ASMAs provide a framework for managing activities so as to improve coordination of different activities and minimize environmental impacts. They may include areas where activities pose risks of mutual interference or cumulative environmental impacts. They may also include places of historical significance. ASMA status is available under Annex V to assist in the coordination of activities and the minimization of environmental impacts for areas of greater activity, or areas where more than one operator is active.

Before the Protocol, international concern regarding seal populations led to the *Convention for the Conservation of Antarctic Seals* (1972), fishing rights and catch sizes led to the *Convention for the Conservation of Antarctic Marine Living Resources* (1980), while possible mineral exploitation led to the *Convention to Regulate Antarctic Mineral Resource Activities* (1988). There was also proposed the idea of a Worldpark, which became a significant Antarctic conservation issue between 1981 and 1984 at successive International Union for Conservation of Nature and Natural Resources meetings with non-governmental organization support (Lucas, 1995). The Worldpark designation would have provided overriding protection of Antarctica, although its failure probably sparked some of the debate that led to the Madrid Protocol.

In addition, many nations who have signatory status in the ATS also have specific domestic laws to regulate their citizen's activities in Antarctica and thus their interactions with wildlife (see Bastmeijer, 2002). Under the auspices of the International Association of Antarctica Tour Operators (IAATO), there are also industry self-regulatory tourism management mechanisms that directly affect wildlife; however, these will be discussed later in relation to tourism activity and tourism management.

Tourism

> We cannot build a barrier around the Antarctic and keep tourists or the science community out. The Antarctic Treaty grants us all freedom of access to Antarctica. With that freedom comes a responsibility which we all share.
>
> (Landau, 2000, p. 15)

Responsibility encompasses proper management practices, creating a balance for tourism and wildlife. Tourism in Antarctica has traditionally been defined to include activities such as:

- Commercial sea-borne operations, accessing coastal sites;
- Private yacht visits;
- Continental overflights;
- Flights to King George Island, Patriot Hills or the South Pole for land-based operations.

However, argument over such definition (either as being to inclusive or too exclusive) does occur in previous research (see Enzenbacher, 1992; Benson, 2000; Bauer, 2001).

While Antarctic tourism is not a recent phenomenon as hundreds of thousands of tourists have visited continuously since 1965; it is still small in scale compared to global tourism. Often portrayed as a new pressure on the southern polar region, it is quite possible that tourism activity had simply been overlooked until the huge growth over the last two decades. Lars-Eric Lindblad began large-scale ship-borne tourism in 1966, but tourists had made landings on subantarctic islands as early as 1882 and by 1933 most large subantarctic islands surrounding the continent had been visited (Headland, 1994).

Landau and Splettstoesser (2007) provide an explicit look at the growth of tourism in Table 16.1, which collates data from IAATO reports to the Antarctic Treaty Consultative Meetings (1992–2006). Remaining fairly constant, however, is the fact that visitors to Antarctica today are from a wide variety of nations, but are still typically first-world citizens. In 2005/06, 39% were from the USA, 15% from the UK, 10% from Germany, 8% from Australia, 6% from Canada, 3% from the Netherlands, 2% from Switzerland, 2% from Japan and the remaining 14% from other countries (all numbers rounded) (IAATO, 2007a). Typically, these tourists are tertiary educated, well travelled, have high disposable incomes and are looking for a unique nature-based experience (Kriwoken and Rootes, 2000).

Geographically, visits to the continent are highly concentrated, with less than 0.5% of the continental area visited; this is an area measuring only 56,000 km², or about the size of the Canadian province of Nova Scotia. Overall, the sites are widely dispersed around the continent, but the Antarctic Peninsula takes 90% of the tourist activity (Cessford, 1997). A comparison of

Table 16.1. Numbers of operators, ships, voyages and passengers involved in Antarctic tourism 1992–2007. (From Landau and Splettstoesser, 2007.)

Year	Operators (Owners and Chartering Companies	Vessels	Voyages	Passengers landed	Passengers on non-landing vessels
1992/1993	10	12	59	6,704	0
1993/1994	9	11	65	7,957	0
1994/1995	9	14	93	8,098	0
1995/1996	10	15	113	9,312	0
1996/1997	11	13	104	7,322	0
1997/1998	12	13[a]	92[a]	9,473	0
1998/1999	13	15[a]	116	9,857	0
1999/2000	17	22	153	14,623	936
2000/2001	15[a]	32	131	12,109	0
2001/2002	19[a]	37	117	11,429	2,029
2002/2003	26	47	136	13,263	2,424
2003/2004	31	51	180	19,369	4,949
2004/2005	35	52	207	22,294	5,027
2005/2006	47	44	249	25,167	4,632
2006/2007	45	50	278	27,575	7,500

2006/07 figures are based on estimates received from operators as of mid-2006.
[a]Small numbers of yachts reported in seasons that are not included in totals.

the 2005/06 season underscores the bias towards peninsula visits; the most visited site in the peninsula region was the Lemaire Channel with 33,535 visits, while the most visited site continentally was Coulman Island with 1189 total tourists (IAATO, 2007a). Headland (1994) estimated that the 'footprint' of tourists is less than 1% that of science programmes in Antarctica, mainly because of the ship as hotel aspect of tourism, and because it is generally a summer-only activity. This percentage is likely to have changed quite drastically since the early 1990s given the overall growth in tourism.

Antarctic tourism is a marine activity, based upon how much of its activity deals with large vessels, yachts and small boats for shore visits. An additional number visit Antarctica on overflights without landings, and a smaller number are flown to the interior for adventure tourism. To some it is quite important to distinguish between the total numbers of tourists who visit Antarctica, and those who actually set foot on land (see Table 16.1).

In any case, most landings occur primarily in coastal areas where marine wildlife and scenery are the major attractions, and the transport of passengers from tour vessels to shore is done by rubber inflatable boat. There is some shore transport done by helicopter, but this is largely dependent on the vessel (only a few tourist vessels carry helicopters), and is also generally done at scientific bases such as those in the Ross Sea and Australian Antarctic Territory. Shore visits normally last for 2–3h, but can be as long as 5–6h in the case of ships with capacities in excess of 500 passengers (IAATO operators conduct shore activities in accordance with IAATO guidelines that permit no more than 100 passengers ashore at any one time). Shore visits usually involve passengers being organized into groups and being led by expedition staff for the purposes of interpretation, and maintaining approved and safe distances from wildlife.

While the numbers who set foot on shore may be one way used to differentiate visitors, it is also meaningful to examine Antarctic tourism divided into the categories of ship-borne, land-based and airborne tourism (Hall and Johnston, 1995), of which as stated earlier only ship-borne tourism is directly linked to marine wildlife. However, given that nearly all Antarctic tourism is solely based on viewing marine wildlife, all three categories will be discussed.

Ship-borne tourism

With closer proximity and less time crossing the Southern Ocean, ship-borne tours from South America to the Antarctic Peninsula are much cheaper and friendlier, in terms of comfort, than those from New Zealand or Australia to the Ross Sea Region (Hall and Wouters, 1995). The ease of transport, distance and a milder marine climate have led scientists to refer to the peninsula as the 'Banana Belt' (Campbell, 1993), and with the build-up of tourism the peninsula had also been dubbed the 'Antarctic Riviera' (Hart, 1988).

To highlight why this is, consider that Cape Horn at the tip of South America extends to approximately 56°S, and the closest destination related to the Antarctic Peninsula (the South Shetland Islands) are only at 62°S. This is in stark contrast to the Ross Sea side of the continent, where Bluff, New Zealand lies at approximately 46°S, and destinations such as Cape Adare (71°S) and Ross Island (77°S) are considerably further away. From Ushuaia, the Antarctic Peninsula can be reached in as little as 48h, whereas from New Zealand and Australia to the Ross Sea Region the voyage may take as long as 10 days. To break up this 10-day voyage operators stop at Australian and New Zealand subantarctic islands along the way (see Fig. 16.2).

Also possible in the peninsula region are private yacht tours, which create a difficult situation for IAATO and the ATS in that numbers are increasing and the activity of yachts is much more difficult to regulate and monitor (Splettstoesser, 1999). Yacht tours will remain popular in Antarctica because of price and flexible schedules, but to many ATS signatories such tours are much more of an environmental threat than any other type of tourism (Splettstoesser, 1999). The precise reason for this concern is perhaps counter-intuitive, in that large vessels would leak more fuel if damaged, but yachts are not nearly as controlled in terms of itineraries and keeping to a set schedule, and are also more susceptible to damage in high seas and around sea ice.

In 1970, Lars-Eric Lindblad built the *Lindblad Explorer*, the first polar vessel constructed specifically for tourist purposes (Benson, 2000). Having gone through various name and ownership changes, the *M/S Explorer* still remains a leader in Antarctic tourism (Headland, 1994). Two other important vessels in ship-borne Antarctic tourism history are the *Bahia Paraiso* and the *Kapitan Khlebnikov* (KK). The *Bahia Paraiso* was an Argentine naval resupply vessel that additionally carried tourists between Ushuaia and King George Island in the South Shetland Islands chain. On January 28, 1989, the *Bahia Paraiso* became grounded in Arthur Harbour near the US Palmer Station, was then abandoned

Fig. 16.2. Tourists landed on Australia's Macquarie Island. (Photograph
P. Maher.)

and eventually sunk (Headland, 1994). The logistics of rescue and tourist management during this incident led to a closer examination of Antarctic tourism
and, in turn, likely spurred the formation of IAATO in 1991 (Splettstoesser,
1999). As a note, recent (2000 and onwards) increases in incidents (e.g. ground-
ings) have led to further debate on the logistics of rescue again.

The KK was the first vessel to circumnavigate Antarctica after 2 months at
sea (Splettstoesser *et al.*, 1997), but more importantly it is among a class of
ex-Soviet vessels that became available for Antarctic tourism charters. Cruises
by these vessels began in the 1992/93 summer and an entirely new dimension
was added to tourism with this new versatility of vessels (icebreakers) and prospects of reaching new areas of Antarctica that were otherwise inaccessible by
tour ships (utilizing on-board helicopters). For example, icebreakers are capable
of breaking ice in most coastal areas throughout the summer (prior to 1992/93
and even today ships travelling to the Ross Sea are routinely held up due to the
highly variable sea ice conditions), and helicopters can provide flight-seeing for
passengers, as well as transport to Emperor Penguin breeding colonies, a bird
species not often seen on tourist cruises by conventional Antarctic vessels.

The KK (see Fig. 16.3) in its 1996/97 circumnavigation also opened up a
new possibility. Given the interest shown by tourists in seeing both the wildlife

Fig. 16.3. The *Kapitan Khlebnikov* (KK) in McMurdo Sound. (Photograph P. Maher.)

of the Antarctic Peninsula and the history of the Ross Sea region, it is likely that circumnavigations of the continent, such as that completed by the KK, or at least partial circumnavigation will increase in popularity.

Land-based tourism

The building of a 1300 m hard runway at the Chilean Tiente Rodolfo Marsh Station on King George Island in 1979/80 signalled the ability for land-based and airborne tourism to be able to operate in the Antarctic (Benson, 2000). On 8 January 1982, a group of 40 tourists flew to Marsh Station to stay prior to boarding a cruise (Swithinbank, 1992). From 1982 to 1992, Chile operated the *Hotel Estrella Polar*, a converted, 80-bed military barrack at Marsh Station, which served as a rest spot for tourists between cruise ships and tourists flights to King George Island (Headland, 1994). Both the Chilean military and commercial operators offered flights to the 'hotel'. Excursions to nearby attractions were conducted as well. Following the cessation of Chile's polar hotel operations, Argentina began flying tourists to its base on Seymour Island, but today all such accommodations have reverted to official use.

Today, land-based tourism in Antarctica generally centres around one particular company, Adventure Network International (ANI). ANI operates a tented summer camp at Patriot Hills in the Ellsworth Mountains, which can accommodate 50 people and takes advantage of a natural blue ice runway to land large Hercules aircraft (Benson, 2000). From Patriot Hills, ANI operates a service using smaller planes to Vinson Massif, the South Pole and numerous glaciers and Emperor Penguin colonies (Benson, 2000; Kriwoken and Rootes, 2000). In 1997/98, ANI carried 131 passengers to Antarctica with eight Hercules flights being made between Punta Arenas and Patriot Hills (Swithinbank, 1998). Four years later, 2001/02, ANI only carried 159 of the total 11,588 tourists who landed in Antarctica that season (IAATO, 2007a).

Airborne tourism

Ship-borne and land-based tourism may include elements of airborne tourism. Air travel from ships is limited to those vessels equipped with helicopters such as the KK. Figure 16.4 shows tourists dropped by helicopter at New Zealand's Scott Base. These tourists were also given flight-seeing tours of Mt Erebus, before being driven over to the US McMurdo Station, and meeting the KK docked nearby. ANI's airborne tourism is primarily a means of transporting visitors and goods rather than offering sightseeing as found on overflights (Benson, 2000).

This category of tourist travel currently consists primarily of continental overflights from Australia or Chile, and in the past from New Zealand. Overflights began in 1956 with LAN Chile flying over the South Shetland Islands (Stonehouse and Crosbie, 1995). No regular flights were made over Antarctica until February 1977, when both Qantas and Air New Zealand began operations (see Swithinbank, 1992; Kriwoken and Rootes, 2000). Both companies flew extensively through 1979 with a total of 11,145 passengers and 43 flights

Fig. 16.4. Tourists at New Zealand's Scott Base. (Photograph P. Maher.)

(Reich, 1980). The journey involved in these overflights was 11 h in duration from New Zealand or Australia; the actual overflight of the continent lasted a total of 90 min (Reich, 1980). Overflights ceased on 28 November 1979, when Air New Zealand flight TE901 crashed into Mt Erebus on Ross Island, killing all 257 passengers and crew aboard (MacFarlane, 1991). Resuming in 1994/95, overflights are now being organized by Croydon Travel in Australia and LAN Chile.

Tourism Management

Today, management of tourism in Antarctica is essentially the combined responsibility of Antarctic Treaty Consultative Parties and IAATO. As all human activities in the Antarctic are subject to the collective ATS, tourism is no different. The ATS provides a complex framework for the governance of Antarctica, south of 60° S. Management of Antarctica, for wildlife or tourism, is thus established for a special kind of global commons. Having discussed the ATS extensively in the previous section on wildlife management, this section will focus on the IAATO side of tourism management.

Formed in 1991, IAATO's mission is 'to advocate, promote and practice safe and environmentally responsible private-sector travel to the Antarctic' (IAATO, 2007b). With the growth of tourism since the late 1980s/early 1990s, management strategies developed by IAATO have enabled its members to be better prepared for their activities. The original seven companies have now expanded to 84 companies from 14 countries (plus the Falkland Islands/Islas Malvinas) (Landau and Splettstoesser, 2007).

The secret to IAATO's success is that as interest in Antarctica has grown and more companies have marketed this remote area of the world, management strategies have also evolved to cope with increasing numbers of ships and tourists. These strategies have required a dynamic, interactive process that includes cooperation between IAATO, Antarctic Treaty Parties and other experts (Landau and Splettstoesser, 2007).

When IAATO was formed the tourism situation was relatively simple, with six ship-borne operators and one land-based operator as members, and only a few vessels conducting cruises in Antarctica each austral summer. Today, tourist activities have evolved and diversified, operators offer much more than traditional cruising with Zodiacs for shore visits, frequently offering kayaking, camping, scuba diving, skiing, mountain climbing and even marathons (IAATO, 2007b).

Continued growth over several years has required additional effort to ensure only minor or transitory impact occurs to the Antarctic environment as well as improved guidelines for wildlife protection, safety at sea, communications, contingency plans and shipping operations (fuel type, ballast restrictions, etc.). A major consideration in response to the growth in the number of vessels and tourist numbers relates to continuing protection of the environment and awareness that the continent is dedicated to the conduct of science. IAATO respects these facts as a visitor to Antarctica and also takes its responsibility for the management of tourism seriously. Education of passengers is a key component of IAATO's objectives for the industry, as well-informed individuals may share the attitude of environmental protection as much as the tour operators do.

IAATO provides a single voice for liaison and dialogue with Antarctic Treaty Consultative Parties and other related organizations; a means of data reporting and collation for use in determining trends; standardized procedures in field operations; and a secretariat (Executive Director) to coordinate activities for all members.

Much of the contention that is attached to any activity in Antarctica is related to environmental issues, and how best to visit and leave no impact. Everyone agrees that there should be little or no impact arising from any activity. However, in today's reality, management of numerous companies, operating 40–45 vessels and carrying more than 30,000 tourists to Antarctica, requires many strategies in order to ensure that all participants are following the same guidelines and continuing to protect the environment. Some management examples utilized by IAATO are given below.

Bylaws

IAATO bylaws provide the framework for operating successfully and also help ensure compliance with treaty practices. These bylaws are amended whenever changes are needed and 2006 bylaws are available online (IAATO, 2007c).

Categories of membership

Up until June 2001, operators of ships carrying more than 400 passengers to Antarctica were denied membership in IAATO. As increasingly larger ships came on the scene, members agreed to develop new categories of membership that would accommodate these new aspects of operations. There are currently seven categories of membership (Landau and Splettstoesser, 2007). During the 2005/06 operating season, the numbers of members in each category are shown in parentheses (numbers from Landau and Splettstoesser, 2007):

1. Organizers of expedition ships or yachts that carry fewer than 200 passengers are required to limit to 100 the number of passengers ashore at any one time (29 members).
2. Organizers of vessels carrying 200–500 passengers who are intending to land passengers. Stringent restrictions on landing activities of time and place could apply. The limit of 100 passengers ashore at one site at one time also applies (4 members).
3. Organizers of cruise ships making no landings (cruise only). Cruise ships carrying more than 500 passengers are not permitted to make landings (3 members).
4. Organizers of land-based operations (2 members).
5. Organizers of over-flight operations (no landings) (2 members).
6. Organizers of the combination of air and cruise operations (1 member).
7. Members consisting of organizations and individuals interested in or promoting travel to Antarctic and who choose to support IAATO objectives (34 members).

The above categories, depending on organizer interests and types of activities, can be further grouped into any of the following types of membership (with further details given by IAATO, 2007c):

1. *Full Members* are experienced organizers who operate travel programmes to the Antarctic and who pledge to abide by IAATO bylaws, agree to the above-mentioned categories and to not have more than 100 passengers ashore at any one site at the same time, and maintain a staff to passenger ratio of 1:20 ashore.
2. *Provisional Members* are organizers who operate travel programmes to the Antarctic who are requesting full membership in IAATO. An observer is required on a voyage in the first season of operation.
3. *Probationary Members* are current or past full or provisional members who have not fully complied with IAATO bylaws or who otherwise are not in good standing in the view of full members.
4. *Associate Members* are other organizations and individuals interested in or promoting travel to the Antarctic and choose to support IAATO objectives.

Site-specific guidelines

As more visits have been made to selected sites that offer numerous attractions for tour operators and their passengers, a need arose for IAATO to ensure that these popular sites had guidelines that pertained to the vulnerability of the wildlife present as well as the vegetation. These were in addition to IAATO's general visitor guidelines (see IAATO, 2007e), as procedures were needed to ensure that over-visitation would be avoided with regard to the carrying capacity of the site. Specific guidelines and monitoring were thus developed for these locations, in discussions with ATS parties and IAATO members (IAATO, 2007f).

Briefing of passengers and crew

IAATO's Code of Conduct dates to 1991 when IAATO was formed and has been adopted and incorporated in Treaty Recommendation XVIII-1, *Guidance for Tour Operators and Visitors to Antarctica*, in order to brief passengers and crew regarding procedures while ashore, including actions around wildlife as one example (see IAATO, 2007e). Because of the diversity of nationalities and languages among tourists now, these guidelines have been printed in nine languages. Printed copies of the guidelines are provided to all passengers and crew prior to start of the itineraries, and are also referred to onboard with a formal PowerPoint or other presentation to emphasize the importance of each part (see IAATO, 2007e).

Marine wildlife watching guidelines

This is a major component of present-day IAATO management procedures, especially with regard to marine wildlife–tourist interaction. In order to ensure protection of wildlife in the marine environment, special guidelines have been developed to provide precautions for the presence of whales, seals and seabirds while on itineraries (IAATO, 2007d; see Fig. 16.5). Attention to the presence of marine mammals and seabirds also provides the advantage for tour naturalists to conduct

IAATO MARINE WILDLIFE WATCHING GUIDELINES
(WHALES, DOLPHINS, SEALS AND SEABIRDS)
FOR VESSEL & ZODIAC OPERATIONS

INTRODUCTION

In 2002, the International Association of Antarctica Tour
Operators (IAATO) initially developed Marine Wildlife
Watching Guidelines to provide guidance to vessel operators
while viewing cetaceans, seals, and birds in their marine
environment.

These guidelines minimize potential environmental impacts to
wildlife and suggest ways to comply with Annex II
(Conservation of Antarctic Fauna and Flora) of the Protocol on
Environmental Protection to the Antarctic Treaty. The
guidelines do not replace any domestic governmental laws, but
provide an additional code of conduct to help reduce potential
disturbance to the marine environment. Some countries have
guidelines or regulations stricter than these which may override
these guidelines. Violation of national regulations may be
punishable by fines, imprisonment and, in extreme cases, seizure
of vessel. Members/operators of IAATO should be aware that
compliance with the IAATO guidelines might be insufficient to
prevent violation of, and penalties resulting from, national laws
and regulations.

Compliance with the International Regulations for Preventing
Collisions at Sea has priority over these guidelines at all times.

A. These Guidelines Provide Standard Operating Procedures which:

1. Are intended to be used by IAATO members operating:

- Any type of vessel (e.g. ship, sailboat, yacht,
  Zodiac, small boats, kayak, etc. Note: The use of jet-
  skis, surfboards or windsurfers should not occur in
  areas of known wildlife);
- By the officers, crew, expedition staff and visitors
  involved in navigating in wildlife-rich areas during
  viewing sessions; and

2. Aim to:

- Minimize wildlife disturbance;
- Protect cetaceans, seals and seabirds while ensuring a
  high quality wildlife-watching experience through
  responsible observation. (Many passengers are
  concerned about the welfare of wildlife and expect high
  standards of conduct by operators);
- Avoid harmful impacts on marine wildlife populations
  by ensuring that the normal patterns of daily and
  seasonal activity of the animals are maintained in the
  short and long term.

*Competent, careful boat handling avoids harming wildlife and
leads to better wildlife watching.*

B. Possible Impacts from Vessels

Possible negative impacts from vessel operations include
physical injury, interference with or disruption of normal
behavior, stress, underwater noise and possibly increased
exposure to predators. In addition, animals could be exposed to
increased levels of environmental contaminants such as oil from
leaking outboard engines and discharged bilges. The
recommended guidelines will help minimize the level of
potential disturbance and should prevent the following from
occurring:

- Displacement from important feeding areas;
- Disruption of feeding;
- Disruption of reproductive and other socially important

Fig. 16.5. IAATO Marine Wildlife Watching Guidelines. (From IAATO, 2007d.)

behaviors;

- Changes to regular migratory pathways to avoid human interaction zones;
- Stress from interaction;
- Injury;
- Increased mortality or decreased productivity/survivorship (and therefore population decline).

C. Approaching Marine Mammals and Seabirds

General Principles

The animal/s should dictate all encounters.

It is very important for vessel operators to be able to evaluate the animal/s' behavioral patterns. This can be difficult in practice and a good reason to have experienced ships officers and naturalists onboard.

The guidelines take into account the approach towards the animals, arrival at and departure from an optimal viewing area, and recommended distances from the animals. Sometimes an animal will approach a vessel. If a marine mammal wants to interact, it may remain with the vessel. The vessel can then drift passively. If the animal is moving away from the vessel, it is choosing not to interact with, or approach, the vessel.

Take all care to avoid collisions. This may include stopping, slowing down, and/or steering away from the animal/s. Do not chase or pursue animals.

The following principles address vessels in general:

1. Cetaceans (Whales, Dolphins, Porpoises)

Cetaceans should never be approached head-on. Ideally, they should be approached from slightly to the side and rear of the animal (see Figure 1). Once traveling with the animal, travel parallel with it/them.

Figure 1.

Correct approach angles

1a. Vessels, Officers, Crew, Expedition Staff:

- Keep a good lookout forward (and ideally on the sides and from the stern) where cetaceans may be present.
- Always give the animals the benefit of the doubt.
- Avoid sudden change in speed and direction (including putting vessel in reverse).
- Avoid loud noises, including conversation, whistling, etc.
- Keep radios on a low volume setting.
- Should a vessel get closer than the recommended minimum distance, withdraw at a constant, slow, no-wake speed, to at least the recommended minimum distance.
- If animals approach the vessel, put engines in neutral and do not re-engage propulsion until they are observed well clear of your vessel. If the animals remain in a local area, and if it is safe to do so, you may shut off the vessel's engine. Some whales will approach a silent, stationary vessel.

(Note: Allowing a vessel to drift within accepted recommended distances could constitute an intentional approach.)

1b. Awareness of the Animal/s' Behavioral Patterns:

Use your best judgment. Animals may alter their behavior if they are disturbed by your activities. When in doubt, err on the side of caution, and give animals time and space. If the cetacean is agitated or no longer interested in staying near the vessel, the following behavioral changes may be observed:

- Changes in traveling direction.
- Regular changes in direction, or speed, of swimming.
- Moving away from the area.
- Apparent general agitation.
- Hasty dives.
- Changes in respiration patterns.

Fig. 16.5. *Continued*

- Increased time spent diving compared to time spent at the surface.
- Changes in acoustic behavior.
- Certain surface behaviors such as tail or pectoral fin slapping or trumpet blows.

1c. General code of conduct around marine mammals

- Do not stay with the animal/s too long, with a suggested maximum time of 1 hour. If signs of disturbance or change in behavior occur at any time during the stay with the animals, retreat slowly and quietly.
- Never herd (circle), separate, scatter, or pursue a group of marine mammals, particularly mothers and young.
- If a cetacean approaches a vessel to bow-ride, maintain a relatively constant course and speed. Do not enter a group of dolphins to encourage them to bow-ride.
- If a cetacean surfaces in the vicinity of your vessel, take all necessary precautions to avoid collisions, while avoiding sudden changes in speed or direction. This may include, slowing down, slowly coming to a stop, and/or steering away from the animal.
- If a cetacean comes close to shore, or your boat, remain quiet.
- Avoid sudden movements that might startle the cetacean.
- In line with Recommendation XVIII-1 and IAATO general codes of conduct never attempt to touch or feed animals.
- Playback of underwater sound of any kind should not occur. This includes recorded whale or dolphin sounds. If hydrophones are used from small boats to listen to the underwater sounds it is preferable to have the engines of the small boats shut down. The sounds can be listened to with headphones or mini speakers and can also be recorded. There are a number of sites on the Internet that offer hydrophones and recording equipment for sale.

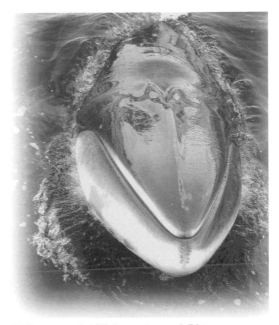

1d. Recommended Minimum Approach Distances:

- No intentional approach within:
  - ~ 30 meters or 100 feet for small boats (including kayaks);
  - 100 meters or 100 feet for small boats (including kayaks) if cetaceans communally feeding;
  - ~ 100 meters or 300 feet for ships,
  - ~ 150m/500 ft. if ship over 20,000 tons;
  - ~ 200m/600 ft. if 2 ships present.

- Helicopters or any aircraft should not approach closer than 300 meters or 1000 feet vertical distance. *Aircraft should cease contact if the animals repeatedly dive or increase speed.*

1e. When Whales Are Sighted:
Approximately 1500 to 3000 meters / one to two miles away

- Reduce speed to less than 10 knots.
- Post a dedicated lookout to assist the vessel operator in monitoring the location of all marine mammals.

1500 to 750 meters / one to one-half mile away

- Reduce speed to 5 knots.

Fig. 16.5. *Continued*

Approximately 750 meters / half a mile or closer

- Reduce speed to less than 5 knots.
- Maneuver vessel to avoid a head-on approach.
- Avoid sudden gear changes (*e.g.* into reverse).

1f. Close Approach Procedure for Vessels and/or Zodiacs:
Approximately 200 meters/600 feet or closer:

- Approach at no faster than 'no-wake' speed or at idle, whichever is slower.
- Approach the animal/s from parallel to and slightly to the rear (at 4 or 8 o'clock to the whales heading 12 o'clock, see Figure 1).
- Never attempt an approach head-on or from directly behind.
- Stay well clear of feeding baleen whales.
- Try to position the vessel downwind of the animals to avoid engine fumes drifting over them.
- Establish communication between vessels, small boats or Zodiacs in multi-vessel approaches to coordinate viewing and to ensure that there is no disturbance or harassment of the animals.
- Radio volume should be kept to the minimum necessary for needed communication.
- Do not 'box-in' cetaceans, create a 'tunnel' of zodiacs or kayaks, or cut off their travel or exit routes. This is particularly important when more than one vessel is present.
- If multiple vessels are watching the animal/s at one time, it is suggested that a *maximum* of two ships or four small craft, should position themselves adjacent to

each other to ensure the cetaceans have large open avenues to depart through.

- Beware of local geography – never 'trap' animals between the vessel and shore. Assess the presence of obstacles such as other vessels, structures, natural features, rocks and shoreline.
- *Remember: Avoid sudden or repeated changes in direction, speed or changing gears when close to marine mammals.*

1g. Close Approach Zone:
(Note: Ideally this should be no more than one vessel at a time)
Approximately 30 meters/100 feet for Zodiacs/ 100 meters/300 feet for ships.

- When stopping to watch cetaceans, put your engines in neutral and allow the motor to idle without turning off; or allow the motor to idle for several minutes before turning off. This prevents abrupt changes in noise that can startle the animals and allows them to become aware of your presence and current location.
- Avoid excess engine use, gear changes, maneuvering or backing up to the animals. These produce sudden, large changes in underwater noise levels, which may startle, agitate or disturb the animals.
- Avoid the use of bow or stern lateral thrusters to maintain position. Thrusters can produce hi-pitched acoustics as well as intensive cavitations (air bubble implosion) underwater.
- Be aware that whales may surface in unexpected locations.
- Breaching, tail-lobbing or flipper slapping whales may be socializing and may not be aware of boats. Keep your distance.

Fig. 16.5. *Continued*

- Feeding humpback whales often emit sub-surface bubbles before rising to feed at the surface. Avoid these light green bubble patches.
- Emitting periodic noise may help to let whales know your location and avoid whale and boat collisions. For example, if the small boat or Zodiac engine is not running, occasionally tap on the engine casing with a hard object (not your radio!).
- If cetaceans approach within 30 meters or 100 feet of your vessel, put engines in neutral and do not re-engage propulsion until they are observed clear of the vessel. On rare occasions, whales have been known to use ships as 'backscratchers', if so, remain drifting as long as safety is not compromised.
- Stay quiet, turn radios down, and restrict passenger movement in small boats or Zodiacs during close encounters.
- Enjoy the experience.

1h. Departure Procedures:

- Move off at a slow 'no-wake' speed to the minimum distance of the close approach zone. Avoid engaging propellers within the minimum approach distance, if possible.
- Always move away from the animals to their rear, *i.e.*, not in front of them.
- Do not chase or pursue 'departing' animals.

2. Seals

2a General Guidelines for viewing hauled-out seals

- Seals hauled out on land, rock or ice are sensitive to boats and human presence. Noises, smells and sights may elicit a reaction. Be aware of seal behavior that indicates a seal has been disturbed. Such behaviors

include, but are not limited to:

- ~ an increase in alert or vigilance
- ~ head turning
- ~ change in posture from lying to erect,
- ~ hurriedly moving away from the approaching vessel,
- · open mouth threat displays (e.g. such as in Leopard Seals on ice, or Elephant seals on land).
- ~ aggressive displays or bluff charges in your direction.
- When viewing seals do not surround or separate them, especially mothers and pups. Stay on the side where they can see you.
- On beaches, avoid getting between seals and the sea, walk 'above' them.
- Try not to break their horizon, or tower over hauled out seals – stay low.
- Similar with all Antarctic species, do not attempt to touch or feed seals.
- Pups are often left alone when the mother is feeding. They are not abandoned and should be left alone and not touched.
- Keep commentary, conversation and engine noise to a minimum and be aware of your radio volume.

Fig. 16.5. *Continued*

- Any seal response other than a raised head should be avoided.
- If an individual or a herd moves towards the water or there is a hurried entry into the water by many individuals, you should retreat slowly and carefully
- Beware that fur seals and sea lions are highly mobile on land and might charge (and potentially bite) you if approached too closely.
- Beware of animals in tussock grass areas. Ideally, staff member should lead, carrying walking stick or equivalent.
- Suggested minimum distances ashore 5-10 meters (25 meters from jousting bulls) –

2b. Viewing seals that are in the water

When observing seals in water, please apply similar principles as outlined for cetaceans (Section 1).

3. Seabirds

3a. Vessel & zodiac operations near birds:

Sometimes spectacular concentrations of seabirds may be found out at sea – rafts of birds either feeding on the surface, diving from it, or simply resting and bathing. Many of these birds may have flown hundreds or thousands of miles, often to find food for their young.

- Stay on the fringes of these concentrations. Ships should stay 100 meters and small boats or Zodiacs 30 meters away.
- Birds such as penguins may be subject to disturbance by Zodiac operations close to landing sites or colonies.

Fig. 16.5. *Continued*

- Approach or depart a landing site or colony slowly to minimize any disturbance.
- Staff/crew should assess the best landing point – ideally as far from the birds as possible. This is particularly important if birds are moulting near the shore.
- Avoid boat operations in waters where birds enter and exit, are bathing, or are feeding close to colonies.
- Be aware of birds in the water, slow down and/or alter course to avoid collision.
- There may be occasions when swimming penguins find themselves in a Zodiac when they 'porpoise', landing on the deck. Occupants should remain quiet and wait for the penguin to find its own way over the side and return to the water, normally by jumping onto the anchor box. It is normally not necessary to assist.

The same advice applies to 'feeding frenzies', which may involve species diving from the air into and under the surface of the sea. Some seabirds may be attracted to drifting vessels.

- Under no circumstances should 'chumming' (depositing fish guts or oil) occur to attract birds south of 60°.
- Never feed wild birds.

3b. Viewing birds ashore:

- Walk slowly and encourage passengers to simply sit and watch the animals.
- Avoid blocking 'walkways' in colonies and water entry and exit points.
- If parent birds are blocked from returning to their nests, increased predation of eggs and chicks may occur by skuas and gulls. In addition, parent birds will waste precious energy by avoiding human obstacles on their way to their nests or being displaced from the shortest access route. Take care in tussock grass where birds

may be nesting, including in burrows under bare earth.

- If skuas (jaegers) or terns start dive-bombing, they are protecting young or nests. Retreat in the direction you approached from. Beware that eggs and young are well camouflaged and might be hidden from your view.
- Recommended approach distances:
- In general, keep 5-10 meters from nesting seabirds.
- Keep 10 meters from nesting, and 25 meters from displaying Albatross on South Georgia.
- Giant Petrels seem particularly prone to disturbance whilst nesting, stay 25-50 meters away, if possible.
- Aircraft (including helicopters) should follow the guidelines laid out in Antarctic Treaty Resolution 2 (2004) '*Guidelines for the Operation of Aircraft near Concentrations of Birds in Antarctica.*'

4. Entanglement and Strandings

- Any animals entangled in fishing equipment etc., should be assisted where possible. Please only use experienced staff/crew for these situations and take the necessary precautions such as protective clothing – seal bites are particularly prone to disease.
- Photographs of the entanglement should be taken. Please complete a report and send it to IAATO.
- Should you not be able to assist, please record details including geographic position (expressed as coordinates in latitude and longitude), species,

and type of entanglement. Please report the event as soon as possible, so assistance may be sought from other vessels with experienced staff onboard.

- Details of dead (floating) animals and 'strandings' (beached) cetaceans should be recorded and reported to IAATO. Where possible, please take photographs recording the front andside of the head of the animal (for species identification). Please include a scale of measurement (e.g., a ruler or Zodiac paddle) in the photographs. If the state of decomposition of the animal allows, please also take photographs of the fluke (tail) and the dorsal fin (if present) to allow recognition of potentially known individuals (i.e. using photo-identification).

5. Identification and Data Collection

Identifying and, in many cases, recording species for the voyage log purposes is part of most onboard naturalists' remit. Logs, which include these records coupled with latitude and longitude of sightings, species identification, and any additional information such as identification photographs, are of immense value. Please send copies to IAATO at iaato@iaato.org

Fig. 16.5. *Continued*

Helpful Hints!

- Reduce pollution from engines – In all close wildlife encounters, please ensure you are using 'clean running' engines, especially on small boats or Zodiacs, and are creating minimum air and water pollution.
- Polarizing sunglasses can considerably enhance viewing of submerged/partially submerged marine animals and birds.
- Encourage the use of binoculars for viewing marine mammals and seabirds.

Recommended Field Guides:

- *Whales, Dolphins and Other Marine Mammals of the World* by Shirihai and Jarrett 2006
- *Birds of Chile, Antarctica and Southern Argentina* by Jaramillo, Burke and Beadle 2003
- *A Complete Guide to Antarctic Wildlife* by Shirihai and Jarrett 2002
- *National Audubon Guide to Marine Mammals of the World* by Folkiens et al. 2002
- *Cetaceans: Whales, Dolphins and Porpoises* by Carwardine & Camm 1995
- *Seabirds: A Photographic Guide* by Peter Harrison, 1987

 The Sea Mammal Research Unit, Getty Marine Laboratory, University of St. Andrews have endorsed these Guidelines.

IAATO Secretariat
P.O. Box 2178
Basalt, Colorado 81621 USA
Tel: 970 704 1047
Fax: 970 704 9660
E-mail: iaato@iaato.org
website: www.iaato.org

Guidelines revised: 2007.

Fig. 16.5. *Continued*

censuses of these animals and report results to the treaty parties and their scientists. Many results are in the public domain, as can be seen in the IAATO Information Paper at the Warsaw Antarctic Treaty Consultative Meeting (IAATO, 2002).

Guidelines for other activities

In addition to specific guidelines for marine mammal watching, numerous guidelines also exist for the variety of tour operator activities, which now offer camping, kayaking, scuba diving, helicopter operations and the likes. Presently, individual IAATO member companies have developed these guidelines in accordance with safety procedures, acceptance of IAATO requirements and treaty party knowledge. Land-based activities also have their own operating guidelines, pertaining to operations centred on the temporary summer camp of ANI/Antarctic Logistics and Expeditions (Landau and Splettstoesser, 2007).

Staff training

A key ingredient for IAATO operations pertains to the training and experience of primary personnel, particularly expedition leaders and Zodiac drivers. Several IAATO member companies have formalized their own staff training programmes, and other plans are being developed to institute standardized training for both expedition leaders and Zodiac drivers in order to provide the safety and operational procedures that will apply to all member operations (Landau and Splettstoesser, 2007).

Conclusions: Pulling It All Together

If we return to the three questions posed in the chapter introduction, how does the situation fare?

1. Are tourism and wildlife compatible in the Antarctic?
2. Can tourism support and conserve Antarctic wildlife?
3. Can Antarctic wildlife support current or increased tourism?

Antarctic wildlife, which is predominantly marine wildlife, is quite bio-dense, but not biodiverse. That is to say there are many healthy populations and large numbers of wildlife, but if something were to happen to a single breeding ground, it would have catastrophic consequences. One breeding ground contains a large proportion of a single species' population.

In terms of wildlife management, the parties of the ATS have put a comprehensive system in place, which over the course of the last 50 years appears to be working. Marine wildlife has not appeared to be in any sort of decline, and as much as we can posit, the wildlife–human interaction appears to be in balance.

Since 1991, numbers have increased nearly each year, with growing global interest in the Antarctic that may increase even further given the 2007–2009 International Polar Year(s). Tourism management being primarily an industry self-regulation affair also appears to be growing in terms of operator member-ship, but at the same time the association (IAATO) appears to be able to evolve and flex appropriately. IAATO has been quite proactive in their management of tourism, having addressed visitor guidelines for many years, which form the basis of the visitor guidelines in all ATS documents. With guidelines for specific wildlife (see Fig. 16.5), IAATO has much more practical and focused efforts than the multi-year debates the ATS has had regarding what they classify together as 'Tourism and Non-governmental Activities'. That said, recently the ATS has also been more open to tourism discussion, and has since 1991 taken IAATO's expertise to heart.

So, are tourism and wildlife compatible? To date the answer appears to be yes. Will there be continued issues and conflict assuming that tourism numbers continue to grow? Again, the answer is probably yes, but it seems that every-one involved has Antarctica and its wildlife's best interests in hand. Can tourism support and conserve wildlife? The answer to this also seems to be yes. Not only have the ATS and IAATO implemented policies to accomplish this, but also the tourist has bought into it. IAATO companies generally support the logistics and movement of scientists who seek to provide sound conservation-minded research (see Stonehouse and Crosbie, 1995; Williams and Crosbie, 2007: as both examples and overviews of such work), but passengers and companies alike donate directly to organizations such as Orca Project, American Bird Conservancy, Oceanites and Birdlife International-Albatross. Donations totalled US$350,000 in the 2005/06 tour season (Landau and Splettstoesser, 2007). In the end, Antarctica is as unique for its legal regimes as it is for its wildlife. As a means and literally grounds for compromise in international rela-tions, 'if we cannot succeed in Antarctica we have little chance of succeeding elsewhere' (Mickleburgh, 1988, p. 7).

References

Antarctic Treaty (2002) Available at: http://www.nsf.gov/od/opp/antarct/anttrty.htm
Bastmeijer, K. (2002) *The Antarctic Environmental Protocol and Its Domestic Legal Implementation*. Kluwer Law International, The Hague, The Netherlands.
Bastmeijer, K. and Roura, R. (2004) Regulating Antarctic tourism and the precautionary prin-ciple. *The American Journal of International Law* 98(4), 763–781.
Bauer, T.G. (2001) *Tourism in the Antarctic: Opportunities, Constraints and Future Prospects*. The Haworth Hospitality Press, New York.
Benson, J. (2000) Tourism in Antarctica: a unique undertaking in development and international environmental management. *Pacific Tourism Review* 4(1), 7–18.
Campbell, D.G. (1993) *The Crystal Desert: Summers in Antarctica*. Minerva, London.
Cessford, G.R. (1997) Antarctic tourism: a frontier for wilderness management. *International Journal of Wilderness* 3(3), 7–11.
Chester, S.R. (1993) *Antarctic Birds and Seals*. Wandering Albatross, San Mateo, California.

Davis, B.W. (1992) Antarctica as a global protected area: perceptions and reality. *Australian Geographer* 23(1), 39–43.

Enzenbacher, D. (1992) Tourism in Antarctica: an overview. In: Kempf, C. and Girard, L. (eds) *Tourism in Polar Areas: Proceedings of the First International Symposium*. IUCN and Ministry of Tourism France, Colmar, France.

Hall, C.M. and Johnston, M.E. (eds) (1995) *Polar Tourism: Tourism in the Arctic and Antarctic Regions*. Wiley, Chichester, UK.

Hall, C.M. and Wouters, M. (1995) Issues in Antarctic tourism. In: Hall, C.M. and Johnston, M.E. (eds) *Polar Tourism: Tourism in the Arctic and Antarctic Regions*. Wiley, Chichester, UK, pp. 147–166.

Hart, P.D. (1988) The growth of Antarctic tourism. *Oceanus* 31(2), 93–100.

Headland, R.K. (1994) Historical development of Antarctic tourism. *Annals of Tourism Research* 21(2), 269–280.

IAATO (International Association of Antarctica Tour Operators) (2002) *Bibliography of Publications by Staff Naturalists/Lecturers Involved in Tour Activities in Antarctica, 1991–2001*. XXV ATCM Warsaw, IP-071.

IAATO (International Association of Antarctica Tour Operators) (2007a) Tourism statistics. Available at: www.iaato.org/tourism_stats.html

IAATO (International Association of Antarctica Tour Operators) (2007b) About IAATO. Available at: www.iaato.org/about.html

IAATO (International Association of Antarctica Tour Operators) (2007c) Bylaws. Available at: www.iaato.org/bylaws.html

IAATO (International Association of Antarctica Tour Operators) (2007d) Marine wildlife watching guidelines. Available at: www.iaato.org/docs/Wildlife_Watching_Guidelines.pdf

IAATO (International Association of Antarctica Tour Operators) (2007e) Visitor guidelines Available at: http://www.iaato.org/docs/English_guidelines.pdf

IAATO (International Association of Antarctica Tour Operators) (2007f) Operational. Available at: http://www.iaato.org/operational.html

Kriwoken, L.K. and Rootes, D. (2000) Tourism on ice: environmental impact assessment of Antarctic tourism. *Impact Assessment and Project Appraisal* 18(2), 138–150.

Landau, D. (2000) Tourism scenarios. In: Antarctica New Zealand (ed.) *Proceedings of the Antarctic Tourism Workshop*. Antarctica New Zealand, Christchurch, pp. 15–18.

Landau, D. and Splettstoesser, J. (2007) Management of tourism in the marine environment of Antarctica: The IAATO Perspective. *Tourism in Marine Environments* (in press).

Lucas, P.H.C. (1995) National parks and protected areas in Polar Regions. In: Martin, V.G. and Tyler, N. (eds) *Arctic Wilderness: The 5th World Wilderness Congress*. North American Press, Golden, Colorado, pp. 162–169.

MacFarlane, S. (1991) *The Erebus Papers*. Avon Press Ltd, Wellington, New Zealand.

Mickleburgh, E. (1988) *Beyond the Frozen Seas: Visions of Antarctica*. Bodley Head, London.

Prosser, R. (1995) Power, control and intrusion with particular reference to Antarctica. In: Cooper, D.E. and Palmer, J. (eds) *Just Environments: Intergenerational, International and Interspecies Issues*. Routledge, London, pp. 108–120.

Reich, R.J. (1980) The development of Antarctic tourism. *Polar Record* 20(126), 203–214.

Shirihai, H. (2002) *The Complete Guide to Antarctic Wildlife*. Princeton University Press, Princeton, New Jersey.

Splettstoesser, J. (1999) Antarctic tourism: successful management of a vulnerable environment. In: Singh, T.V. and Singh, S (eds) *Tourism Development in Critical Environments*. CCC, New York, pp. 137–148.

Splettstoesser, J.F., Headland, R.K. and Todd, F. (1997) The first circumnavigation of Antarctica. *Polar Record* 33(3), 244–245.

Stonehouse, B. and Crosbie, K. (1995) Tourist impacts and management in the Antarctic Peninsula area. In: Hall, C.M. and Johnston, M.E. (eds) *Polar Tourism: Tourism in the Arctic and Antarctic Regions*. Wiley, Chichester, UK, pp. 217–233.

Swithinbank, C. (1992) Airborne tourism in the Antarctic. In: Kempf, C. and Girard, L. (eds) *Tourism in Polar Areas: Proceedings of the First International Symposium*. IUCN and Ministry of Tourism France, Colmar, France.

Swithinbank, C. (1998) Non-government aviation in Antarctica 1997/98. *Polar Record* 34, 249.

Waterhouse, E. (ed.) (2001) *Ross Sea Region 2001: A State of the Environment Report for the Ross Sea Region of Antarctica*. Antarctica New Zealand, Christchurch.

Williams, R. and Crosbie, K. (2007) Antarctic whales and Antarctic tourism. *Tourism in Marine Environments* (in press).

IV Marine Wildlife and Tourism Management

17 Managing the Whale- and Dolphin-watching Industry: Time for a Paradigm Shift

R. Constantine and L. Bejder

Watching whales, dolphins and porpoises in the wild (in this chapter commonly referred to as whale watching) is a rapidly growing commercial industry that includes land-, boat- and aircraft-based activities (Hoyt, 2001). Unfortunately, management of this industry still ranges from complex and difficult to implement, to inadequate, or indeed completely lacking despite commercial whale watching having been engaged in for over 50 years. Initially, little attention was paid to the potential impacts of commercial whale-watching tours, most operators being pleased to take tourists to see whales in the wild and offer an alternative to the commercial whale hunts that were destroying great whale stocks throughout many of the worlds' oceans. With whale hunting as a basis for comparison, whale watching was never really considered an activity likely to cause harassment or disturbance to wild cetaceans. However, whale-watching tourism targets specific communities of animals that are repeatedly sought out for prolonged, close-up encounters. Since the early 1990s, concerns over the potential for detrimental consequences to targeted animals have been raised (e.g. IWC, 1996; Samuels et al., 2003; Corkeron, 2004). Repeated disruptions of breeding, social, feeding and resting behaviour have long been speculated to result in deleterious effects on reproductive success, health, ranging patterns and availability of preferred habitat. Emergent research has now revealed that dolphin watching can cause biologically significant impacts on targeted communities, notably by displacing dolphins from critical habitats and reducing their reproductive success (Lusseau, 2005; Bejder, 2005; Bejder et al., 2006a).

The seriousness of documented impacts on dolphins exposed to dolphin-watching tourism in Shark Bay, Western Australia (WA), was acknowledged by the first decision by a government agency in any country to reduce the number of commercial dolphin-watching licenses from two to one (Ministry Media Statement, 2006). The presiding minister noted that the withdrawal of one license was in the interest of the welfare of the local dolphin population and a necessary sacrifice for the long-term sustainability of the area. In other areas where whale watching has been shown to impact on dolphin behaviour, changes in the growth and operation of the local industry have been made. In Kaikoura, New Zealand (NZ), a moratorium on new permits was put in place after research revealed changes in dolphin and whale behaviour in the presence of boats (Barr and Slooten, 1999; Richter *et al.*, 2006). Also, in Kaikoura and Northland, changes have been made to dolphin-watching operators' permits by creating periods of time when boats are not allowed to interact with dolphins in order to minimize disturbance. These managerial changes are indicative of a growing awareness that whale- and dolphin-watching tourism, whilst once viewed as benign, can have biologically significant impacts that require thoughtful mitigation strategies if the industry is to move towards sustainability. However, these examples are also the exception rather than the norm when it comes to management intervention based on high-quality research on impact assessment of whale watching on target animals. For example, no management action has been taken despite strong evidence of population-level effects on bottlenose dolphins exposed to unsustainable dolphin-watching tourism practices in Fiordland, New Zealand (Lusseau *et al.*, 2006).

The International Whaling Commission (IWC) has maintained support for the development of the whale-watching industry, conditional on this development being in a manner that minimizes the risk of adverse impacts and is, ultimately, sustainable (IWC, 1996). During the 2006 annual IWC meeting, the Scientific Committee concluded that there was sufficient evidence that whale watching can have population-level impacts and can endanger the viability of small coastal populations of whales and dolphins (IWC, 2006).

This recent recognition of the extent of impacts of whale and dolphin watching, and the resultant instances of management changes, represent a paradigmatic shift and, accordingly, calls for a revised and more stringent approach to management of the industry to ensure its long-term sustainability.

Economic Value and Growth

Whale watching was first established in the USA in the 1950s. It became a popular tourism activity targeting grey whales along the west coast and humpback whales along the east and west coasts, as well as in Hawaii (Hoyt, 2002). Industry growth was facilitated by the readily accessible populations of grey whales migrating close inshore to calving grounds in Mexico, and humpbacks migrating to winter breeding grounds of Hawaii and summer feeding grounds off the New England coast. Other than the initial expense of investing in or seasonally leasing boats, whales were a reliable, natural resource that tour operators did not need to pay for. In the 1980s, many other countries began whale-watching tours and it was around this time that dolphin watching and swim-with-dolphin tours were also being established (Samuels *et al.*, 2003). This resulted in the transformation of many small towns and communities as tourists spent increasing amounts of money to travel often great distances to interact with whales and dolphins in the wild.

Globally, the direct income from ticket sales and indirect income from accommodation, food and souvenirs were estimated to be US$14 million with approximately 400,000 participants in 1981 (Hoyt, 2001). In the most recent worldwide survey, Hoyt (2001) estimated that over US$1 billion was spent on whale watching in over 87 countries and territories in 1998. This estimate is now almost a decade old, but with recent trends of 11% growth per annum from 1998 to 2004 in New Zealand (IFAW, 2005) and 15% growth per annum from 1998 to 2003 in Australia (IFAW, 2004), the industry would appear to be worth well in excess of US$1 billion per annum today. Whale watching is thus big business, with the potential to contribute significant financial gains for small towns (e.g. Shark Bay, WA; Kaikoura, NZ) and produce large tourism revenue for otherwise remote countries and communities (e.g. Vava' u, Tonga; Zanzibar, East Africa; Baja Peninsula, Mexico).

With whale watching bringing literally millions of dollars into an ever-growing number of regions, great challenges for governments lie in balancing ecological values and the viability of cetacean populations with the economic and social benefits whale watching brings to the community. That the economic gains and social benefits to communities are significant is doubtless, however, recent research findings indicate that enforced management is imperative to minimize impacts and ensure the long-term sustainability of the industry (Lusseau, 2003b; Scarpaci *et al.*, 2003; Constantine *et al.*, 2004). There are many development models (e.g. government legislation, voluntary guidelines, codes of conduct and the designation of marine protected areas (MPAs)) available to help newly forming whale-watching operations avoid the pitfalls of a short-sighted approach to business growth; these will be discussed in more detail below. If animals, the very resource upon which these businesses depend avoid boats or swimmers, or become displaced from preferred habitats, the industry will not be able to provide the experience that tourists seek and will fail in the medium to long term.

Legislation

There is a variety of codes of conduct, guidelines and regulations designed to manage the whale-watching industry (see review by Carlson, 2004). The USA was the first country to develop legislation to protect whales from harassment; the Marine Mammals Protection Act (MMPA) 1972. This legislation was designed primarily to minimize harassment and disturbance and required permits for 'takes', e.g. by-catch in fisheries and hunting of whales and dolphins. The MMPA provided the impetus and guidance for other nations to develop their own protection laws. New Zealand adopted their MMPA in 1978 and, with the development of sperm whale watching and dusky dolphin watching in Kaikoura in the 1980s, the Marine Mammals Protection Regulations were adopted in 1989. These regulations were advanced for their time and, to this day, provide the legal basis for issuing permits to run commercial tours to interact with marine mammals and prescribe appropriate operational behaviour for all vessels around marine mammals. These advances allowed management

Fig. 17.1. Commercial and recreational dolphin watching in Panama City Beach, Florida, USA. (Photograph L. Bejder.)

agencies in New Zealand to have some measure of control over the growth of the whale-watching industry. Nevertheless, this legislation is only as good as its enforcement and recent research findings point towards an urgent need for regulation changes to increase their efficacy (e.g. Constantine, 2001; Lusseau, 2003b, 2005; Constantine *et al.*, 2004). The two examples above represent the legislative approach. Many other variations in management techniques have been adopted throughout the world, from simple voluntary guidelines (e.g. New Caledonia, Zanzibar) to industry codes of conduct and creating MPAs (e.g. Brazil), and even banning certain types of whale watching (e.g. only land-based whale watching is allowed in the Cook Islands). Important issues facing managers throughout the world include when to adopt legislation and how to adequately enforce legislation. There are a number of examples in which blatant disregard has been shown for the law and, indeed, little or no response by the relevant authorities – due to either inadequate resources, lack of legal security or poor leadership. Such examples include feeding dolphins at Panama City Beach, Florida (Samuels and Bejder, 2004) (see Fig. 17.1) and swim-with-dolphin activities in Port Phillip Bay, Victoria (Scarpaci *et al.*, 2003) and Zanzibar (Stensland and Berggren, 2007). It has become apparent that appropriate legislative controls and enforcement are crucial in allowing the industry to develop sustainably, especially in light of long-term data revealing significant impacts on dolphin populations exposed to tourism. What is clear is that there is no 'one-size-fits-all' solution and that each country, territory or even local region must decide upon what is an appropriate management regime.

Evaluating Impact: Migratory Versus Non-migratory Species

Research on the effects of both commercial and recreational whale watching began in the 1980s (e.g. Salden, 1988; Baker and Herman, 1989) and increased considerably in the late 1990s. In most cases, studies have been undertaken some years after the establishment of tourism industry activities, providing challenges to scientists when interpreting changes in behaviour or habitat use in the absence of baseline, or 'pre-impact', data (Bejder and Samuels, 2003). Nevertheless, significant impacts on whales and dolphins as a direct result of whale watching are being proven and managers are struggling to mitigate against impacts on target populations.

While watching whales has been occurring for several decades more than dolphin watching, most recent research has focused on the potential impacts on coastal dolphin populations (see below for research examples). There are distinct differences between the effects of whale watching on non-migratory coastal populations of whales or dolphins, and those on the migratory populations of primarily great whales. Coastal, resident populations of dolphins are likely to be exposed to year-round, daily dolphin watching (see Fig. 17.2) or swim-with-dolphin tours, e.g. bottlenose dolphins in the Bay of Islands, New Zealand (Constantine, 2001; Constantine *et al.*, 2004), Zanzibar, Africa (Stensland and Berggren, 2007), Port Stephens, New South Wales (Allen *et al.*, in press a) and Shark Bay, WA (Bejder *et al.*, 2006a,b); killer whales in the waters off Vancouver Island, Canada (see Fig. 17.3, Williams *et al.*, 2002) and dusky and common dolphins in Patagonia, Argentina (Coscarella *et al.*, 2003) and Kaikoura, NZ (Würsig *et al.*, 1997).

Occasionally, portions of a whale population are coastal residents and are the subject of year-round whale-watching tours, for example, sperm whale watching in Kaikoura, New Zealand (Gordon *et al.*, 1992; Richter *et al.*, 2006). This may place greater pressure on the population than that experienced by migratory species. For populations which range over many hundreds or thousands of kilometres of coastline (e.g. the bottlenose dolphins of the

Fig. 17.2. Boat-based dolphin watching in Bunbury, Western Australia. (Photograph Bunbury Dolphin Discovery Centre.)

Fig. 17.3. Boat-based killer whale watching in British Columbia. (Photograph S. Allen.)

north-east coast of New Zealand), home ranges include several locations where whale-watching tours occur (Constantine *et al.*, 2003). The majority of these tours are thus targeting the same population of dolphins. This cumulative effect is more difficult to assess than that in areas where there is extreme site fidelity and small home ranges (e.g. Doubtful Sound, NZ and individuals in the Port Stephens, NSW and Shark Bay, WA populations). In all cases, however, disturbance to behavioural patterns, group structure, habitat use and/or reproductive success has been documented.

In the case of migratory species, the effects of whale-watching tourism are generally isolated to a particular season, e.g. humpback whales on the winter breeding grounds in Hawaii and Tonga, or the summer feeding grounds off New England and Alaska; grey whales in the winter breeding grounds of the Baja lagoon system in Mexico; and southern right whales on their summer breeding grounds in South Africa (Hoyt, 2002). Occasionally, tourism also occurs along the whales' migratory path, e.g. humpback whale watching off the east Australian coast (IFAW, 2004). There are concerns over the potential for cumulative effects of whale watching since whales may first approach land near south-eastern Australia, then continue migrating north along the New South Wales and Queensland coasts (National Parks and Wildlife Service, NSW Amendment (Marine Mammals) Regulation 2004). Some individuals continue further on to breeding grounds in New Caledonia where a rapidly developing, unregulated whale-watching industry has developed (C. Garrigue, Operation Cetaces, 2006, personal communication). Exposure to whale watching then continues on their return migration south. This could result in 7 months of exposure to boat-based tourism for some individuals, and while it is still uncertain whether the same individuals pass close enough to shore to be targeted by boats on the northern and southern migration, in the absence of these data the precautionary principle should be applied to protect this recovering population (IWC, 1996).

Both migratory and non-migratory populations of cetaceans present unique management challenges as the implications of chronic, year-round tourism differ from the acute, seasonal bursts of tourism activity. In many places where whale-watching tours operate, there is additional vessel activity that causes changes in cetacean behaviour, e.g. Moray Firth, Scotland (Janik and Thompson, 1996; Hastie *et al.*, 2003), Panama City Beach, USA (Samuels and Bejder, 2003) and Bay of Islands, New Zealand (Constantine *et al.*, 2004). This must also be carefully considered when managing the growth of an industry. In addition, research has shown that there are inter-species differences in responses to particular types of boat handling or swimmer placement (e.g. Constantine and Baker, 1997) and these results need to be incorporated into management plans. Well-designed, quantitative studies are vital and the collection of long-term data is important in providing an accurate picture of the effect of the industry on target animals. The complexity of managing tourism requires caution and the use of an adaptive management plan where all stakeholders are involved in the process, i.e. managers, researchers and tour operators.

Research into Short- and Long-term Impacts

In almost all situations, the lack of pre-tourism data on target animals' behaviour, habitat use and fecundity, the often urgent need for information as a new whale-watching industry is being developed, and the fact that cetaceans are long-lived, slow-breeding animals makes for a difficult research environment. Even with these limitations, however, many studies have shown clear short-term changes in behaviour in the presence of tour vessels. Documented short-term impacts include changes in habitat use (Salden, 1988; Lusseau, 2005), behaviour (Corkeron, 1995; Würsig, 1996; Lusseau, 2003a,b; Constantine *et al.*, 2004; Allen *et al.*, in press a), swimming speed and direction (Kruse, 1991; Williams *et al.*, 2002; Scheidat *et al.*, 2004), inter-animal distance (Blane and Jaakson 1995; Bejder *et al.*, 1999, 2006b; Allen *et al.*, in press a) and vocal communication (Scarpaci *et al.*, 2000). Animals are also exposed to vessel traffic not involved in whale watching in many cases (e.g. Janik and Thompson, 1996; Allen and Read, 2000; Nowacek et al., 2001; Van Parijs and Corkeron, 2001; Hastie et al., 2003), which can add extra pressures on populations exposed to tourism.

In some cases, the use of several short-term studies on the same population can provide longer-term information useful to managers employing an adaptive management plan. Short-term studies may be misleading (see Bejder *et al.*, 2006b) however, if repeated over a longer time period these can provide useful measurements of behavioural change. One example of this was the comparison of bottlenose dolphin responses to swim attempts in the Bay of Islands, New Zealand over two time periods, 1994–1995 and 1997–1998 (Constantine, 2001). Even in this relatively short time period, there was a significant increase in avoidance behaviour, which was attributed to swimmer placements that left the dolphins no choice of whether to approach the swimmers or not. In response, the New Zealand Department of Conservation (management agency) made placing swimmers line abreast of the dolphins' path of travel the only option for swimmer placement and included this as a condition of the operators' permits in Northland.

Ideally, long-term data sets that include a period of pre-tourism data collection provide the most useful information about the effects of whale watching on the target population. Such data sets are rarely available. The most comprehensive example to date exists in Shark Bay, WA, where data were available both before and during vessel-based dolphin-watching tourism and at two tourism levels (Bejder, 2005; Bejder *et al.*, 2006a). Furthermore, there were subsets of the population with very different levels of exposure to tour vessels – hence providing both before/after and control/impact comparisons. Based on decades of detailed behavioural records, dolphin abundance was compared within adjacent tourism and control sites over three consecutive 4.5-year periods wherein research activity was relatively constant, but tourism levels increased from zero to two operators.

At this location, when comparing periods in which there was no tourism and then one operator within the tourism site, there was no change in dolphin abundance per square kilometre; however, as tour operator numbers increased from one to two, there was a significant average decline of 14.9% in dolphins per square kilometre, approximating a decline of one per seven individuals. Concurrently, within the control site there was a non-significant average increase of 8.5% in dolphins per square kilometre. While acknowledging that research vessels are likely to have contributed to the documented effects, it was concluded that, given the substantially greater presence and proximity to dolphins of tour vessels relative to research vessels, tour vessel activity was identified as the most significant contributor to declining dolphin numbers within the tourism site (Bejder *et al.*, 2006a).

While few study sites have such extensive data sets, the benefits of long-term data collection on cetacean populations are numerous. Recent work on bottlenose dolphins in Fiordland, New Zealand, revealed significant differences between the ways in which male and female dolphins respond to boat disturbance, and that these responses were exacerbated for female dolphins when boats violated regulations prescribing appropriate boat handling around dolphins (Lusseau, 2003a,b). It is likely that the energetic demands for female dolphins are different from those of male dolphins, especially when accompanied by calves. Research on the bottlenose dolphin populations in Fiordland was initiated in Doubtful Sound in 1991, with data being collected on population size and ranging behaviour (e.g. Williams *et al.*, 1993; Schneider, 1999; Lusseau, 2003b, 2005). There has been a decrease in dolphin population size in Doubtful Sound from 67 to 56 individuals between 1997 and 2005 which is of concern for such a small population (Lusseau *et al.*, 2006). With demonstrated impacts of boating traffic on dolphin behaviour (Lusseau, 2003b), an increase in boat traffic operating in the narrow fjord system and no changes to tourism management in response to research findings and recommendations (Lusseau and Higham, 2004; Lusseau *et al.*, 2006), this does not represent a picture of long-term sustainability.

A Case Study of Unsustainable Dolphin-watching Tourism Management

Fiordland is a popular tourism destination in New Zealand. Scenic cruises are one of the main activities visitors can experience there, and those tours interact on a daily basis with bottlenose dolphins in three fjords. In Milford Sound, one of those three fjords, the dolphins leave the area altogether during cruising peak times as the disturbance from boat traffic is too high (Lusseau, 2005). With Milford Sound becoming too crowded, many tourists are attracted towards other fjords and the visitor volume to Doubtful Sound is now increasing significantly (D. Lusseau, Dalhousie University, 2006, personal communication). From 1999 to 2002, a study by Lusseau (2003a) showed that the levels of tourism activities in Doubtful Sound were putting significant biological stress on bottlenose dolphins residing there. Dolphins tended to avoid interactions with boats and the energetic costs of these avoidance strategies were greater for females, especially mothers who already had an added energetic burden due to calf dependence (Lusseau, 2003b). These findings presented a warning that tourism could affect the reproductive success of this population and potentially trigger a decline in abundance. Managers, the New Zealand Department of Conservation, were advised to create a multi-level reserve, establishing no-go areas to eliminate interactions with tour boats during the more sensitive times for dolphins and still allowing for further growth of the industry (Lusseau and Higham, 2004). Despite this advice, no management actions were taken and the industry was allowed to grow unchecked. This inaction has been followed by a decline in the abundance of the dolphin population (Lusseau *et al.*, 2006). In mid-2007, managers finally produced a discussion document after having the research findings for four years and the 2006 advice from the IWC urging New Zealand to act speedily to increase the protection of the Doubtful Sound bottlenose dolphin population (IWC, 2006). Lusseau *et al.* (2006) warn of the likelihood that the population will become extinct within the next 50 years if no action is taken. It is hoped that action will soon be taken to afford greater protection to this vulnerable species.

Management Directions

So what should managers do? We suggest that there needs to be a widespread paradigm shift away from attempting to demonstrate impact subsequent to the establishment of a regional whale-watching industry. It is time to shift the burden of proof on to the industry to show that their activities are sustainable and not detrimental to the health of target populations (Mangel *et al.*, 1996). The current systems of either reactive (constantly playing 'catch-up') or proactive management that attempt to minimize impact until research can prove that the industry is not causing biologically significant changes is unrealistic and does not afford the animals the protection they require (Corkeron, 2004). There are so few studies that have the longevity to demonstrate biological impact that managers must pay attention to those that exist and draw inference from these studies to design an adaptive management system. This, of course, represents an enormous challenge, but considering the financial and educational gains that a sustainable whale-watching industry can contribute over the long term, this also presents an enormous opportunity. Small, isolated populations of cetaceans are most vulnerable to disturbance (evidence from the Shark Bay, WA, and Fiordland, NZ studies) and this can lead to biologically significant impacts on the population. Obviously, long-term studies provide us with the most informed management advice, but the ever-increasing volume of studies demonstrating short-term impacts can provide useful proxies.

Management for sustainability requires the introduction of whale-watching 'no-go' protected areas based on research that demonstrates preferred habitats in which behaviours vulnerable to disturbance occur (such as resting or foraging; e.g. Lusseau, 2003b; Constantine et al., 2004). A system of multi-levelled marine mammal sanctuaries has been proposed for the tourism-impacted dolphin population in Fiordland, allowing for a continuing dolphin-watching industry while also allowing the dolphins, the resource upon which the industry depends, critical habitat free of boating activity (Lusseau and Higham, 2004; Lusseau et al., 2006).

Other management options include creating 'no-go' times of day in which interactions with cetaceans are prohibited, greater minimum approach distances, capping the number of boats allowed within a particular distance of target animals and prohibiting or restricting intrusive activities such as swimming with cetaceans. Similar stipulations are already in place in various parts of the world, either as voluntary codes, guidelines or governmental legislation, but they are frequently rendered ineffective by a lack of monitoring or enforcement (Samuels and Bejder, 2003; Scarpaci et al., 2003; Allen et al., in press b). The whale-watching industry is yet to demonstrate effective self-regulation and infringements of codes and laws are reported even when operators are aware of some assessment of their activities (Scarpaci et al., 2003; Allen et al., in press b).

Whale watching has been occurring for over half a century, yet even the highest-quality research, designed specifically to provide managers with potential solutions to resource management challenges, is rarely acted upon. In only one case has a permit been withdrawn due to demonstrated impacts on the target animals – that being Shark Bay, Australia (Ministry Media Statement, 2006). In a few other cases, amendments have been made to whale-watching tour operating permits such that disturbance was minimized (Bay of Islands and Kaikoura, NZ). While these instances represented steps in the right direction, the reality is that little monitoring or enforcement occurs even in these areas and that the industry continues to grow on a broad scale. As an example, permits continue to be issued to watch and swim with humpback whales in Tonga despite a distinct lack of data on the quality of operations and the potential impacts on the whales. The industry is vital to this small community (Orams, 1999), but the increase in the number of permitted operators without monitoring this recovering population of whales on their calving grounds, and no guideline enforcement, cannot be described as 'sustainable development'. Humpback whale watching also continues to spread on the New South Wales coast of Australia without a permit system and the potential for a cumulative impact is of concern (IFAW, 2004).

Conclusion

The whale-watching industry has enormous potential to educate people and provide them with a once-in-a-lifetime experience, which hopefully will encourage people to protect cetaceans and their environment. Bejder and Samuels (2003) urged for more high-quality, quantitative research to evaluate the poten-

tial impacts of whale watching to help managers ensure the long-term sustainability of the industry. We are now beginning to see the results of these studies
and the concern raised by scientists is being echoed by the IWC and some of the
non-governmental organizations (such as the Whale and Dolphin Conservation
Society). The IWC is urging member states to take the threats to populations
impacted by whale watching seriously and, in some places, this has occurred.
Sadly though, despite all the potential economic, educational and conservation
benefits, many places are failing to protect their cetaceans. If the burden of
proof is shifted on to the industry and managers to show sustainability, then the
industry will grow cautiously. As Corkeron (2004) pointed out, it is time to raise
the question as to when and where whale watching should not occur; with the
evidence we have in hand it is time for a paradigm shift.

References

Allen, M.C. and Read, A.J. (2000) Habitat selection of foraging bottlenose dolphins in relation
    to boat density near Clearwater, Florida. *Marine Mammal Science* 16, 81–84.
Allen, S., Constantine, R., Bejder, L., Waples, K. and Harcourt, R. (in press a) 'Can't sleep, can't
    eat – let's split': Indo-Pacific bottlenose dolphin responses to tour boats in Port Stephens,
    Australia. *Journal of Cetacean Research and Management.*
Allen, S., Smith, H., Waples, K. and Harcourt, R. (in press b) The voluntary code of conduct for
    dolphin-watching in Port Stephens, Australia: is self-regulation an effective management
    tool? *Journal of Cetacean Research and Management.*
Baker, C.S. and Herman, L.M. (1989) *Behavioral Responses of Summering Humpback Whales
    to Vessel Traffic: Experimental and Opportunistic Observations.* Report to National Parks
    Service, US Department of the Interior, NPS-NR-TRS-89-01.
Barr, K. and Slooten, L. (1999) *Effects of Tourism on Dusky Dolphins at Kaikoura.* Conservation
    Advisory Science Notes 229, Department of Conservation, Wellington, New Zealand.
Bejder, L. (2005) Linking short- and long-term effects of nature-based tourism on cetaceans. PhD
    thesis, Dalhousie University, Nova Scotia, Canada.
Bejder, L. and Samuels, A. (2003) Evaluating the effects of nature-based tourism on cetaceans.
    In: Gales, N., Hindell, M. and Kirkwood, R. (eds) *Marine Mammals: Fisheries, Tourism and
    Management Issues.* CSIRO Publishing, Collingwood, Australia, pp. 229–256.
Bejder, L., Dawson, S.M. and Harraway, J.A. (1999) Responses by Hector's dolphins to
    boats and swimmers in Porpoise Bay, New Zealand. *Marine Mammal Science* 15,
    738–750.
Bejder, L., Samuels, A., Whitehead, H., Gales, N., Mann, J., Connor, R., Heithaus, M., Watson-
    Capps, J., Flaherty, C. and Krützen, M. (2006a) Decline in relative abundance of bottlenose
    dolphins exposed to long-term disturbance. *Conservation Biology* 20, 1791–1798.
Bejder, L., Samuels, A., Whitehead, H. and Gales, N. (2006b) Interpreting short-term behavioural
    responses to disturbance within a longitudinal perspective. *Animal Behaviour* 72, 1149–1158.
Blane, J.M. and Jaakson, R. (1995) The impact of ecotourism on the St. Lawrence beluga
    whales. *Environmental Conservation* 21, 267–269.
Carlson, C. (2004) *A Review of Whale Watch Guidelines and Regulations around the
    World.* Report to the International Whaling Commission Meeting Scientific Committee,
    International Fund for Animal Welfare, Yarmouth Port, Massachusetts.
Constantine, R. (2001) Increased avoidance of swimmers by wild bottlenose dolphins (*Tursiops
    truncatus*) due to long-term exposure to swim-with-dolphin tourism. *Marine Mammal
    Science* 17, 689–702.

Constantine, R. and Baker, C.S. (1997) *Monitoring the Commercial Swim-with-Dolphin Operations in the Bay of Islands*. Science for Conservation No.56, Department of Conservation, Wellington, New Zealand.

Constantine, R. Brunton, D.H. and Baker, C.S. (2003) *Effects of Tourism on Behavioural Ecology If Bottlenose Dolphins of Northeastern New Zealand*. DoC Science Internal Series 153, Department of Conservation, Wellington, New Zealand.

Constantine, R., Brunton, D.H. and Dennis, T. (2004) Dolphin watching tour boats change bottlenose dolphin (*Tursiops truncatus*) behaviour. *Biological Conservation* 117, 299–307.

Corkeron, P.J. (1995) Humpback whales (*Megaptera novaeangliae*) in Hervey Bay, Queensland: behaviour and responses to whale-watching vessels. *Canadian Journal of Zoology* 73, 1290–1299.

Corkeron, P.J. (2004) Whale watching, iconography, and marine conservation. *Conservation Biology* 18, 847–849.

Coscarella, M.A., Dans, S.L., Crespo, E.A. and Pedraza, S.N. (2003) Potential impact of unregulated dolphin-watching activities in Patagonia. *Journal of Cetacean Research and Management* 5, 77–84.

Gordon, J., Leaper, R., Hartley, F.G. and Chappell, O. (1992) *Effects of Whale-watching Vessels on the Surface and Underwater Acoustic Behaviour of Sperm Whales off Kaikoura, New Zealand*. Science and Research Series No. 52, Department of Conservation, Wellington, New Zealand.

Hoyt, E. (2001) *Whale Watching 2000: Worldwide Tourism Numbers, Expenditures, and Expanding Socioeconomic Benefits*. Report to the International Fund for Animal Welfare, Yarmouth Port, Massachusetts.

Hoyt, E. (2002) Whale watching. In: Perrin, W.F., Würsig, B. and Thewissen, J.G.M. (eds) *Encyclopedia of Marine Mammals*. Academic Press, San Diego, California, pp. 1305–1310.

IWC (1996) Report of the Scientific Committee. International Whaling Commission Meeting 48, Washington, DC.

IWC (2006) Report of the Scientific Committee. International Whaling Commission Meeting 58, Washington, DC.

IFAW (International Fund for Animal Welfare) (2004) *The Growth of Whale Watching Tourism in Australia*. A report for IFAW – the International Fund for Animal Welfare, Yarmouth Port, Massachusetts.

IFAW (International Fund for Animal Welfare) (2005) *The Growth of the New Zealand Whale Watching Industry*. A report for IFAW – the International Fund for Animal Welfare, Yarmouth Port, Massachusetts.

Janik, V.M. and Thompson, P.M. (1996) Changes in the surfacing patterns of bottlenose dolphins in response to boat traffic. *Marine Mammal Science* 12, 597–602.

Hastie, G., Wilson, B., Tufft, L.H. and Thompson, P.M. (2003) Bottlenose dolphins increase breathing synchrony in response to boat traffic. *Marine Mammal Science* 19, 74–84.

Kruse, S. (1991) The interactions between killer whales and boats in Johnstone Strait, B.C. In: Pryor, K and Norris, K.S. (eds) *Dolphin Societies: Discoveries and Puzzles*. University of California Press, Berkeley, California, pp. 149–159.

Lusseau, D. (2003a) The effects of tourism activities on bottlenose dolphins (*Tursiops* spp.) in Fiordland, New Zealand. PhD thesis, University of Otago, Dunedin, New Zealand.

Lusseau, D. (2003b) Effects of tour boats on the behaviour of bottlenose dolphins: using Markov chains to model anthropogenic impacts. *Conservation Biology* 17, 1785–1793.

Lusseau, D. (2005) Residency pattern of bottlenose dolphins, *Tursiops* spp., in Milford Sounds, New Zealand, is related to boat traffic. *Marine Ecology Progress Series* 295, 265–272.

Lusseau, D. and Higham, J.E.S. (2004) Managing the impacts of dolphin-based tourism through the definition of critical habitats: the case of bottlenose dolphins (*Tursiops* spp.) in Doubtful Sounds, New Zealand. *Tourism Management* 25, 657–667.

Lusseau, D., Slooten, L. and Currey, R.J.C. (2006) Unsustainable dolphin watching tourism in Fiordland, New Zealand. *Tourism in Marine Environments* 3, 173–178.

Mangel, M., Talbot, L.M., Meffe, G.K., Agardy, M.T., *et al.* (1996) Principles for the conservation of wild living resources. *Ecological Applications* 6, 338–362.

Ministry Media Statement (2006) Available at: http://www.mediastatements.wa.gov.au/media/media.nsf/news/958A19167C70F7934825719900206D69. West Australian Government.

Nowacek, S.M., Wells, R.S. and Solow, A.R. (2001) Short-term effects of boat traffic in bottlenose dolphins, *Tursiops truncatus*, in Sarasota Bay, Florida. *Marine Mammal Science* 17, 673–688.

Orams, M.B. (1999) *The Economic Benefits of Whale Watching in Vava'u, The Kingdom of Tonga.* Centre for Tourism Research, Massey University at Albany, North Shore, New Zealand.

Richter, C., Dawson, S. and Slooten, E. (2006) Impacts of commercial whale watching on male sperm whales at Kaikoura, New Zealand. *Marine Mammal Science* 22, 46–63.

Salden, D.R. (1988) Humpback whale encounter rates offshore of Maui, Hawaii. *Journal of Wildlife Management* 52, 301–304.

Samuels, A. and Bejder, L. (2004) Chronic interaction between humans and free-ranging bottlenose dolphins near Panama City Beach, Florida, *Journal of Cetacean Research and Management* 6, 69–77.

Samuels, A., Bejder, L., Constantine, R. and Heinrich, S. (2003) Swimming with wild cetaceans, with a special focus on the Southern Hemisphere. In: Gales, N., Hindell, M. and Kirkwood, R. (eds) *Marine Mammals: Fisheries, Tourism and Management Issues.* CSIRO Publishing, Collingwood, Australia, pp. 277–303.

Scarpaci, C., Bigger, S.W., Corkeron, P.J. and Nugegoda, D. (2000) Bottlenose dolphins (*Tursiops truncatus*) increase whistling in the presence of 'swim-with-dolphin' tour operations. *Journal of Cetacean Research and Management* 2, 183–185.

Scarpaci, C., Nugedoga, D. and Corkeron, P. (2003) Compliance with regulations by 'swim-with-dolphins' operations in Port Phillip Bay, Victoria, Australia. *Environmental Management* 31, 342–347.

Scheidat, M., Castro, C., Gonzalez, J. and Williams, R. (2004) Behavioural responses of humpback whales (*Megaptera novaeangliae*) to whale watch boats near Isla de la Plata, Machalilla National Reserve, Ecuador. *Journal of Cetacean Research and Management* 6(1), 63–68.

Schneider, K. (1999) Behaviour and ecology of bottlenose dolphins in Doubtful Sound, Fiordland, New Zealand. PhD thesis, University of Otago, New Zealand.

Stensland, E. and Berggren, P. (2007) Behavioural changes in female Indo-Pacific bottlenose dolphins in response to boat-based tourism. *Marine Ecology Progress Series* 332, 225–234.

Van Parijs, S.M. and Corkeron, P.J. (2001) Boat traffic affects the acoustic behaviour of Pacific humpback dolphins, *Sousa chinensis. Journal of the Marine Biology Association, UK* 81, 1–6.

Williams, J.A., Dawson, S.M. and Slooten, E. (1993) The abundance and distribution of bottlenosed dolphins (*Tursiops truncatus*) in Doubtful Sound, New Zealand. *Canadian Journal of Zoology* 71, 2080–2088.

Williams, R.M., Trites, A.W. and Bain, D.E. (2002) Behavioural responses of killer whales (*Orcinus orca*) to whale-watching boats: opportunistic observations and experimental approaches. *Journal of Zoology, London* 256, 255–270.

Würsig, B. (1996) Swim-with-dolphin activities in nature: weighing the pros and cons. *Whalewatcher* 30, 11–15.

Würsig, B., Cipriano, F., Slooten, E., Constantine, R., Barr, K. and Yin, S. (1997) Dusky dolphins (*Lagenorhynchus obscurus*) off New Zealand: status of present knowledge. *Report for the International Whaling Commission* 47, 715–722.

18 Managing Marine Wildlife Experiences: The Role of Visitor Interpretation Programmes

M. Lück

Interacting with marine wildlife is an increasingly popular tourism activity (Orams, 1994; Schänzel, 1998; Mayes et al., 2004). These interactions range from viewing marine wildlife at aquaria and marine parks or in the wild to feeding, swim-with activities and fishing. Orams (1999) identifies four main management domains for marine (wildlife) tourism: regulatory, physical, economic and educational. Andersen and Miller (2006) suggest that, for example, in a whale-watching management context, marine environmental education can complement regulation with the goal of changing human conduct and attitudes. Most marine wildlife tourism activities have the potential to incorporate interpretive components. In fact, it is widely accepted that education and interpretation should be fundamental parts of tourism, in particular, at heritage sites and in the natural environment (Uzzell, 1989; Ham, 1992; Hall and McArthur, 1996; Uzzell and Ballantyne, 1998; Fennell, 1999; Newsome et al., 2002). Many regulatory bodies make an educational component a condition of granting licenses for tour operators. For example, New Zealand's Marine Mammals Protection Regulations (1992) state in Section 6(h) 'that the commercial operation should have sufficient educational value to participants or to the public'. In the wider context, national parks in general are for the 'benefit, education and enjoyment of the people of Canada' (Government of Canada, 1990, in Butler and Hvenegaard, 2002, p. 184). Simonds (1990) suggests that through nature-based tourism, ecological awareness can be increased when nature education is incorporated. Probably most important for the participants in nature-based and/or wildlife tours is a high level of satisfaction. Interpretation may help to increase the satisfaction level and make the visit a richer and more enjoyable experience (Weiler and Davis, 1993). It is surprising that not more researchers have taken the lead of the work of, for example, Forestell and Kaufman (1990) and Orams (1995). In this light, this chapter reviews the nature of education and interpretation, with a particular focus on an outdoor setting, and marine wildlife viewing.

Education and Interpretation

The terms 'education' and 'interpretation' are often used synonymously in a tourism context. However, despite often going hand in hand, there are distinct differences between these two. According to Hammitt (1984), education usually involves more formal approaches, whereas interpretation is usually a more informal tool. Thus, education takes place in a formalized setting, such as classrooms, and involves 'repeat' students as part of a captive audience. In contrast, interpretation commonly involves a 'first time' audience in a non-captive setting (Hammitt, 1984). Also, the motivation for participation can vary significantly between captive and non-captive audiences. The motivations of the first are usually driven by a certain tangible result, such as grades, diplomas, certificates, money or career advancement. In contrast, participants in non-captive settings are commonly more intrinsically motivated, for example, they want to have fun, or are seeking entertainment, self-enrichment or a better life, or are just passing time (have nothing better to do) (Plummer, 2005). Interpretation takes education in a slightly different direction, and builds a bridge between education and leisure activities (Plummer, 2005). It takes tourists from 'passive appreciation to exciting understanding of the natural and cultural environments' (Knudson *et al.*, 1999, p. 13). Interpretation is designed to stimulate interest and enthusiasm and provide an educational aspect. It thus has a pedagogic and an entertainment role (Moscardo and Pearce, 1986). One of the most recognized definitions of interpretation dates back to 1957, when Tilden (1957, p. 8) wrote:

> Interpretation is an educational experience which aims to reveal meaning and relationships through the use of original objects, by first hand experience, and by illustrative media, rather than simply to communicate factual information.

It is now recognized that education and entertainment are the two important components of interpretation. Newsome *et al.* (2002) contend that interpretation incorporates an educational, a recreational and a conservational component. By referring to parks and protected areas in Canada, Butler and Hvenegaard (2002, pp. 185–186) suggest that interpretation includes the following attributes:

1. It is on-site, emphasizing first-hand experiences with the natural environment (e.g. a park interpretive centre will introduce, clarify and direct the visitor to the outdoors for direct interactions between visitors and the park environment, unlike a museum in the city, which functions as the destination).
2. It is an informal form of education (i.e. interpretation does not employ a rigid, classroom-style approach).
3. It deals with a voluntary, non-captive audience (i.e. visitors participate by choice during their leisure time).

4. Visitors normally have an expectation of gratification (i.e. they want to be rewarded or to have a need or want satisfied).

5. It is inspirational and motivational in nature (i.e. interpretation does not present just factual information).

6. Its goals are expansion of knowledge, shifts in attitude and alterations in behaviour of visitors (i.e. visitors should increase their understanding of and their appreciation and respect for the park environment).

7. It is an extrinsic activity (e.g. an interpretive sign or exhibit along the trail side, which is not natural or innate to the setting), based on the intrinsic values of the landscape (e.g. a waterfall, a calypso orchid or the song of a red-eyed vireo). Interpretation facilitates understanding, appreciation and protection of the park's intrinsic landscape values.

The basis of all interpretation is communication (Plummer, 2005). Communication can involve various techniques, such as simple signs and plates (Fig. 18.1), audio tours (Fig. 18.2), video screenings and personal information provided by a tour guide (Fig. 18.3).

Communication involves a sender and a receiver. Often, the tour guide (or audio tape or a sign) is the sender, who conveys a certain message to the tourist. However, the roles can easily change, and with this change comes a dynamic process, which enables the tourist to be active part of the communication process. Such a change of role is illustrated in the example in Box 18.1.

Fig. 18.1. Interpretative mural at Marineland, Niagara Falls, Canada. (Photograph M. Lück.)

Fig. 18.2. Audio tours and signs at Rainbow Springs Nature Park, Rotorua, New Zealand. (Photograph M. Lück.)

Fig. 18.3. Personal interpretation by a guide (left) on a dolphin-watching tour in Hong Kong. (Photograph M. Lück.)

Box 18.1. The Communication Process. (After Plummer, 2005, pp. 279–280.)

Communication is at the basis of all types of interpretation. The basic process of communication is illustrated in Fig. 18.4 and is explained using a situation in which two individuals are penguin watching. The process of communication is initiated when an individual (encoder) has a concept that he wants to express to someone else.

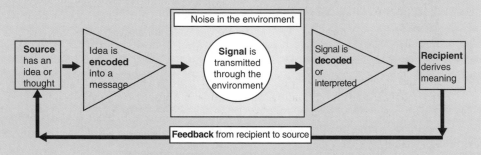

Fig. 18.4. The communication process. (From Plummer, 2005, p. 280.)

In this case, one of our penguin watchers, Neil, is very excited because he has just seen what he believes to be a rare penguin. The first step in the communication process involves the individual taking the image that he saw (rare penguin) and encoding it. This may occur in any number of forms; a lengthy description, a series of hand gestures and/or a specific name. Once the individual has encoded the concept, he projects the message. In our penguin example, our inexperienced watcher, Neil, tries to be very quiet, so as to not disrupt any of the penguins. He whispers the words 'yellow-eyed' and points to the location of the penguin in question. His companion, Anna, an experienced wildlife enthusiast, receives the message and decodes it. In decoding the message, the recipient attempts to make sense of the intended idea. In our example, the diction with which the penguin's name was said and accompanying hand gesture lead Anna to decipher that Neil was making a query regarding the nature of the penguin in question. Anna takes a look at the penguin and shakes her head side to side with a smile. She whispers the words 'little blue' back to Neil with a big smile on her face thereby providing feedback.

The Importance of Interpretation and Education

A variety of benefits of marine wildlife tours, including visitor benefits (educational benefits and psychological benefits) and conservation benefits, have been introduced by Heather Zeppel and Sue Muloin in Chapter 2. Thus, in this chapter only two examples shall suffice in order to highlight the importance of interpretive programmes on marine wildlife tours.

Example 1: Swim-with dolphin programmes in New Zealand

Although there has been some criticism about interpretation and education of tourists (Butler, 1990; Wheeller, 1991, 1994; Pleumarom, 1993), Aldridge

(1989, p. 64) suggests that 'interpretation is the art of explaining the significance of a place to the people who visit it, with the object of pointing a conservation message'. Many (eco)tour operators recognized that interpretation can be a powerful tool for conservation. However, the question is whether tourists indeed *want* to be educated. McKercher (1993) argues that tourists are nothing but consumers, and as such want to be entertained. In contrast, MacCannell (1976) notes that tourists are searching for the truth, the meaning and authenticity. Markwell and Weiler (1998) take it a step further and claim that it is the (eco)tourist's *commitment* to act environmentally and ecologically friendly, and that this can be supported through interpreted experiences. However, it appears that in many cases, the desires of the tourists are not recognized, and that there is a disequilibrium between the importance of interpretation objectives from a management's point of view and the tourist's point of view (McArthur and Hall, 1996). McArthur and Hall (1996) argue that the objectives are practised by heritage managers in the opposite order of importance to how they are perceived by visitors.

Lück (2003a,b) conducted research on swim-with-dolphins tours in Akaroa, Kaikoura and Paihia (New Zealand). One part of the survey was a number of questions about education and interpretation on the tours. A total of 79.7% (n = 733) of the respondents strongly agreed or mildly agreed that their dolphin tour was an educational experience, and 63.3% strongly agreed or mildly agreed that they had the feeling that on this tour they learned a lot about dolphins. However, when asked if they had learned a lot about other marine life, only 21.6% strongly agreed or mildly agreed. In an open-ended section of the survey, many respondents commented on the interpretation, complimenting the programmes and the enthusiastic crews on the tours. But they also mentioned that they would have loved to learn more about other marine wildlife in the area (whales, penguins, seals, etc.), about the role of the Department of Conservation (the Crown's conservation body in New Zealand) and about environmental threats to the marine wildlife in general, and the dolphins in particular (Lück, 2003a,b).

Example 2: Dolphin-watching at Tangalooma, Australia

Orams (1997) reports on a study undertaken at Tangalooma on Moreton Island, Australia, a site where dolphin feeding is allowed. Orams undertook a study on the changes after the implementation of a structured interpretive programme, which was introduced at the Tangalooma Dolphin Feeding Programme in 1994. Orams interviewed two visitor groups: first, a group of visitors before the interpretive programme was in place (the control group) in 1993, and a year later a group of visitors who were exposed to the new interpretive programme (experimental group). The new interpretive programme was quite extensive. First, a new Dolphin Education Centre was constructed, which contained a small library, posters, displays and a small video theatre. It was open every afternoon before the actual feeding. Second, a public address system was constructed at the actual feeding site. Presentations concentrated on educating tourists about various issues about the dolphins, for example, aspects of their behaviour and biology, while visitors could feed the dolphins. During this commentary, visitors were also encouraged to become more environmentally responsible (Orams, 1997). The results were astonishing (standardized to 100 feed events): The number of touches dropped from 6.7 in the control group to 1.2 in the experimental group. The number of staff cautions also dropped from 2.6 in the control group to 1.2 in the experimental group. Finally, the number of inappropriate behaviours dropped from 3.2 (control) to 1.1 (experimental). It also became clear that tourists felt that they had learned more after the interpretive programme had been established. A total of 32.4% of the control group agreed with the statement 'It was good, but I would have liked to have learned more', while this figure dropped to 11.6% after the implementation of the programme (Orams, 1997).

The two above studies illustrate two important points:

1. Marine wildlife tourists desire interpretation, often to a greater extent than what they experienced on their respective tours. Of Orams' (1997) control group, 32.4% would have loved to learn more, and during Lück's (2003a,b) study, only 21.6% strongly agreed or mildly agreed that they had learned a lot about other marine life.

2. Orams' study also clearly proved that well-planned, structured interpretive programmes do have an impact on the participants' learning and their attitudes towards conservation, including the intentions of being more environmentally sensitive.

Structured Interpretation Programmes

The previous sections attempted to define interpretation and education, and showed why interpretation is important as an agent for conservation. This section will look at interpretation concepts. There is considerably more experience about interpretation in terrestrial settings, but Gubbay (1989) contends that the planning process in coastal/marine settings is very similar. The various stages of such interpretation programmes are illustrated in Fig. 18.5. Gubbay uses the example of the UK, and suggests that the Marine Conservation Society regards the promotion of conservation as one of the main goals of marine park wardens. She also identifies four different target audiences, each of whom is likely to need customized interpretation programmes:

1. Those with a managerial interest in the coast;
2. Those with a recreational interest in the coast;
3. Those with a commercial interest in the coast;
4. Those with an educational interest in the coast.

Once the target audience has been identified, the following step includes the identification of the 'messages' that are to be conveyed. These can be split into three themes, including 'appreciation', the 'need for concern' and 'how you can help' (Gubbay, 1989).

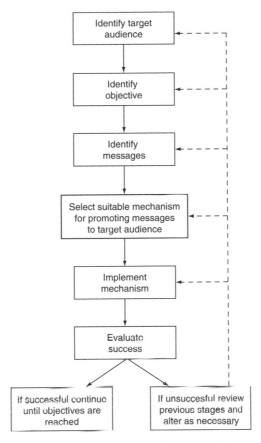

Fig. 18.5. Stages in promoting marine conservation at coastal sites. (From Gubbay, 1989, p. 171.)

In the next step, Gubbay identified a large number of possible techniques to convey the chosen messages. They can be shore-based, sea-based or remote. Some examples for shore-based techniques include seashore walks, leaflets, self-guided walks, events, touch tanks, games, videos and more. Sea-based techniques may include boat trips, underwater nature trails, viewing tunnels, glass-bottom boats, waterproof leaflets, etc. Finally, remote interpretation techniques can include displays, slide shows, films, aquaria, games, formal education, demonstrations, mass media and many more. Depending on the target audience, and the message to be brought across, interpretation techniques can vary considerably. Decisions as to what technique should be used in a specific context interpreters have to look at the capability of the respective technique to convey the message, and the practical requirements of the chosen technique. Also, Gubbay (1989, p. 177) recommends choosing techniques that have the following characteristics:

1. Uses *active involvement*;
2. Shows the *relevance* of the information;
3. Makes the experience *enjoyable*;

4. Generates curiosity and interest;

5. Uses personal contact.

The probability that interpretation programmes will be successful, i.e. that tourists enjoy themselves, while at the same time learn about the respective topic, is greatest if as many of these characteristics as possible are achieved for the chosen technique. There is evidence that people are more likely to retain information if it is presented in a way that includes the characteristics above (Gubbay, 1989; McArthur and Hall, 1996).

While Gubbay's (1989) model provides a clear structure for the implementation of interpretive programmes, there is still a lack of understanding when it comes to the specific content of these programmes. In a marine tourism context, the work of Forestell and Kaufman (1990), Forestell (1991) and subsequently Orams (1995) has been the most recognized in the academic literature, stimulating further research serving this field (e.g. Finkler, 2001; Lück, 2003a). Forestell and Kaufman based their model on the concept of cognitive dissonance, and Orams expanded this model into five major steps.

The Forestell and Kaufman Model

Forestell and Kaufman (1990) reviewed literature on cognitive psychological theory for the development of their model for effective interpretation, based on whale-watching tours in Hawaii. A key principle of their model is that a 'direct guided experience' is more effective than just either a 'guided experience' or a 'direct experience'. With 'direct experience', they refer to a real-life situation, for example, on a whale-watching tour without a guide, whereas a 'guided experience' is the exposure to a knowledgeable guide, but not in a real-life situation. Both concepts combined lead to a guided, real-life situation, which is the most effective form. Forestell and Kaufman's (1990, p. 404) model is based on a three-point approach:

1. Creating a perceived need for information;

2. Providing the needed information in an informed and interesting manner;

3. Facilitating participation in follow-up activities, which incorporate the new information into a changed behavioural repertoire.

They argue that a whale-watching tour can be divided into three different stages, each of which bears different information needs (Fig. 18.6).

During the pre-contact stage, tourists are excited about the coming experience and have the need for information regarding their safety, the surroundings and

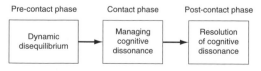

Fig. 18.6. Forestell and Kaufman's interpretation model. (From Orams, 1995, p. 85.)

their following encounter with whales. The contact phase is a time when tourists are interacting with whales. During this stage, they have specific questions about the mammals and their behaviour, as well as about the knowledge of the guides. The final, post-contact stage is a time of personal validation, in which participants compare knowledge and expectation with the just-experienced encounter. Forestell and Kaufman observed that during the post-contact phase, whale-watchers are very receptive to environmental issues in general. In this stage, they often reconsider global environmental threats and habitat degradation. Since they just encountered marine wildlife, these threats are not abstract issues far away from their home, but very tangible ones that are affecting the whales they have just encountered. Forestell and Kaufman conclude that interpretation was most effective if a final stage would be added. The proposed follow-up activities would include lobby activities, calls for signing petitions and making information material available to participants. Although they suggest that there is no scientific data available to support this proposition, Forestell and Kaufman stress the significant opportunities of this model, including the chance to change the participants' behaviour even for other marine activities in the future, such as snorkelling, nature cruises or diving trips.

The Orams Model

Forestell and Kaufman's (1990) model was the starting point for Orams (1993, 1994, 1996, 1997) to further develop this model. Orams suggests a model, which is based on five major steps, as illustrated in Fig. 18.7. The design of the interpretation programme includes both theories of cognitive dissonance and the affective domain (Festinger, 1957, 1964; Fishbein, 1967; Piaget and Inhelder, 1969; Fishbein and Ajzen, 1975; Ajzen and Fishbein, 1977, 1980; Piaget *et al.*, 1977). An interpretation programme should offer a variety of interesting questions, so that participants become curious and develop a cognitive dissonance between the questions and their knowledge. With stories about the animals encountered, for example, marine mammals, the affective domain shall be addressed through the involvement of participants' emotions. A state of cognitive dissonance is meant to motivate and provide an incentive to act. Orams suggests that the interpreter should address specific environmental problems and issues, and offer solutions for each participant to act (Fig. 18.7).

Ideally, participants are given concrete opportunities to act during the experience, such as petitions to sign, signing up for membership of an environmental organization or products to purchase that support environmental research. Orams, like Forestell and Kaufman (1990), stresses the importance of this stage, because tourists are highly motivated after the experience and more likely to act than they would be once they are back at home. The final stage is crucial for the design of programme changes. Feedback and assessment are indicators for the success of the programme and should include observation, interviews of participants or questionnaires. In order to investigate the long-term effects of the educational programme, follow-up surveys should be undertaken (Orams, 1993, 1996).

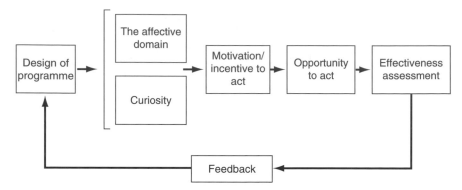

Fig. 18.7. Interpretation techniques (features of an effective interpretation programme). (From Orams, 1997, p. 297.)

Conclusion

This chapter addresses the differences between interpretation and education, with a focus on marine wildlife tourism. It highlights the importance of interpretation on marine wildlife tours, using two case studies from Australia and New Zealand. A framework to support structured interpretation programmes, with the goal of promoting conservation, is then considered. Finally, the two frameworks for successful interpretation, by Forestell and Kaufman (1990) and Orams (1995), are discussed. The discussion in this chapter contends that interpretation and education of (marine) wildlife tourists, but also of guides and members of other stakeholder groups, can play a crucial role in the endeavour for conservation of marine resources. Research clearly indicates that not only do tourists learn if well-organized, structured interpretive programmes are in place (see Orams, 1997), but also that tourists increasingly expect and demand a certain depth of interpretation during their marine wildlife experience (see Orams, 1997; Lück, 2003a,b). However, there still appears to be a lack of understanding of the desires of the tourists, i.e. what exactly they expect from the tours and guides, and what kind of specific additional information they would like to receive during the onsite experience. Recent research suggests that there is significant tourist demand for information about the wider marine environment, threats to marine wildlife and the role of conservation bodies (Lück, 2003a,b), but there remains a need for structured and well-planned research into this area.

References

Ajzen, I. and Fishbein, M. (1977) Attitude-behaviour relations: a theoretical analysis and review of empirical research. *Psychological Bulletin* 84(5), 888–918.

Ajzen, I. and Fishbein, M. (1980) *Understanding Attitudes and Predicting Social Behavior.* Prentice-Hall, London.

Aldridge, D. (1989) How the ship of interpretation was blown off course in the tempest: some philosophical thoughts. In: Uzzell, D. (ed.) *Heritage Interpretation Volume 1: The Natural and Built Environment*. Belhaven Press, London, pp. 64–87.

Andersen, M.S. and Miller, M.L. (2006) Onboard marine environmental education: whale watching in the San Juan Islands, Washington. *Tourism in Marine Environments* 2(2), 111–118.

Butler, J.R. and Hvenegaard, G.T. (2002) Interpretation and environmental education. In: Dearden, P. and Rollins, R. (eds) *Parks and Protected Areas in Canada: Planning and Management*, 2nd edn. Oxford University Press, Oxford, pp. 179–203.

Butler, R.W. (1990) Alternative tourism: pious hope or Trojan horse? *Journal of Travel Research* 28(3), 40–45.

Fennell, D.A. (1999) *Ecotourism: An Introduction*. Routledge, London/New York.

Festinger, L. (1957) *A Theory of Cognitive Dissonance*. Tavistock Publications, London.

Festinger, L. (1964) *Conflict, Decision & Dissonance*. Tavistock Publications, London.

Finkler, W. (2001) The experiential impact of whale watching: implications for management in the case of the San Juan Islands, USA. MSc thesis, University of Otago, Dunedin.

Fishbein, M. and Ajzen, I. (1975) *Belief, Attitude, Intention and Behaviour: An Introduction to Theory and Research*. Addison-Wesley, Reading, Massachusetts.

Fishbein, M.E. (1967) *Readings in Attitude Theory and Measurement*. Wiley, New York/London/Sydney.

Forestell, P.H. (1991) *Marine Education and Ocean Tourism: Replacing Parasitism with Symbiosis*. Paper presented at the 1990 Congress on Coastal and Marine Tourism: A Symposium and Workshop on Balancing Conservation and Economic Development, Honolulu, Hawaii.

Forestell, P.H. and Kaufman, G.D. (1990) *The History of Whale Watching in Hawaii and Its Role in Enhancing Visitor Appreciation for Endangered Species*. Paper presented at the 1990 Congress on Coastal and Marine Tourism: A Symposium and Workshop on Balancing Conservation and Economic Development, Newport, Oregon.

Gubbay, S. (1989) Interpreting the United Kingdom's marine environment. In: Uzzell, D. (ed.) *Heritage Interpretation. Volume 1: The Natural and Built Environment*. Belhaven Press, London, pp. 170–178.

Hall, C.M. and McArthur, S. (eds) (1996) *Heritage Management in Australia and New Zealand*. Oxford University Press, Oxford.

Hammitt, W.E. (1984) Cognitive processes involved in environmental interpretation. *Journal of Environmental Education* 15(4), 11–15.

Ham, S.H. (1992) *Environmental Interpretation: A Practical Guide for People with Big Ideas and Small Budgets*. North American Press/Golden, New York.

Knudson, D.M., Cable, T.T. and Beck, L. (1999) *Interpretation of Cultural and Natural Resources*. Venture Publishing, State College, Pennsylvania.

Lück, M. (2003a) Environmental education on marine mammal tours as agent for conservation – but do tourists want to be educated? *Ocean and Coastal Management* 46(9 & 10), 943–956.

Lück, M. (2003b) Environmentalism and on-tour experiences of tourists on wildlife watch tours in New Zealand: a study of visitors watching and/or swimming with wild dolphins. PhD thesis, University of Otago, Dunedin, New Zealand.

MacCannell, D. (1976) *The Tourist: A New Theory of the Leisure Class*. The Macmillan Press, London.

Markwell, K. and Weiler, B. (1998) Ecotourism and interpretation. In: Uzzell, D. and Ballantyne, R. (eds) *Contemporary Issues in Heritage and Environmental Interpretation*. The Stationery Office, London, pp. 98–111.

Mayes, G., Dyer, P. and Richins, H. (2004) Dolphin–human interaction: pro-environmental attitudes, beliefs and intended behaviours and actions of participants in interpretation programs: a pilot study. *Annals of Leisure Research* 7(1), 34–53.

McArthur, S. and Hall, C.M. (1996) Visitor management: principles and practice. In: Hall, C.M. and McArthur, S. (eds) *Heritage Management in Australia and New Zealand.* Oxford University Press, Oxford, pp. 37–51.

McKercher, B. (1993) Some fundamental truths about tourism: understanding tourism's social and environmental impacts. *Journal of Sustainable Tourism* 1(1), 6–16.

Moscardo, G. and Pearce, P.L. (1986) Visitor centres and environmental interpretation: an exploration of the relationship among visitor enjoyment, understanding and mindfulness. *Journal of Environmental Psychology* 6, 89–108.

Newsome, D., Moore, S.A. and Dowling, R.K. (2002) *Natural Area Tourism.* Channel View Publications, Clevedon, UK.

Orams, M.B. (1993) *The Role of Education in Managing Marine Wildlife–Tourist Interaction.* Paper presented at the 7th Annual Marine Education Society of Australasia Conference, Brisbane, Australia.

Orams, M.B. (1994) Creating effective interpretation for managing interaction between tourists and wildlife. *Australian Journal of Environmental Education* 10(September), 21–34.

Orams, M.B. (1995) Using interpretation to manage nature-based tourism. *Journal of Sustainable Tourism* 4(2), 81–94.

Orams, M.B. (1996) An interpretation model for managing marine wildlife–tourist interaction. *Journal of Sustainable Tourism* 4(4), 81–95.

Orams, M.B. (1997) The effectiveness of environmental education: can we turn tourists into 'Greenies'? *Progress in Tourism and Hospitality Research* 3, 295–306.

Orams, M.B. (1999) *Marine Tourism: Development, Impacts and Management.* Routledge, London.

Piaget, J. and Inhelder, B. (1969) *The Psychology of the Child.* Routledge & Kegan Paul, London.

Piaget, J., Grize, J.-B., Szeminska, A. and Bang, V. (1977) *Epistemology and Psychology of Functions,* Vol. 83. D. Reidel Publishing, Dordrecht, The Netherlands.

Pleumarom, A. (1993) What's wrong with mass ecotourism? *Contours* 6(3–4), 15–21.

Plummer, R. (2005) *Outdoor Recreation: An Interdiscplinary Perspective.* Kendall/Hunt Publishing, Dubuque, Iowa.

Schänzel, H.A. (1998) *Wildlife-Viewing Tourism on the Otago Peninsula.* DipTour dissertation, University of Otago, Dunedin.

Simonds, M.A.C. (1990) *Natural Encounters: Adventure, Learning, and Research Programs for the Marine Environment.* Paper presented at the 1990 Congress on Coastal and Marine Tourism: A Symposium and Workshop on Balancing Conservation and Economic Development, Honolulu, Hawaii.

Tilden, F. (1957) *Interpreting Our Heritage.* The University of North Carolina Press, Chapel Hill, North Carolina.

Uzzell, D. (ed.) (1989) *Heritage Interpretation Volume 1: The Natural and Built Environment.* Belhaven Press, London.

Uzzell, D. and Ballantyne, R. (eds) (1998) *Contemporary Issues in Heritage and Environmental Interpretation.* The Stationery Office, London.

Weiler, B. and Davis, D. (1993) An exploratory investigation into the roles of the nature-based tour leader. *Tourism Management* 14(2), 91–98.

Wheeller, B. (1991) Tourism's troubled times: responsible tourism is not the answer. *Tourism Management* 12(2), 91–96.

Wheeller, B. (1994) Ecotourism: a ruse by any other name. In: Cooper, C.P. and Lockwood, A. (eds) *Progress in Tourism, Recreation and Hospitality Management* Vol. 6. Wiley, Chichester, UK, pp. 3–11.

19 Marine Wildlife Viewing: Insights into the Significance of the Viewing Platform

J.E.S. HIGHAM AND W.F. HENDRY

Introduction

Marine wildlife observation takes place from 'viewing platforms', which are designed to temporarily accommodate visitors at locations where interactions with wild animals are relatively predictable. At the most simple level, marine wildlife viewing typically takes place from either land-based observatories (e.g. to view colonial nesting seabirds in island or mainland colonies), boat based (e.g. viewing marine mammals) or airborne (e.g. viewing whales from fixed-wing aircraft or helicopter) viewing platforms. All these platforms are, for example, used to view whales in different parts of the world (Higham and Lusseau, 2004). As marine wildlife viewing has increased in popularity and expanded in scale, so the range and locations of viewing platforms have diversified. Thus marine wildlife species can be viewed on foot (e.g. walking among nesting penguins and breeding sea lions in the peri-Antarctic islands), in kayaks, on inflatable and highly manoeuvrable rubber boats (IRBs), on four wheel motorbikes, in retired army transport vehicles, submarines and submerged cages designed to withstand shark attack, in remote locations via closed circuit television and even (if one takes the definition of viewing platform to an extreme) in the water if one chooses to 'swim with dolphins'. Such a diversity of viewing platforms and marine wildlife-viewing contexts raises a plethora of questions that, if unanswered, leave sustainable tourist–wildlife interactions an improbability.

In 1990, Duffus and Dearden identified three key elements of wildlife-viewing experiences: site users, the focal wildlife population and the wider ecology of the site at which the experience takes place. In this chapter it is suggested that all three of these key elements are directly influenced by the design, location and management of the viewing platforms from which people observe animals, either individual animals or local populations of animals, in the wild. It is argued here that the experiences of site users, the responses of wild animals being observed and the impacts of tourists upon site ecologies, are all largely determined by the viewing platform that is accessed by visitors, and at which the visitor experience takes place. This chapter begins by reviewing what is known or, perhaps more accurately, what is *not* known about wildlife-viewing platforms. In beginning to address some important aspects of the human dimensions of viewing marine wildlife, the results of a study of whale watching in the San Juan Islands are reviewed, and conclusions drawn, before considering questions for future researchers to address, so as to provide a much needed and long overdue understanding of the viewing platform and its relevance to sustainable marine wildlife tourism.

Wildlife-viewing Platforms

Little research that specifically addresses the viewing platform has been published in the field of wildlife tourism. Much of what is known about wildlife-viewing platforms is largely anecdotal. Viewing platforms concentrate visitors in well-defined locations where interactions with wildlife are predictable or quite constant (Whittaker, 1997). Viewing platforms can be either static or mobile. The location of static viewing platforms is typically determined by the presence, predictability and/or abundance of resident wildlife. The location of static viewing platforms is no doubt also dictated by factors such as visitor convenience and (wherever possible) accessibility in relation to tourist routes, infrastructure and services. As a consequence, it is probable that the location and design of wildlife-viewing platforms is determined by human comfort and convenience rather than wildlife impact mitigation. Viewing platforms are also likely to be located in critical and/or important wildlife habitats (e.g. breeding, socializing and feeding habitats), where sightings are most consistent and/or spectacular behaviours (e.g. adolescent displaying, courtship and socializing behaviours) are most likely to be observed. As demand increases it is typical that existing viewing platform facilities are expanded or multiplied. Concern for the biological consequences of such developments is usually disregarded or, at best, clouded due to ignorance.

It is evident, at least in New Zealand, that a recent trend has been the proliferation of viewing platforms. While most commercial wildlife-viewing sites once supported only static land-based viewing platforms, numerous variations now exist, including mobile (e.g. vehicle-based) platforms, aircraft and boat-based platforms and remote (e.g. closed circuit television) platforms. A further variation of the viewing platform is to actually engage directly with a focal animal (or animals) such as is the case with the 'swim-with-dolphins' phenomenon where the viewing platform, as such, actually involves being in the water with marine mammals. Protective cages (e.g. shark and polar bear viewing in different parts of the world) are, again, variations of such platforms. In some cases, such as viewing Royal Albatross at the Taiaroa Head Colony in Dunedin (land- and boat-based) and whale watching in Kaikoura (air- and boat-based), multiple platforms may exist at the same site. This raises questions of symmetric or asymmetric social impacts between visitors who are visible and/or able to interact with each other at different viewing platforms. The behaviours exhibited by visitors (as well as guides and drivers) at these viewing platforms may be important factors in terms of both wildlife impacts and visitor experiences.

Mobile viewing platforms present a range of additional management challenges. One significant challenge relates to the duration and frequency of contact between visitors and focal animals. This issue is particularly acute in cases where increasing demand leads to constant interaction with focal animals. Others include intrusion into critical habitats (knowingly or otherwise) and irresponsible manoeuvring (including relentless pursuit) when in close contact with focal animals. Evidence from numerous dolphin-watching locations from around the world confirms a widespread inability to protect dolphins, either spatially or temporally, from interaction with tourist vessels (Würsig, 1996; Constantine, 1999; Driscoll-Lind and Östman-Lind, 1999; Lusseau, 2002; Bejder *et al.*, 2006), despite the existence of research that confirms the urgent need for such protection (Lusseau, 2002; Lusseau and Higham, 2004). Indeed the reverse is commonly the case as dolphin-watching permits have continued to be issued to commercial operators, and non-permitted viewing has proliferated, as visitor numbers to dolphin-watching sites have continued to increase (Lusseau *et al.*, 2006a,b).

Apart from the impacts of viewing platform design and location on wildlife, some researchers have addressed elements of the human dimensions of wildlife-viewing platforms. Whittaker (1997) assessed viewers' capacity norms on bear-viewing platforms in Alaska and found that there were upper limits for platform capacities irrespective of platform size and design. The works of Findlay (1997), Kind-Keppel *et al.* (1999) and Malcolm *et al.* (2002) also bear much relevance to viewing platforms. These studies confirm the dearth of social science research that examines the human dimensions of visitor experiences achieved at wildlife-viewing platforms.

Marine Wildlife-viewing Platforms: The Case of Whale Watching

There are three main platforms from which whales in the wild can be watched: land-based, boat-based and air-based. Land-based platforms are located in areas where whales frequently appear close to the shore. These platforms are common in California (USA), Quebec (Canada), South Africa and Australia (Hoyt, 1994). Boat-based platforms include a multitude of vessels such as kayaks, sailboats, small- and medium-sized powerboats and large cruise ships (International Fund for Animal Welfare, World Wildlife Fund, Whale & Dolphin Conservation Society, 1999). Air-based whale watching occurs from helicopters and small planes and can be found in areas such as Hawaii and New Zealand.

The growth and diversification of whale watching has raised concerns over the planning and management of whale-watching platforms. The long-term survival of whale populations is central to these concerns (Beach and Weinrich, 1989; Amante-Helweg, 1995). The management of whale-watching platforms varies considerably around the world (Orams, 2000). However, in most cases whale-watching management is based on regulatory strategies such as:

approach speed, direction and minimum distance of boats; duration of interaction with whales; and manoeuvring of vessels in close proximity to whales (Baxter, 1993; Baird et al., 1998). Industry self-regulation through 'codes of ethics' or 'codes of practice' guidelines is also common (McKegg, 1996). Overall, whale watching continues to grow, whereas its management remains largely uncontrolled (Spalding and Blumenfeld, 1997; Baird et al., 1998; Corkeron, 1998).

The potential consequences of impacts from whale watching were classified into three levels by Duffus and Dearden (1993). First, *immediate consequences* of whale disturbance are caused by such events as whale–vessel collisions possibly resulting in reduced population numbers and/or impacts on the breeding success of whales. Second, *short-term consequences* describe interference with important behaviours such as feeding, resting, courtship and the care of juveniles. This may lead to energetic imbalances, reduced foraging success and interference with acoustic processes. Third, changes in ecological and energetic systems and breeding success may lead to *long-term consequences* and the reduction of fitness on a local or regional basis. The cumulative impacts over time may contribute to long-term biological impacts, giving rise to wider population and ecological change.

An increasingly comprehensive literature concerning these possible impacts on cetaceans in different regions and contexts has emerged (Osborne, 1986, 1991; Finley et al., 1990; Briggs, 1985; MacGibbon, 1991; Gordon et al., 1992; Phillips and Baird, 1993; Corkeron, 1995; Williams et al., 2002). Increasingly, studies are based on long-term observations, or otherwise designed to capture a temporal element of analysis (Barber, 1993; Würsig, 1996; Constantine, 1999). Much of this research relates to the impacts of boat-based whale watching rather than land-based or airborne tourist activity. Little is known about how the physical impacts of tourism vary between whale-watching platforms.

Even less is known about the human dimensions of whale watching and remarkably little is known about the perceptions of whale watchers although the need for such information is now well recognized (Findlay, 1997; Muloin, 1998; International Fund for Animal Welfare, World Wildlife Fund, Whale & Dolphin Conservation Society, 1999; Kind-Keppel et al., 1999; Orams, 2000). More specifically, little is known about how whale watchers differ with respect to their expectations and satisfaction between different whale-watching platforms (e.g. land-based, boat-based).

Over the last decade, some notable contributions to the human dimensions of whale watching have emerged (Duffus and Dearden, 1992; Duffus and Baird, 1995; Duffus, 1996). For example, Malcolm et al. (2002) provide a comprehensive analysis of ecotourism development and management strategies relating to whale watching. They observed that social science research has recently become an important focus for developing management options at the increasingly diverse range of whale-watching sites. Their sample of 1617 whale watchers drawn from three different locations in British Columbia, Canada, highlighted the importance of site-specific education programmes as a strategy for managing whale watching. In addition, Malcolm *et al.* (2002) showed that whale watchers have different expectations and levels of experience. This, in turn, influenced visitor satisfaction ratings at each location.

Findlay's (1997) study of land-based tourists viewing southern right whales (*Eubalaena australis*) in South Africa revealed resident support for land-based whale watching, and opposition to the development of boat-based whale watching due to concerns regarding whale disturbance. Kind-Keppel *et al.* (1999, p. 18) suggested that:

> Perhaps whale watchers on commercial boats feel the same sense of stewardship towards the Killer Whale as do observers on the shoreline, but now they find their sense of stewardship being compromised, given the close proximity to whales that their captain has placed them in. This, plus the fact that they have just paid a sizable fee ($35–$60) to be on the boat, makes it much easier for them to change their attitudes.

Understanding humans and their activities at wildlife-viewing platforms is as important as understanding the responses of whales to those activities (Orams, 2000). It is absurd to consider the management of whale viewing without consideration of the human component of the activity, given that management responses are focused mainly on people, not whales (Forestell and Kaufmann, 1993; Orams, 2000). The following sections, therefore, provide insights into the human dimensions of whale watching in the San Juan Islands, focusing on the perceptions and experiences of whale watchers at different viewing platforms.

Whale-watching Platforms: The Case of the San Juan Islands

The San Juan Islands are located in Washington State, USA, in close proximity to the Canadian border. Southern resident killer whales (*Orcinus orca*) annually forage the waters around the San Juan Islands from April to November (Kind-Keppel *et al.*, 1999). Lime Kiln Point State Park, located on the western side of San Juan Island, serves as the most popular shoreline whale observation area in these islands (i.e. land-based platform). This park attracts approximately 200,000 visitors annually (Hoyt, 2000). This area also attracts many boat-based whale watchers. The number of whale-watching boats in this area increased from an average of 4.4 boats per day in 1990 to 26.2 boats per day in 1996 (Kind-Keppel *et al.*, 1999). On one occasion in 1997, 108 boats (commercial and private) were recorded following a group of 25 whales (Kind-Keppel *et al.*, 1999). This study site represents a unique place to conduct research on human–whale interactions, as it offers both land-based and boat-based whale-watching opportunities.

Data collection at land- and boat-based viewing platforms

The methodology included self-completed visitor surveys and an observational checklist (Finkler, 2001). The survey was pilot tested at the study site in May

2000. The main data collection occurred from May to August 2000. The sample population comprised people participating in whale-watching trips at one of the two locations. First, visitors were surveyed on four different commercial whale-watching boats in the San Juan Islands (boat-based platform or BB platform). People were surveyed at the completion of the whale-watching experience during the return to port. Second, people watching whales from the shore within Lime Kiln State Park (land-based platform or LB platform) were approached after their whale-watching experience. The study employed a systematic sampling procedure to eliminate researcher bias. Every fourth person over 16 years of age was approached and invited to participate in the study. All surveys were completed on site.

In total, 327 LB and 306 BB whale watchers (n = 633) completed the survey. The researcher spent a total of 64 days at the land-based whale-watching site and made field observations; whales were visible at this site on 47 of these days. BB participants were sampled during 16 days on a total of 26 individual boat trips. Using a checklist, observational data were also collected and recorded for each whale-watching occasion. The checklist comprised of the number and type of boats (e.g. powerboats, sailboats, kayaks), the number of people on land or in boats, the number of whales seen and the total time that whales were present.

The rating systems used for most closed-response questions in the surveys were 5-point Likert-type scales, which allowed participants to respond to a range of variables related to the visitor experience. Respondents indicated their post-experience perceptions of items including levels of tourist presence, vessel activity and noise. Independent sample t-tests and chi-square (χ^2) tests were used to compare data among respondents at the LB and BB whale-watching platforms (Babbie, 1995). In addition, effect size measures were reported where appropriate (Cohen, 1988; Vaske et al., 2002).

Results

On-site observations
On-site field observations revealed that on each occasion when whale watching was taking place, between four and 120 people were watching from the shoreline (M = 43.8) (Table 19.1). At these times, the number of boats travelling with whales ranged from one to 64 (M = 24.8). On average, there were 16.5 powerboats, 6.8 kayaks and 1.7 sailboats accompanying the whales when they were being viewed from the shore. From the LB platform, an average of 20.6 whales were visible for an average duration of 36.7 min on each occasion. During the surveying period, the number of people onboard whale-watching boats ranged from six to 53 people (M = 28.0). The number of boats with whales during each whale-watching occasion ranged from three to 36 boats (M = 18.8).

On average, there were 16.5 powerboats, 1.4 kayaks and 0.7 sailboats travelling with the whales during each BB encounter. The average number of whales visible from the boats was 22.5 and the average time spent with whales

Table 19.1. Summary of boat- and whale-related variables for land- and boat-based viewing platforms.

	Land-based				Boat-based					
	Min	Max	Mean	SD	Min	Max	Mean	SD	t-value	p-value
Number of people in group	4	120	43.8	28.0	6	53	28.0	13.0	9.14	0.001
Total number of boats	1	64	24.8	17.0	3	36	18.8	9.1	5.55	0.961
Number of powerboats	1	36	16.5	9.8	3	33	16.5	7.6	0.049	0.001
Number of kayaks	0	28	6.8	7.9	0	10	1.4	2.8	11.55	0.001
Number of sailboats	0	5	1.7	1.6	0	2	0.7	0.8	9.66	0.001
Time spent with whales (min)	10	122	36.7	27	30	110	70.9	18	18.55	0.001

during a BB whale-watching encounter was 70.9 min. There were significant differences in the viewing circumstances experienced at each platform. In general, LB whale watchers viewed whales in larger groups ($M = 43.8$ people) than their BB counterparts ($M = 28.0$) ($t = 9.14$, $df = 471.4$, $p < 0.001$). Furthermore, whales that were being watched from land were generally accompanied by more kayaks ($t = 11.55$, $df = 413.5$, $p < 0.001$) and sailboats ($t = 9.66$, $df = 482.0$, $p < 0.001$). Differences in levels of powerboat traffic between viewing platforms were statistically insignificant ($t = 5.55$, $df = 611.8$, $p = 0.961$). The time spent viewing whales was significantly ($t = 18.55$, $df = 570.8$, $p < 0.001$) longer for BB visitors ($M = 70.9$ min) compared to LB watchers ($M = 36.7$ min). These results suggest that LB whale-watching within Lime Kiln Point takes place in larger groups and for shorter durations than BB whale-watching in the area.

Respondent profile

The majority of whale watchers at each platform lived in the USA (LB = 93% and BB = 94%), while 3% of the LB whale watchers were from Canada (Table 19.2). The rest were from Europe, Australia, Israel and Mexico. A high proportion of the US whale watchers on each platform came from Washington State (LB = 52% and BB = 23%). The difference in residency between the two platforms was statistically significant ($\chi^2 = 12.0$, $df = 4$, $p = 0.017$), however, using the guidelines suggested by Cohen (1988) and Vaske *et al.* (2002), this difference can be considered as 'small' or 'minimal', respectively ($V = 0.138$).

Female whale watchers predominated at both platforms although this result was not statistically significant ($\chi^2 = 0.668$, $df = 1$, $p = 0.414$). The age profile could be described as predominantly in the 20–49 age range at the LB platform, and 30–59 age range at the BB platforms ($\chi^2 = 30.2$, $df = 5$, $p < 0.001$,

$V = 0.221$). LB whale watchers were more likely to hold college and university degrees although they were less likely to hold graduate degrees ($\chi^2 = 13.4$, $df = 4$, $p = 0.01$, $V = 0.149$). The effect sizes reported in Table 19.2 range from 0.1 to 0.3, which may be described as 'small' to 'medium' (Cohen, 1988) or 'minimal' to 'typical' (Vaske et al., 2002). This suggests that although the results are statistically significant, the effect sizes indicate that practical significance may be negligible.

Whale watchers' perceptions of the visitor experience

LB and BB respondents reported concerns about the manoeuvring of power-boats into the path of killer whales and the noise from powerboats as causes of whale disturbance. The majority of LB whale watchers (74%) expressed concerns about the noise from powerboats impacting the whales' behaviour (Table 19.3). Fewer people (54%) watching from boats were concerned about this possible negative impact. This difference was statistically significant ($\chi^2 = 25.7$, $df = 2$, $p < 0.001$, $\phi = 0.212$). LB whale watchers also expressed concern

Table 19.2. Summary of origin, gender, age range and highest level of education.

	Land-based (%)	Boat-based (%)	χ^2	df	p-value	Cramer's V or Phi ϕ
Visitor origin						
USA	93	94	12.0	4	0.017	0.138
Canada	3	–				
UK	2	3				
Europe	2	3				
Other	1	1				
Gender						
Female	59	56	0.668	1	0.414	0.330
Male	41	44				
Age range (years)						
16–19	4	6	30.2	5	0.001	0.221
20–29	26	12				
30–39	27	26				
40–49	20	24				
50–59	17	19				
Over 60	6	14				
Highest level of education						
High school	16	22	13.4	4	0.010	0.149
Technical school degree	5	4				
Community college degree	10	5				
University/ college degree	44	35				
University graduate degree	23	30				

Table 19.3. Summary of visitor responses to impact issues at each platform.[a]

	Land-based (%)	Boat-based (%)	χ^2	df	p-value	Phi ϕ
Noise disturbance by powerboats impacting upon the whales' behaviour	74	54	25.7	2	0.001	0.212
Powerboats impacting on killer whale safety by positioning themselves in the path of the whales	73	56	19.7	2	0.001	0.188
Disturbance of the whales' behaviour by powerboats	69	52	18.2	2	0.001	0.179
Disturbance of your visual experience by powerboats	53	28	39.7	2	0.001	0.262
Noise of powerboats disturbing your personal experience	47	27	26.4	2	0.001	0.215
Kayaks impacting on killer whale safety by positioning themselves in the path of the whales	27	18	6.9	2	0.031	0.118

[a]Cell entries represent the % of respondents at each platform specifying this as a 'concern'.

about powerboats compromising whale safety by being positioned in the path of whales (73%). Significantly fewer BB whale watchers (56%) expressed concerns with this issue ($\chi^2 = 19.72$, $df = 2$, $p < 0.001$, $\phi = 0.188$). This potential impact was clearly a concern for LB whale watchers, whereas their BB counterparts were somewhat divided on this issue. The effect size, however, was 'small' to 'medium' (Cohen, 1988).

The third most common concern for LB whale watchers was that of general disturbance to whale behaviour by powerboats (69%), followed by the disturbance to their own visual experience by powerboats (53%). The latter issue only concerned 30% of the BB whale watchers ($\chi^2 = 39.70$, $df = 2$, $p < 0.001$, $\phi = 0.262$). More than half of the BB whale watchers (52%) expressed concerns regarding whale disturbance by other powerboats, as opposed to the boat that they were on (32%) ($\chi^2 = 18.2$, $df = 2$, $p < 0.001$, $\phi = 0.179$). Nearly half of the LB whale watchers (47%) expressed concerns regarding the noise of powerboats disturbing their own personal experience, while fewer BB whale watchers (27%) were concerned about noise ($\chi^2 = 26.4$, $df = 2$, $p < 0.001$, $\phi = 0.215$). Whale watchers reported relatively little concern with kayaks moving into the path of killer whales. More LB (27%) than BB (18%) whale watchers expressed concerns regarding this potential impact ($\chi^2 = 6.9$, $df = 2$, $p = 0.031$, $\phi = 0.118$). Disturbance of the visual experience of LB whale watchers and general concern for noise impacts disrupting both the visitor experience and whale behaviour emerged as perhaps the most significant impact issues.

Open-ended questions provided respondents with the opportunity to express their views regarding aspects of their whale-watching experience. Qualitative statements provided by respondents indicated that whale watchers at both platforms enjoyed their experience and responded to it with positive feelings and emotions. People stated that they were very satisfied but, none the less, felt the desire to encounter whales from the closest possible proximity. Whale watchers from both platforms, however, expressed concern about adverse impacts on whales. Comments included: 'I was thinking about what kind of impact we are having on the whales while we were out watching; it's great to watch them, but I don't think it should be at the expense of the whales' and 'I often feel torn between experiencing the Killer Whales and disturbing them'.

In total, over one-third of LB whale watchers (39%) mentioned 'impacts on the whales', particularly in relation to the boats getting too close to the whales, as the least enjoyable aspect of their experience. By comparison, 21% of BB whale watchers mentioned this as the least enjoyable aspect of their trip. Comments, for example, included: 'when too many boats go too fast or too close to the whales it really makes me angry' and 'versus boat-based whale watching, I liked being far enough away on the land, so that I was not disturbing the whales'.

Discussion

LB whale watchers appeared to be significantly more concerned than BB respondents about the disturbance of whales caused by visitor activities, particularly from noise and boat position. In addition, these visitors were more concerned about the disturbance of their own experiences as a result of the presence of too many powerboats. These results are consistent with the findings of Kind-Keppel et al. (1999) who identified a widespread perception, from both tourists and host community residents, that boats harass whales. This perception appears to be greater for LB than BB observers. Kind-Keppel et al. (1999) suggest that such differences in perception may relate to the different visual perspectives between LB and BB viewing platforms. To watch interactions from the shore might exacerbate perceptions of impacts, as it is difficult to estimate vessel–whale distance from this position (Phillips and Baird, 1993).

Observers on boats may not experience this crowding effect as substantially or frequently.

Whale watchers at both platforms enjoyed their experience and reported positive feelings and emotions. Many visitors, however, stated that they felt the desire to encounter killer whales at a closer proximity, but acknowledged the potential for adverse impacts on the whales to be a consequence. Blewett (1993, p. 1) recognized this dilemma and referred to it as 'the whale watcher's paradox'. This describes the desire of whale watchers to get as close as possible to whales, while simultaneously recognizing the possibility that their desired experience may impact the whales. In addition, the results of this study are consistent with Lindberg and McKercher (1997), who argued that although tourists might enjoy themselves and respond with high satisfaction rankings, they also desire improvements and changes in activities and/or management conditions.

This study confirms that whale watchers from LB and BB platforms recognize the issue of whale disturbance and see their own participation as being a possible negative impact. Many respondents expressed concerns about negative impacts on the whales as the least enjoyable aspect of their whale-watching experience. This suggests that the rigorous and comprehensive management of commercial whale-watching activities and site (e.g. platform)-specific education programmes may help to enhance the whale-watching experience (Malcolm *et al.*, 2002).

Conclusion

In the marine tourism context, the diversity of wildlife-viewing platforms is noteworthy, as are the experiential, impact and management challenges associated with different platforms. As such, manifold research questions remain unanswered as to the impacts and management challenges associated with wildlife-viewing platforms in different marine tourism contexts. Visitor perceptions of the impacts of wildlife-based tourism from LB and BB platforms are not well understood; so, too, the physical impacts of wildlife-based tourism as they relate to the design of, and management of visitors at, different viewing platforms.

Research conducted at BB and LB whale-viewing platforms in the San Juan Islands (USA) demonstrated that visitors at both platforms were concerned about adverse impacts on whales due to whale-watching activities. LB whale watchers, however, were significantly more concerned about the presence of boats disturbing not only the whales, but also their own experiences. In particular, the noise and manoeuvring of whale-watching boats were identified as issues of concern. Concern for the welfare of whales was shown to detract from the visitor experience, especially for LB whale watchers.

The findings suggest that site-specific strategies for managing whale watching are necessary. Possible strategies include limiting the number of boats and the frequency and duration of boat interactions with whales. Furthermore, platform-specific visitor education programmes may be useful for helping to address viewers' concerns. Codes of ethics, codes of best practice and self-regulation may be an ineffective management response if they take place in the absence of regulations governing levels of visitor activity. This suggests that the effective regulation of the development and management of wildlife-viewing platforms is critical to the sustainability of marine wildlife viewing. An adequate understanding of the social (human) and biological (wildlife) dimensions of wildlife viewing at specific sites, and from particular viewing platforms, is a challenging but important first step in this direction. Positioning the viewing platform as the central element in understanding sustainable wildlife tourism in the marine context provides much scope for further research serving this field.

Acknowledgement

The San Juan Islands case material presented in this chapter was originally published in *Human Dimensions of Wildlife* 9(1), 103–117 (2004).

References

Amante-Helweg, V.L. (1995) Cultural perspectives of dolphins by ecotourists participating in a 'swim with dolphins' programme in the Bay of Islands. MA thesis, University of Auckland, New Zealand.

Babbie, E. (1995) *The Practice of Social Research*. Wadsworth Press, Belmont, California.

Baird, R., Otis, R. and Osborne, R. (1998) *Killer Whales and Boats in the Haro Strait Area: Biology, Politics, Aesthetics and Human Attitudes*. Abstracts of the Whale-Watching Research Workshop of the World Marine Mammal Science Conference, January 18, Monaco.

Barber, A. (1993) The Hawaiian spinner dolphin (*Stenella longirostris*): effects of human activities. MSc thesis proposal, Texas A&M University, College Station, Texas.

Baxter, A. (1993) Management of whale and dolphin-watching, Kaikoura, New Zealand. In: Postle, D. and Simmons, M. (eds) *Encounters with Whales'93*. Great Barrier Reef Marine Park Authority, Townsville, Queensland, pp. 108–120.

Beach, D.W. and Weinrich, M.T. (1989) Watching the whales: is an educational adventure for humans turning out to be another threat for endangered species? *Oceanus* 32(1), 84–88.

Bejder, L., Samuels, A., Whitehead, H., Gales, N., Mann, J., Connor, R., Heithaus, M., Watson-Capps, J., Flaherty, C. and Kruetzen, M. (2006) Decline in relative abundance of bottlenose dolphins (*Tursiops* sp.) exposed to long-term disturbance. *Conservation Biology*. DOI: 10.1111/j.1523-1739.2006.00540.x

Blewett, C. (1993) *A Survey of Orcinus Orca Whale-Watching in Haro Strait and Possible Applications in the Rewriting of Federal Whale-Watching Guidelines*. Unpublished paper presented at the Killer Whale Ecology Summer Satellite Course, School for Field Studies, Northeastern Universities, Boston, Massachusetts.

Briggs, D.A. (1985) Report on the effects of boats on the Orcas in the Johnstone strait from July 11, 1984–September 1, 1984. UCSC, Santa Cruz, California.

Cohen, J. (1988) *Statistical Power Analysis for the Behavioural Sciences*, 2nd edn. Lawrence Erlbaum Associates, Hillsdale, New Jersey.

Constantine, R. (1999) *Effects of Tourism on Marine Mammals in New Zealand*. Science for Conservation Report 106, Department of Conservation, Wellington, New Zealand.

Corkeron, P. (1995) Humpback whales (*Megaptera novaeangliae*) in Hervey Bay, Queensland: behaviour and responses to whale-watching vessels. *Canadian Journal of Zoology* 73, 1290–1299.

Corkeron, P. (1998) *Whale-Watching – Management and Research in Australia*. Abstracts of the Whale-Watching Research Workshop of the World Marine Mammal Science Conference, January 18, Monaco.

Driscoll-Lind, A. and Östman-Lind, J. (1999) Harassment of Hawaiian spinner dolphins by the general public. *MMPA Bulletin* 17, 8–9.

Duffus, D.A. (1996) The recreational use of grey whales in southern clayoquot sound, Canada. *Applied Geography* 16(3), 179–190.

Duffus, D.A. and Baird, R.W. (1995) Killer whales, whale-watching and management: a status report. *Whalewatcher* 29(2), 14–17.

Duffus, D.A. and Dearden, P. (1990) Non-consumptive wildlife-oriented recreation: a conceptual framework. *Biological Conservation* 53(3), 213–231.

Duffus, D.A. and Dearden, P. (1992) Whales, science, and protected area management in British Columbia, Canada. *The George Wright Forum* 9(3–4), 79–87.

Duffus, D. and Dearden, P. (1993) Recreational use, valuation and management of killer whales (*Orcinus orca*) on Canada's Pacific Coast. *Environmental Conservation* 20(2), 149–156.

Findlay, K. (1997) Attitudes and expenditures of whale watchers in Hermanus, South Africa. *South African Journal of Wildlife Research* 27(2), 57–62.

Finkler, W. (2001) *The Experiential Impact of Whale-Watching: Implications for Management in the Case of the San Juan Islands, USA*. MSc. thesis, Department of Marine Science, University of Otago, New Zealand.

Finley, K., Miller, G. and Davis, R. (1990) Reactions of Belugas (*Delphinapterus leucas*), and Narwhals (*Monodon monocerus*), to ice-breaking ships in the Canadian high arctic. *Canadian Bulletin of Fisheries and Aquatic Science* 224, 97–117.

Forestell, P.H. and Kaufmann, G.D. (1993) Resource managers and field researchers: allies or adversaries? In: Postle, D. and Simmons, M. (eds) *Encounters with Whales '93*. Great Barrier Reef Marine Park Authority, Townsville, Queensland, pp. 17–26.

Gordon, J., Leaper, R., Hartley, G.F. and Chappell, O. (1992) *Effects of Whale-Watching Vessels on the Surface and Underwater Acoustic Behaviour of Sperm Whales off Kaikoura, New Zealand*. Department of Conservation, Wellington, New Zealand.

Higham, J.E.S. and Lusseau, D. (2004) Ecological impacts and management of tourist engagements with cetaceans. In: Buckley, R. (ed.) *Environmental Impacts of Ecotourism*. CAB International, Wallingford, UK, pp. 173–188.

Hoyt, E. (1994) Whale-watching worldwide: an overview of the industry and the implications for science and conservation. In: Evans, P.G.H. (ed.) *Proceedings of the 8th Annual Conference of the European Cetacean Society*. Cambridge, pp. 24–29.

Hoyt, E. (2000) *Whale-Watching 2000: Worldwide Tourism Numbers, Expenditures, and Expanding Socioeconomic Benefits*. International Fund for Animal Welfare, Crowborough, UK.

International Fund for Animal Welfare, World Wildlife Fund, Whale & Dolphin Conservation Society (1999) *Report of the Workshop on the Socioeconomic Aspects of Whale-Watching, Kaikoura, New Zealand, 8–12 December 1997*. Yarmouth Port, Massachusetts.

Kind-Keppel, J., Nikolay, A., Muloin, S. and Otis, R. (1999) *Whale Watchers Attitudes Towards Boats Accompanying Killer Whales* (Orcinus orca). Paper presented at the 1999 Northeast Regional Animal Behaviour Society Annual Meeting, C.W. Post Campus, Long Island University, New York.

Lindberg, K. and McKercher, B. (1997) Ecotourism: a critical overview. *Pacific Tourism Review* 1, 65–79.

Lusseau, D. (2002) The effects of tourism activities on bottlenose dolphins (*Tursiops* spp.) in Fiordland. PhD thesis, University of Otago, Dunedin, New Zealand.

Lusseau, D. and Higham, J.E.S. (2004) Managing the impacts of dolphin-based tourism through the definition of critical habitats: the case of bottlenose dolphins (*Tursiops* spp.) in Doubtful Sound, New Zealand. *Tourism Management* 25(5), 657–667.

Lusseau, D., Slooten, E. and Currey, R.J. (2006a) Unsustainable dolphin-watching activities in Fiordland, New Zealand. *Tourism in the Marine Environments* 3(2), 173–178.

Lusseau, D., Slooten, E. and Currey, R.J. (2006b) Unsustainable dolphin-watching activities in Fiordland, New Zealand. Paper SC/58/WW6, 58th meeting of the International Whaling Commission Scientific Committee, St Kitts and Nevis, June 2006, 13 pp.

Malcolm, C., Duffus, D.A. and Rollins, R.B. (2002) The case for site-specific education strategies in ecotourism management: whale-watching on Vancouver island, British Columbia. In: E. Jackson (ed.) *Proceedings of the Tenth Canadian Congress on Leisure Research (CCLR-10)*. University of Alberta, Edmonton, Alberta, Canada, pp. 211–213.

MacGibbon, J. (1991) Responses of Sperm Whales (Physeter macrocephalus) to Commercial Whale Watching Boats off the Coast of Kaikoura. Department of Conservation, Wellington, New Zealand.

McKegg, S. (1996) Marine tourism in New Zealand: environmental issues and options. MSc thesis, University of Otago, Dunedin.

Muloin, S. (1998) Wildlife tourism: the psychological benefits of whale watching. *Pacific Tourism Review* 2, 199–213.

Orams, M.B. (2000) Tourists getting close to whales, is it what whale-watching is all about? *Tourism Management* 21, 561–569.

Osborne, R.W. (1986) A behavioural budget of puget sound killer whales. In: Kirkevold, B. and Lockard, J.S. (eds) *Behavioural Biology of Killer Whales*. Alan Liss Publishing, New York, pp. 211–249.

Osborne, R. (1991) *Trends in Killer Whale Movements, Vessel Traffics, and Whale-Watching in Haro Strait*. Puget Sound Research '91, Seattle, Washington.

Phillips, N. and Baird, R. (1993) Are killer whales harassed by boats? *The Victoria Naturalist* 50(3), 10–11.

Spalding, M. and Blumenfeld, J. (1997) *Legal Aspects of Whale-Watching in North America*. Unpublished paper presented at the symposium on the Legal Aspects of Whale-Watching, Punta Arenas, Chile, 17–20 November.

Vaske, J.J., Gliner, J.A. and Morgan, G.A. (2002) Communicating judgments about practical significance: effect size, confidence intervals and odds ratios. *Human Dimensions of Wildlife* 7, 287–300.

Whittaker, D. (1997) Capacity norms on bear viewing platforms. *Human Dimensions of Wildlife* 2(2), 37–49.

Würsig, B. (1996) Swim-with dolphin activities in nature: weighting the pros and cons. *Whalewatcher* 30(1), 11–15.

20 New Frontiers in Marine Wildlife Tourism: An International Overview of Polar Bear Tourism Management Strategies

R.H. LEMELIN AND M. DYCK

Introduction

From cultural conflicts between ecotourists and Inuit guides (Weaver, 2002), to aggressive interactions with humans in protected areas (Clark, 2003), and proposed legislative changes to the status in North America (USA and Canada), polar bears (*Ursus maritimus*) have become synonymous, at least from a media perspective, with increasing human–polar bear interactions and climate change (Slocum, 2004). This growing media awareness has in some cases resulted in increasing concerns for polar bears, while for others, it has piqued interest, and consequently stimulated a demand to view polar bears in their natural environment (Lemelin, 2005; Lemelin and Smale, 2006).

Wildlife viewing, also referred to as wildlife tourism, is considered one of the fastest growing outdoor activities in the world (Higginbottom, 2005; Newsome *et al.*, 2005). The demand for bear-viewing programmes as a specialized form of wildlife tourism has gained wide acceptance, and since the mid-1990s numerous bear-viewing sites managed by various wildlife agencies (i.e. viewing brown bears at Brooks Falls in Katmai National Park, Alaska (managed by the US National Parks Service); Knight Inlet, British Columbia (managed by the BC Ministry of Water, Land and Air Protection)) have been established (Brown, 2006). The growth of polar tourism and polar bear tourism (PBT) and the subsequent changes in the Arctic will require adaptive, multidisciplinary and comprehensive management strategies (Kaltenborn, 1997; Stewart *et al.*, 2005; Johnston, 2006). Understanding this growth, while also acknowledging the potential for impacts of tourism on local communities, both indigenous and non-indigenous have often been overlooked by researchers (Smith, 1997; Nuttall, 2005; Notzke, 2006). That said, research examining the human dimensions of PBT is emerging (see Clark, 2003; Lemelin, 2005; Dyck and Baydack, 2006; Lemelin and Smale, 2006).

Through a closer examination of existing management strategies and research on the human dimensions of PBT, this chapter attempts to provide an overview of five international PBT destinations, including: Barrow and Kaktovik, Alaska, USA; Polar Bear Provincial Park (PBPP), Ontario, Canada; the Svalbard Archipelago (SA), Norway; Ukkusiksalik National Park (UNP), Nunavut, Canada; and Wrangel Island, Russia. These destinations were selected because the frequency and the reliability of polar bear sightings have, to some extent, promoted PBT. Particular attention will also be paid to the types of PBT opportunities currently offered and the visitor management strategies that are implemented at these locations. Churchill, Manitoba, Canada, also known as the 'polar bear capital of the world', will not be addressed in this chapter because an extensive overview of this particular PBT destination is provided in Chapter 5 of this volume.

Tourism impacts can be especially acute for mammals and the coastal areas that these animals frequent (Cessford and Dingwall, 1994; Forbes *et al.*, 2001). The latter impacts are the focus of this chapter, especially as they relate to human dimensions and visitor management frameworks, and these will be highlighted through an examination of existing or non-existing management plans. As will be demonstrated, information pertaining to human–polar bear management is quite variable, with extensive details available for some destinations and virtually no data available for others.

Two caveats are required to explain our approach. First, the authors are relatively aware of current ongoing polar bear research and, in fact, one of the authors is a polar bear biologist. However, this chapter only examines the human dimensions of polar bear management and research dealing strictly with PBT. Second, while some work has been conducted on consumptive approaches to polar bears (e.g. community-based polar bear trophy hunts – see Freeman and Wenzel, 2006), we define PBT in this context as viewing, photographing and otherwise interacting with polar bears in their natural environment without an intent to consume (i.e. not killing and eating the animal). This is, to our understanding, the first attempt at providing an international overview of the PBT from a social perspective. The analysis is based on a literature review and on information collected through correspondence with resource managers. These sources have been supplemented by the personal experience and knowledge of the authors.

A Historical Overview

In the past, northern expeditions undertaken by explorers, whalers and others while travelling in the Arctic relied greatly on polar bears for sustenance. Indigenous people also supplemented their diet and income by selling polar bear meat and pelts (Honderich, 1991; Ovsyanikov, 1998). The unrestricted harvest of polar bears continued well into the 1960s.

> [W]hen trophy hunting became fashionable and technological progress provided the means for people to move around in the High Arctic. About 70 percent of all polar bears killed in Alaska in the 1950s were shot from airplanes; in the 1960s

this figure increased to 90 percent. In Norway, small ships brought hunters to shoot polar bears in the Barrents Sea. During this twenty-year period, a total of 1,200 polar bears were killed worldwide every year.

(Ovsyanikov, 1998, pp. 65–66)

The most vulnerable segment of a polar bear population, females with dependent young (i.e. cubs-of-the-year and yearlings), was hunted mostly at dens (Honderich, 1991; Scott and Stirling, 2002). For example, over three-quarters of all polar bears killed on Wrangel Island were females with cubs (Ovsyanikov, 1998). At times, cubs were also captured and sold for entertainment (Honderich, 1991). While some hunting bans were established to minimize these impacts, it was not until the 1970s when the Convention on International Trade in Endangered Species (CITES; signed in 1973), and the International Agreement on the Conservation of Polar Bears (IACPB; signed by Canada, Denmark, Norway, the USA and the former USSR in 1976) specifically addressed international polar bear research, protection and hunting quotas (Lunn *et al.*, 1998). The latter event, for all intents and purposes, made the management of polar bears an international issue. Since then, meetings of the Polar Bear Specialist Group (PBSG), operating under the auspices of the IUCN/SSC (The World Conservation Union/ Species Survival Commission), examine and review current polar bear population status, research, management guidelines and international legislation. However, while some concerns regarding polar bear–human interactions have been noted, very little research addressing PBT specifically has emerged from these meetings. This is somewhat surprising considering a growing media attention and increasing availability of PBT opportunities in the north. Polar bear–human interactions in the context of tourism and research are discussed next.

Incidents of human–wildlife encounters in the north, more specifically polar bear–human encounters, have been reported by researchers (see Stenhouse *et al.*, 1988; Clark, 2003; Dyck, 2006) and reporters (Struzik, 2004a,b). Although no one has been killed by a polar bear in Canadian National Parks, the number of close calls has been increasing (Struzik, 2004a). In 2002, paddlers camping in Ivvavik National Park, Yukon, were visited by a polar bear (Struzik, 2004a). Although no one was hurt in this encounter, the inquisitiveness of the polar bear combined with ill-prepared visitors (i.e. the pepper spray was left in the kayaks not in the tents) indicates how important education and safety training are in these regions. Examples of polar bear–human encounters in the last 5 years include three people who were attacked and injured by polar bears in two separate incidents involving researchers and tourists (Clark, 2003). In another incident, a hiker in Auyuittuq National Park, Nunavut, was slightly injured by a polar bear in 2000 (Clark, 2003). Until then, this was the 'only injury by a polar bear in a national park' (Lunn *et al.*, 2002, p. 50). In 2001, a polar bear injured two canoeists in Katannilik Territorial Park Reserve in Nunavut. Three years later, a researcher was attacked and injured by a polar bear in Wapusk National Park (Northern Manitoba). In 2005, three European researchers were 'rescued' from polar bears in the Svalbard Archipelago. It should also be noted that these examples only depict encounters between polar bears and non-locals. If we were to include polar bear and local resident encounters, this number would increase significantly.

These polar bear–human encounters should be of concern to wildlife managers and PBT proponents alike, for without proper recourse (i.e. proper deterent and safety training), these encounters may eventually harm tourists, locals (i.e. guides, interpreters) and/or the wildlife. It may also question the management strategies which are supposedly mandated to legally protect these individuals and the wildlife (Stewart *et al.*, 2005).

The Present

A number of researchers have examined tourism and sustainability issues in a northern context. Canadian examples include the need to foster community economic and social well-being in the development of a national park on Banks Island in the Northwest Territories (Wright and McVetty, 2000); the tension between conservation, tourism and community development in the northern Yukon Territory (Marsh and Johnston, 1983); and the challenge on Ellesmere Island of balancing preservation and access requirements for national parks (England, 1982). International examples include Olsen (2006) and Viken's (2006) examination of Sami nomads and global tourism, and Kaltenborn's (1997, 2000) overview of the SA's management plan for tourism and outdoor recreation.

For a number of these locations, a combination of factors, such as improved accessibility (i.e. new transportation routes), increased technology (e.g. helicopters, ice-breaker cruise ships), changes in consumer preferences and the never-ending search for new tourism destinations, has made the Arctic a highly attractive travel destination during the last two decades (Johnston, 1997, 2006; Smith, 1997; Notzke, 2006). Although the numbers of tourists visiting the Arctic are still low compared to those of other tourist destinations, hundreds of thousands of visitors come to northern circumpolar regions every year (Stewart *et al.*, 2005). For example, in Canada, it is estimated that between 33,000 and 40,000 leisure travellers visited the Northwest Territories (Government of Northwest Territories, 2006) in recent years, the Yukon Territory received about 32,000 tourists in 2002 and the Nunavut Territory reported 12,000 visitors (Pagnan, 2003). Svalbard reported about 20,000–25,000 ship-based and 1500–3000 airborne tourists (Kaltenborn, 2000). Discussion with park managers has indicated that visitation to Canadian national parks located in polar regions denotes some variability in visitation patterns with some parks experiencing little growth (e.g. Aulavik, Auyuittuq, Quttinirpaaq), some like Ivvavik, and Wapusk experiencing modest growth, while Sirmilik National Park underwent a significant increase. On the long term, all of the national parks in Nunavut (Auyuittuq, Quttinirpaaq, Simirlik) have projected growth from multiple users including cruise-ship tourism, kayakers, hikers, skiers and researchers (Parks Canada, Communications Manager, Iqaluit Field Unit, communication with the author, March 2007).

Tourism holds promise for communities that might have no other way to generate revenue, or that might or have been losing other primary industries (e.g. mining in Svalbard and Nunavut) (Kaltenborn and Emmelin, 1993). Yet, despite the expectation of local people and governments that tourism will benefit communities and regions, promote conservation and encourage cultural exchange, the potential negative impacts associated with wildlife tourism are numerous (e.g. disruption to subsistence activities, loss of autonomy), and some are more problematic in small, remote Arctic settlements than in larger road-accessible locations (Johnston, 1997, 2006; Smith, 1997; Stewart *et al.*, 2005). Another noted downside of polar tourism is economic leakage where payments for transportation and package tours accrue to the airlines and tour operators, which are usually located outside the tourism region (Reid, 2003).

The following section examines specific destinations offering PBT opportunities and reviews current management strategies. As stated earlier, these areas were selected because of the possibility of viewing polar bears, the availability of information and the authors' familiarity with these sites (Fig. 20.1).

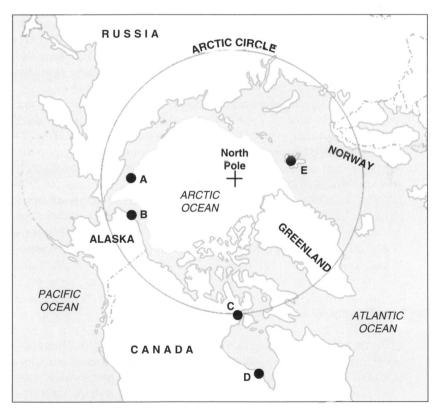

A. *Wrangel Island Zapovednik*, Russia
B. *Kaktovik, Barrow,* Alaska, USA
C. *Ukkusiksalik National Park*, Nunavut, Canada
D. *Polar Bear Provincial Park*, Ontario, Canada
E. *Svalbard Archipelago*, Norway

Fig. 20.1. Northern destinations offering polar bear tourism (PBT) opportunities. (From Chapin, 2004.)

Case Studies

Barrow and Kaktovik, Alaska, USA

The two Alaskan communities of Barrow and Kaktovik offer opportunities to view polar bears. Yet, despite being featured in *Forbes* and *Outside Magazine*, the village of Barrow, with a direct regional aviation link to larger Alaskan cities, offers only a limited number of PBT opportunities. According to local informants, the optimal bear-viewing times are in the spring and fall whaling seasons, where whale carcasses may attract polar bears (George *et al.*, 2004; Norton and Gaylord, 2004; Gearheard *et al.*, 2006). During fall, the presence of polar bears largely depends on the formation of the sea pack ice (Travel Alaska, 2006). While official figures were not available, professional estimates put the figure at 100 tourists per year visiting Barrow (US Fish and Wildlife Services representative, communication with the author, May 2006).

Featured in the Treasure America Project (TAP) and by *Travel Alaska* (see website Travel Alaska, 2006), there is currently no organized PBT in the small, remote subsistence community of Kaktovik, although there are rumours that a commercial viewing operation may start up in the future (US Fish and Wildlife Services representative, communication with the author, May 2006). To date, only a small number of visitors (approximately 10–20) are estimated to visit the village to see polar bears.

Although polar bears are protected under the Marine Mammals Protection Act (1972, revised in 1994), managed by the US Fish and Wildlife Service, there are currently no polar bear-viewing guidelines in Alaska. However, the US National Parks Service is proactively examining the human dimensions of polar bear management by funding a project to collect traditional ecological knowledge of polar bear habitat use in the Chukotka area (Schliebe *et al.*, 2006). In addition, the US Fish and Wildlife Service is attempting to minimize polar bear–human encounters by encouraging indigenous hunters to properly dispose of bowhead whales harvested by subsistence hunters (US Fish and Wildlife Services representative, communication with the author, May 2006).

Polar Bear Provincial Park, Ontario, Canada

This non-operational (i.e. non-operating parks charge no fees, have no on-site staff and only limited infrastructures) provincial park, administered by the Ontario Ministry of Natural Resources (OMNR), was established in 1970 and includes the majority of Ontario's tundra region, and over a third of Ontario's Arctic coastline (Prevett, 1989; Beechey and Davidson, 1999). An important staging and denning area for polar bears, and a Wetlands of International Importance under the Ramsar Convention (1987), PBPP is Ontario's largest provincial park (2,355,200 ha or 24,087 km²), and one of the Canada's largest protected areas (Usher, 1993; Obbart and Walton, 2004; Portman, 2004).

According to the earliest management plans, recreational and tourism opportunities were secondary goals to environmental protection (Brunelle, 1970; OMNR, 1977a). Established in 1970, the initial master plan recognized that the park was of scientific importance particularly with regard to permafrost, Arctic tundra, waterfowl, shore birds and polar bears (Ontario Parks, 2005). The initial goals for PBPP as identified in the park master plan were threefold: (i) to protect its environment for the benefit of present and future generations from significant alterations by humans; (ii) to provide quality, low-intensity wilderness recreational opportunities; and (iii) to provide opportunities for scientific research (OMNR, 1980). Of particular importance to the local communities is the recognition of consultation with the First Nations and the Wabusk Co-Management Agreement within the plan.

In terms of polar bear protection, the park protects two critical habitat elements for Southern Hudson Bay polar bear population (estimated at about 1000 in total population): coastal summer retreat habitat (e.g. Cape Henrietta Maria) used by all classes of bears and inland maternity denning habitat used by pregnant females (i.e. along the Winisk River). That said, some of the identified dens are outside of the PBPP, and females with cubs and bears in dens are not specifically protected in Ontario, but the dens are protected, and only persons with treaty rights can legally hunt or trap polar bears (Environment Canada, 2006). The background document for PBPP (1977) estimated that about 10–30 polar bears are harvested annually by Crees from Fort Severn, Winisk and Attawapiskat. Concerns pertaining to polar bear harvest within and around PBPP have been voiced in 1979 and again in 1994. These concerns regarding the 'over-harvest' of polar bears (for sustenance and/or in protection of property) may have been somewhat excessive since it is, according to one band member, the Weenusk First Nation, only a few polar bears that are actually shot or harvested.

Polar bears in this region of the province are protected through federal and provincial legislation. Additional protection for polar bears was added through Ontario's new Fish and Wildlife Conservation Act (FWCA) (Statutes of Ontario, 1997, Chapter 5) replacing the Game and Fish Act. Under the FWCA, polar bears are prescribed as furbearing mammals by regulation (Ont. Reg. 669/98) (Lunn *et al.*, 2002). Under this Act, there is no hunting season for polar bears by non-Aboriginal people; however, authorization is given to some native trappers in possession of valid trapping licenses to harvest (no more than 30) and subsequently report a limited number of polar bears (Lunn *et al.*, 2002).

Recreational and tourism opportunities were addressed in the original park planning proposal of 1977 (OMNR, 1977b), and subsequently reviewed in the 1980 Master Plan, and the Management Plan Review of 1994. Existing archeological sites and facilities in PBPP include prehistoric sites at Brant River, Hudson Bay Company and Révillon Frères trading outposts site 415 (a series of radar stations spread out along the 55th parallel, otherwise known as the Mid-Canada Line), and several goose-hunting camps (OMNR, 1977a; Usher, 1993). Most of these camps were established in the 1960s, and all are owned by Cree entrepreneurs located in the Attawapiskat, Peawanuck and Fort Severn First Nations. All three First Nations are recognized within the management plan of the PBPP (Usher, 1993). Some camps operate on a seasonal basis (i.e. during the hunting season), and all are small and rustic.

Viewing of the world's most southerly population of polar bears (about 1000 polar bears are estimated in the Southern Hudson Bay population, see Calvert *et al.*, 1995) in northern Ontario is limited to a few entrepreneurs (Canoe Frontiers and Wild Wind Tours; Polar Bear Park Expeditions and Sutton River Lodge). Canoe Frontiers offers both canoe and kayak expeditions up the Winisk River and eco-cultural tours in PBPP through Wild Wind Tours; Polar Bear Park Expeditions (formerly known as Ice Bear Tours) offers polar bear-viewing excursions out of an ecotourism camp located near the mouth of the Sutton River within PBPP. Both operators operate out of the Peawanuck First Nation. In the Fort Severn First Nation, polar bear-viewing opportunities are offered by one local operator. Concerns over perceived decreasing sightings of polar bears have been raised in both communities (Researcher, communication with the author, March 2007).

Management of PBPP is provided through the management plan and the subsequent revisions of this plan. The park management plan presently identifies five access zones where aircraft landing is permitted. The OMNR requires that aircraft landing permits must be obtained for any visitors in PBPP. Generally these permits are obtained by outfitters who are transporting visitors, although a few private individuals do obtain permits each year. Currently, the park does not have a tracking method to determine how many trips are made each year (Ontario Parks management personnel, communication with the author, June 2006). No permits are required for water-based vehicles (i.e. kayaks, canoes).

Usher (1993) estimated that as many as 1300 tourists were visiting the area in the late 1980s and early 1990s, a majority of these visitors being hunters, anglers and recreationists (e.g. canoeists, adventurers). These estimates are quite high compared to those figures provided by the OMNR, which suggest that annual visitation ranges from fewer than 50 people per year (OMNR, 1977a), to 350 visitors per year (OMNR, 1994). The dormancy of the goose-hunting camps in the early 21st century may have resulted in visitation declines in PBPP. Current estimates provided by Canoe Frontiers/Wild Wind Tours indicate that in 2005, 38 specialized adventurers undertook excursions offered in PBPP. If we combine these numbers with other visitors, there are still only a few hundred people visiting the park on an annual basis (Tourism outfitter, communication with the author, December 2006).

The Svalbard Archipelago, Norway

More than 10° N of the Arctic Circle, the archipelago of Svalbard (62,160 km²), which includes the larger islands of Spitsbergen, Nordaustlandet, Barentsoya, Edgeoya, Kong Karls Land, Prins Karls Forland and Bjornoya and hundreds of smaller islands, is Norwegian territory. With the exception of some coastal lowlands, the landscape is mountainous with many large and small glaciers (Humlum, 2005). The SA is home to enormous concentrations of seabirds, including large colonies of dovekies, black-legged kittiwakes, thick-billed murres and northern fulmars. In addition, the Norwegian Polar Institute estimates that there are over 2000 polar bears in the SA. Two national parks, 15 bird sanctuaries and two plant reserves protect nearly half of the SA's landmass. In 1998, Norway requested that the North-east Svalbard Biosphere Reserve, no longer able to meet the criteria established by the Man and Biosphere programme, be removed. Intentions to propose a larger biosphere reserve in the future were mentioned (UNESCO, 1998).

The SA is largely uninhabited except for the mining settlements at Longyearbyen and Barentsburg, and the international research stations at Ny-Alesund. Nearly 1420 persons live in Longyearben, the northernmost town on the planet and the largest settlement on Svalbard, followed by Barentsburg (about 850 inhabitants) (Humlum, 2005). With a declining mining industry, tourism now represents an important means of income for the population of the SA, as does the university in Longyearbyen. There is also some trapping, although the latter is of little economic importance (Humlum, 2005).

Media attention created by the numerous expeditions to the North Pole, combined with luxury travel stimulated early travels, and tourism in the SA (Kaltenborn and Emmelin, 1993). At the beginning of the 21st century, Svalbard's tourism infrastructure consists of an international airport, harbour, several modern hotels and other lodging facilities, restaurants, pubs, shops, indoor sport and swimming facilities, museums, galleries and art shops (Humlum, 2005).

Kaltenborn (2000, p. 29) states that the SA 'presently accounts for about one-fourth of all tourism in the Circumpolar High Arctic'. Cruise-ship tourists constitute the largest group with 20,000–25,000, which is about 80% of the annual total number of visitors that come to Spitsbergen (Kaltenborn, 2000; Humlum, 2005). According to Kaltenborn and Emmelin (1993), cruise-ship travellers spend less time outside the ships. An additional 1500–3000 persons travel around on land in the SA outside the settlements during the summer (Kaltenborn and Emmelin, 1993, p. 45).

Despite the fact that numerous operators advertise PBT in their promotional material, and one operator formerly offering wildlife tourism opportunities in Churchill (see Travel Wild Expeditions, 2006), in direct competition with Churchill, Manitoba, even calls the SA 'the new polar bear capital of the world', the SA's wildlife management policy does not specifically advocate PBT. In fact, correspondence with management representatives indicated that attracting, pursuing or otherwise actively seeking out polar bears is punishable by law. Furthermore, special rules apply to tour organizers and tourist vessels, making the organizers responsible for the safety and for ensuring that visitors are informed of and comply with rules and regulations. Moreover, individual travellers and tour organizers must report their programmes to the wildlife management agency prior to the new season. They must also have appropriate insurance to cover rescue operations and similar assistance (SA management representative, communication with the author, February 1999).

While having these strategies in place, growing concerns over the rapid development of tourism and increasing human–polar bear encounters warranted the re-examination and subsequent development of a comprehensive management plan for the SA. In 1995, a comprehensive macro-level plan (also known as the White Paper) for the entire SA was implemented. The plan became the SA's Environmental Act in June 2001. The main objective of the Act, as it pertains to tourism, is to facilitate tourism and outdoor recreation within limits set by natural, cultural and historical resources in such a way that the wilderness character of the environment is preserved (Kaltenborn, 2000).

> The plan operates according to a zoning system (nature reserves; national parks; outdoor recreation and excursions areas) dividing Svalbard into management areas. For each area or zone, specific goals, resource conditions, management actions, and acceptable activities are described. Thus, the plan becomes a management tool for identifying the amount and type of facilities in each zone as well as access to and restrictions on use. Recreational values are integrated into land management planning through explicit management objectives related to environmental and social conditions in the different areas.
>
> (Kaltenborn, 2000, p. 32)

Since its inception, the plan has been revised and input has been obtained from stakeholders (Kaltenborn, 2000).

Ukkusiksalik National Park, Nunavut

Sila Lodge, located in UNP, Nunavut, is renowned for its wildlife-viewing opportunities, especially swimming polar bears. Located just south of the community of Repulse Bay, the National Park surrounds Wager Bay, a 100 km long saltwater inlet on the north-western coast of Hudson Bay. Established in 2003, UNP includes 20,500 km^2 of eskers, mudflats, cliffs and tundra banks (Portman, 2004). According to Portman (2004), each year about 300 people travel to Sila Lodge, a small eco-lodge constructed on the shores of the Sila River in UNP, to view wildlife. Discussions with park officials (i.e. National Parks Canada) could not corroborate these numbers since official visitation figures for UNP are not, as of yet, available (Parks Canada, Communications Manager, Iqaluit Field Unit, communication with the author, March 2007).

Although a barren-land grizzly was shot and killed at Sila River, polar bears remain the primary concern for Parks Canada management in UNP (Struzik, 2004b). An estimated 130 polar bears are thought to be living and using the areas near the lodge, yet no one really knows how many bears are in this new park, or how park management will deal with the increasing polar bear–human issues brought about through tourism. As Jane Chisholm, the park ecologist explains hikers, kayakers and wildlife tourists 'are going to want to come here. But without a gun to protect themselves, they'll be nothing but bait for the bears. Right now, the only people who can legally carry a gun in a national park are the Inuit and park wardens' (Struzik, 2004b). What the followng observation reveals is that management strategies in the Canadian Arctic can also be hindered by federal rules and regulations, which have little if any, contextual relevance. That said, the territorial government of Nunavut has implemented a number of policies and regulations aimed at wildlife management.

On 21 December 2004, after extensive consultation with affected communities, hunters' and trappers' organizations, regional wildlife organizations and other stakeholders, Olayuk Akesuk, the Minister of Environment (DoE), accepted the decision of the Nunavut Wildlife Management Board (NWMB) regarding polar bear management in the territory of Nunavut. Following this decision, a Memorandum of Understanding (MOU) regarding research and the management for each polar bear population located in Nunavut was created (note that similar MOUs were in place before the Northwest Territories divided into two territories in 1999). Combining *Inuit Qaujimajatuqangit*, otherwise known as IQ or indigenous knowledge, with scientific knowledge, these MOUs are an attempt to regulate the polar bear harvest and to sustain the species' stock. While these MOUs address research on polar bears and the harvest, they do not specifically address PBT.

Wrangel Island *Zapovedniks* (Nature Reserve): Biosphere Reserve and World Heritage Site, Russia

Named after Baron Ferdinand Wrangel (1797–1870), the ownership of the Wrangel Island Zapovedniks (WIZ) was the subject of international dispute between the former USSR and various countries including the USA and Canada, until the former Soviet government forcibly removed a small colony of settlers in the 1930s and established the settlement that survives to this day (Razzhivin, 2005). Permanent residents on the island number four families, who live in the village of Ushakovskoe. Other residents include 12 employees at the meteorological station, four rangers, six frontier post personnel and up to ten rotating resource management agency personnel, plus visiting scientists. No infrastructure catering to tourism exists at present (Razzhivin, 2005).

Wrangel and Herald Islands (*Vrangelya Ostrov*) are located on the border of the East Siberian and Chukchi seas (between 70° N and 177° W), when combined with smaller islands found in the area, the total area of the WIZ encompasses more than 1,430,000 ha. The aquatic and terrestrial areas of Wrangel and Herald islands provide optimum foraging and reproductive habitats for a number of species, some endemic to this area (PBI, 2006). In years with normal ice conditions, a number of polar bears of all ages and both genders come ashore for brief periods of time. During ice-free seasons, however, tens to hundreds of polar bears stand on the shore for weeks, waiting for the ocean to freeze (PBI, 2006). In addition, approximately 350–500 (or 80%) of pregnant female polar bears construct their maternity dens in the WIZ every autumn. Some areas 'support 6–12 bears/km², and are reported to have one of the highest densities of polar bear dens in the world' (Razzhivin, 2005, p. 2197).

Protection of the area first occurred when the WIZ was designated a nature reserve by the former Soviet Government (Razzhivin, 2005). Much like nature reserves elsewhere, scientific research is encouraged while exploration and development are curtailed (UNESCO, 2006). International recognition of this important ecosystem first occurred in 1976 when the area was designated an international biosphere and again in 2004, when it was proclaimed the northernmost World

Heritage Site (Razzhivin, 2005). Additional management strategies aimed at minimizing anthropogenic impacts include the creation of a 13km^2 'general use' zone around the village of Ushakovskoe. The rest of the nature reserve can only be visited with permission from the management agency administered by the community of Mys Shmidta on the Russian mainland (UNESCO, 2006).

Numerous attempts aimed at generating funds for research and increasing tourism in the area have, according to Girard (1996), met with limited success. According to one unofficial source, in the late 1990s some 200 tourists guided by scientists visited the island's coastline on an annual basis. However, between 2000 and 2003 only one group of six tourists visited the site. Other estimates have placed the numbers from eight to 50 annually (ENGO representative, communication with the author, May 2006 – NB: Caution is advised regarding these numbers since none has been confirmed by official sources). Further challenges to PBT in the WIZ include intermittent access to the island, currently provided by international ice-breaker cruise ships, helicopters and fixed-wing aircrafts from the mainland villages of Schmidt and Pevek in Russia. Rudimentary infrastructures and few facilities (i.e. dormitories, construction modules), accommodating at best 40 people, further limit travel opportunities on the WIZ (UNESCO, 2006).

A general lack of information pertaining to PBT is largely symptomatic of polar bear management in Russia.

> With the advent of economic and political reforms, polar bear protection has dropped dramatically throughout the Russian Arctic. The new market economy in Russia has created a demand for polar bear parts-skins and gall bladders-and food shortages in remote Arctic villages and polar weather stations have provoked the killing of bears for meat. Today, poaching is the primary cause of polar bear mortality in Russia, and the extent of the loss is unknown.
>
> (Ovsyanikov, 1998, p. 68)

In addition, polar bears are threatened by planned oil and gas exploration on the continental shelf and by the opening of the North-east Passage to commercial shipping. While tourism is not seen particularly as a threat yet, some concerns regarding potential anthropogenic impacts from tourism in sensitive areas were noted by the World Heritage Committee (UNESCO, 2006).

Case Studies Overview

The polar bear is often perceived as an important component of conservation efforts. Indeed, the animal is often referred to as the 'flagship species' for climate change and an icon for tourism strategies (Slocum, 2004; Lemelin, 2005). Thus, a number of policies and protected area strategies have been established throughout the circumpolar world to protect polar bear staging and denning areas (i.e. from a nature reserve, biosphere/World Heritage Site in Wrangel Island, Russia, to national, provincial and regional parks in Svalbard, Norway and Canada (Ontario, Manitoba, Nunavut) to co-management in PBPP (Lyster, 1985)). Yet, the mere designation of an area as protected may provide a false sense of security where biodiversity or ecological diversity is ensured through some research and legislation. However, of the five case studies, only the SA's comprehensive management strategy specifically addresses the human dimensions of PBT. This is perhaps due in part to the long history of research and tourism in this area. The four other sites are relatively new to PBT, and some like the WIZ and UNP are experiencing socio-political challenges, which can often exacerbate issues associated to wildlife management. Yet even in the WIZ and PBPP where polar bear research is occurring, it continues to be dominated by a positivistic natural science approach with little to no connection to social sciences or local ecological knowledge (LEK) (Kaae, 2002). That said, LEK and IQ are being incorporated in some aspects of polar bear management in Alaska (Barrow and Kaktovik) and Nunavut (UNP), and could be integrated in the Wabusk Co-Management Agreement.

A general lack of information pertaining to the human dimensions of PBT is disconcerting, especially for ecosystem and cooperative approaches to management. Indeed, very few of the case studies highlighted here demonstrated multidisciplinary research approaches that examine the possible environmental, economic and cultural impacts of PBT. A case example of the latter was demonstrated in the 1992 Arctic expedition sponsored by Ecosummer Canada Expeditions (ECC) and the World Wildlife Fund (WWF), when the local guides shot five polar bears to feed the sled dogs (Weaver, 2002). The public outcry regarding this incident caused the WWF to remove their support from the ECC, and also demonstrated potential cultural clashes over wildlife uses. What this particular example and the five case studies indicate is that adaptable management structures and policies are required if PBT is to proceed in the best interest of polar bears, their environment, visitors and local communities (Nuttall, 2005).

Discussion

The mining of diamonds, oil exploration and development in the Arctic pose a wide array of threats to polar bears ranging from oil spills to increased human–bear interactions (Schliebe *et al.*, 2006). Over-harvest, both legal and illegal, is an ongoing concern for some polar bear populations, particularly in areas where there is no information on shared populations (e.g. Québec, east Greenland and the Chukchi Sea) and hunting quotas (Schliebe *et al.*, 2006). Additional threats to polar bears include increasing interactions with human beings (local and non-local), and potential coastline changes or habitat loss due to climatic changes and biomagnifications (Ziaja, 2004).

The Circumpolar North, especially the PBT case studies highlighted in this chapter, contains a unique community of individual countries sharing similar biophysical and socio-cultural features. Some of the described case studies have established the protection of wildlife (i.e. Alaska, USA) and/or their habitat (i.e. the WIZ) through legislation or administrative policies. Others such as the PBPP and the UNP have implemented protected area strategies, and collaborative management strategies while the SA has implemented all three (Alessa and

Watson, 2002). Further, recent development in Alaska (Barrow, Kaktovik) and Nunavut (UNP) regarding local ecological knowledge, co-management agreements and greater involvement of local people and key stakeholders will assist in the management of polar bears in these areas. However, as demonstrated in this chapter, a general lack of knowledge pertaining to the human dimensions of PBT, a factor noted by the chair of the IUCN's polar bear working group in a paper separate from the PBWC publications (see Derocher, 2005), exists within the literature. This is further compounded by limited on-site information. While this should come as no surprise since Stewart *et al.* (2005) also noted numerous gaps in their macroanalysis of polar tourism, it should not be assumed that simply because polar bears are protected through various treaties (regional, national, international), protected areas and management guidelines, that this is sufficient. In fact, legislation and protected areas may not be enough to protect these wandering animals, nor may they provide protection from pollutants and biomagnifications (Lyster, 1993) and anthropogenic disturbances (Forbes *et al.*, 2001).

Currently, there is an intense focus on natural scientific research in polar bear management with few resources allocated to social or tourism research. Given the rapid transitions of some polar communities and the growth of tourism in the north, it appears timely to give higher priority to research of multidisciplinary, socio-cultural and tourism issues (Kaae, 2002). While some of the research highlighted in these case studies may examine some aspects of polar bear management, only one was multidisciplinary. Indeed, as demonstrated in the SA's comprehensive management strategy, some of the impacts, especially those derived by tourism, can be minimized somewhat through increasing understandings in human dimensions of wildlife management, multidisciplinary research and visitor management frameworks aimed at incorporating local ecological knowledge, establishing protected areas, zoning and limiting certain anthropogenic uses.

Since the inception of the IPPSC, Norway and Russia have banned polar bear hunts, although it is believed that up to 200 bears are still being illegally harvested in Russia each year. In Alaska, some 100–200 bears are hunted annually while 500–600 are harvested in Canada. A majority of these animals are harvested in Nunavut and some polar bears have been killed in Northern Ontario. Other governments and wildlife agencies have attempted to manage polar bears through various policies, legislations and wildlife management systems. Some agencies (see Nunavut and Alaska), in an attempt to address the tensions between harvesting practices and polar bear management, have implemented co-management strategies based upon indigenous and scientific knowledge.

While studies examining the human dimensions of polar bear management (i.e. community-based polar bear trophy hunts, see Freeman and Wenzel, 2006) have been conducted, understanding and implementing management approaches in PBT will require greater attention to local–non-local interactions. For example, tensions between local and non-local values were highlighted when 'wildlife tourists' undertaking a polar bear expedition were outraged when their Inuit guides shot three bears to feed the sled dogs (Weaver, 2002). As some researchers and managers have noted, PBT appears to represent a new relationship with polar bears that does not historically reflect Inuit relationships with these animals. Therefore, PBT could have consequences upon indigenous communities that are not readily apparent (Nunavut manager, wildlife services, communication with the author, May 2006). In addition, closer scrutiny pertaining to polar bear–human interactions and conflicts may provide some insights as to why and where these interactions occur. Overall, a need exists to determine the most effective models for interaction between wildlife, local people, tourists and managers. Specifically, there is a need to understand what policy mechanisms most appropriately incorporate the range of views concerning what to protect and how to develop implementation strategies that best accomplish all envisioned goals. Moreover, these considerations must be extended to marine coastal environments (Alessa and Watson, 2002).

In addition, there is an acute need to understand the ways in which these forces impact coastal regions both socioculturally and biophysically, and how these constraints can be mitigated through adaptive management and the incorporation of local stakeholders, so that the protection of coastal environments and the conservation of wildlife are assured (Alessa and Watson, 2002). Since a number of polar bear–human interactions are often instigated by non-local residents (i.e. tourists, scientists), various measures should be implemented to reduce these encounters. They can range from fines, education, training, certification, enforcement and using local guides. Long-term management measures will require multidisciplinary approaches to PBT.

Conclusion

While by no means being exhaustive of all PBT management strategies, this overview provides some interesting insights as to what is, and what is not, being done vis-à-vis the human dimension of polar bear management. The management of polar bears has traditionally been largely dominated by a particular western approach to wildlife management, with certain actors and stakeholders consciously or unconsciously vying for control. For example, polar bear management has tended to emphasize either a conservationist utilitarian approach to game species management or a preservationist approach emphasizing the need for protection of endangered species. The latter often excluding or reducing traditional rights and practices (i.e. Inuit harvest). These somewhat Euro-centric approaches to understanding wildlife management, while effective in the past, may be somewhat myopic given the recognition of 'newer' approaches to wildlife management, which include multidisciplinary research, collaborative management approaches, the interplay between IQ, LEK and the sciences, as well as the emergence of new stakeholders in these socio-political arenas (see Gearheard *et al.*, 2006). If polar bear management is to benefit from these new opportunities, then a reflexive re-examination of the dominant wildlife management paradigm will be required. The acquiescence of co-management policies, IQ and LEK by some management agencies represents a small shift in this direction. However, recognition of the interplay between social values and research will only occur when all crucial stakeholders and their knowledge systems are recognized, and when polar bear management incorporates multidisciplinary approaches (i.e. ecosystem sciences and social sciences) with IQ and LEK into the International Agreement on the Conservation of Polar Bears.

References

Alessa, L. and Watson, A.E. (2002) Growing pressures on circumpolar north wilderness: a case for coordinated research and education. In: Watson, A.E., Alessa, L. and Sproull, J. (eds) *Proceedings of the Wilderness in the Circumpolar North: Searching for Compatibility in Ecological, Traditional, and Ecotourism Values*. USDA Forest Service, Rocky Mountains Research Station, Fort Collins, Colorado, pp. 133–142.

Beechey, T.J. and Davidson, R.J. (1999) The Hudson Bay lowland. In: Labatt, L. and Littlejohn, B. (eds) *Ontario's Parks and Wilderness: Islands of Hope*. Firefly Books/Willowdale Books, Ontario, Canada, pp. 197–202.

Brown, C.E. (2006) An overview of agency-managed bear viewing sites in Alaska: a diversity of agencies, opportunities, and visitor management strategies. *Proceedings of the International Symposium for Society and Resource Management. Global Challenges, Local Responses*. ISSRM, Vancouver, British Columbia.

Brunelle, R. (1970) Polar bear provincial park. *Conservation Biology* 2(2), 147–149.

Calvert, W., Taylor, M., Stirling, I., Kolenosky, G.B., Kearney, S., Crete, M. and Luttich, S. (1995) Polar bear management in Canada 1988–1992. In: Oystein, W., Born, E.W. and Garner, G.W. (eds) *Polar Bears: Proceedings of the 13th Working Meeting of the IUCN/ SSC Polar Bear Specialist Group, 23–28 June 2001, Nuuk, Greenland*. Occasional Paper of the IUCN Species Survival Commission No. 10, Gland, Switzerland, pp. 61–79.

Cessford, G.R. and Dingwall, P.R. (1994) Tourism on New Zealand's sub-Antarctic islands. *Annals of Tourism Research* 21(2), 318–332.

Chapin, M. (2004) A challenge to conservationists. *World Watch* (November/December), pp. 17–30.

Clark, D. (2003) Polar bear–human interactions in Canadian national parks, 1986–2000. *Ursus* 14(1), 65–71.

Derocher, A. (2005) Polar bear. In: Nuttall, M. (ed.) *Encyclopedia of the Arctic*. Routledge, New York, pp. 1656–1658.

Dyck, M.G. (2006) Characteristics of polar bears killed in defense of life and property in Nunavut, Canada, 1970–2000. *Ursus* 17, 52–62.

Dyck, M.G. and Baydack, R.K. (2006) Human activities associated with polar bear viewing near Churchill, Manitoba, Canada. *Human Dimensions of Wildlife* 11, 143–145.

England, J. (1982) Tourism on Ellesmere: what's inside the package? *Northern Perspectives* 10(4), 2–7.

Environment Canada (2006) Polar bears. Available at: http://www.speciesatrisk.gc.ca/search/ speciesDetails_e.cfm?speciesID=167#protection

Forbes, B.C., Ebersole, J.J. and Strandberg, B. (2001) Anthropogenic disturbance and patch dynamics in circumpolar arctic ecosystems. *Conservation Biology* 15, 954–969.

Freeman, M.M.R. and Wenzel, G.W. (2006) The nature and significance of polar bear conservation hunting in the Canadian Arctic. *Arctic* 59(1), 21–30.

Gearheard, S., Matumeak, W., Angutikjuaq, I., Maslanik, J., Huntington, H.P., Leavitt, J., Kagak, M.,Tigullaraqm, G. and Barry, R.G. (2006) It's not that simple: a collaborative comparison of sea ice environments, their uses, observed changes, and adaptations in Barrow, Alaska, USA, and Clyde River, Nunavut, Canada. *Ambio* 35(4), 203–211.

George, J.C., Huntington, H.P., Brewster, K., Eicken, H., Norton, D.W. and Glenn, R. (2004) Observations on shorefast ice failures in Arctic Alaska and the responses of the Inupiat hunting community. *Arctic* 57, 363–374.

Girard, L. (1996) Tourism in the Arctic: blessing or curse. *Ecodecision* 20, 69.

Government of Northwest Territories (2006) Tourism research and statistics. Available at: http://www.iti.gov.nt.ca/parks/tourism/research_and_statistics.html

Higginbottom, K. (ed.) (2005) *Wildlife Tourism: Impacts, Management and Planning.* Sustainable Tourism Cooperative Research Centre (STCRC), Gold Coast, MC, Queensland, Australia.

Honderich, J.E. (1991) *Wildlife as a Hazardous Resource: An Analysis of the Historical Interaction of Humans and Polar Bears in the Canadian Arctic 2,000 BC to AD 1935.* MA thesis, University of Waterloo, Waterloo, Ontario.

Humlum, O. (2005) Svalbard. In: Nuttall, M. (ed.) *Encyclopedia of the Arctic.* Routledge, New York, pp. 1975–1982.

Hummel, M. (1999) Polar bear park. In: Labatt, L. and Littlejohn, B. (eds) *Ontario's Parks and Wilderness: Islands of Hope.* Firefly Books/Willowdale Books, Ontario, Canada, pp. 204–207.

Johnston, M.E. (1997) Polar tourism regulation strategies. *PolarRecord* 33(184), 13–20.

Johnston, M.E. (2006) Impacts of global environmental change on tourism in the polar regions. In: Gössling, S. and Hall, C.M. (eds) *Tourism and Global Environmental Change.* Routledge, New York, pp. 37–53.

Kaae, B.C. (2002) Nature and tourism in Greenland. In: Watson, A.E., Alessa, L. and Sproull, J. (eds) *USDA Forest Service Proceedings RMRS-P-26:Wilderness in the Circumpolar North: Searching for Compatibility in Ecological, Traditional, and Ecotourism Values.* Rocky Mountains Research Station, Fort Collins, Colorado, pp. 43–53.

Kaltenborn, B.P. (1997) Tourism in Svalbard: planned management or the art of stumbling through. In: Price, M.E. and Smith, V.L. (eds) *People and Tourism in Fragile Environments.* Wiley, New York, pp. 89–108.

Kaltenborn, B.P. (2000) Arctic-alpine environments and tourism: can sustainability be planned? Lessons learned on Svalbard. *Mountain Research and Development* 20(1), 28–31.

Kaltenborn, B.P. and Emmelin, L. (1993) Tourism in the high north: management challenges and recreation opportunity spectrum planning in Svalbard, Norway. *Environmental Management* 17, 41–50.

Lemelin, R.H. (2005) Wildlife tourism at the edge of chaos: complex interactions between humans and polar bears in Churchill, Manitoba. In: Berkes, F., Huebert, R., Fast, H., Manseau, M. and Diduck, A. (eds) *Breaking Ice: Renewable Resource and Ocean Management in the Canadian North.* University of Calgary Press, Calgary, Alberta, Canada, pp. 183–202.

Lemelin, R.H. and Smale, B.J.A. (2006) Effects of environmental context on the experience of polar bear viewers in Churchill, Manitoba. *Journal of Ecotourism* 5(3), 176–191.

Lunn, N.J., Taylor, M., Calvert., Stirling, I., Obbard, M., Elliot, C., Lamontagne, G., Schaeffer, J., Atkinson, S., Clark, D., Bowden, E. and Doidge, B. (1998) Polar bear management in Canada 1993–1996. In: Derocher, A.E., Garner, G.W., Lunn, N.J. and Wiig, O. (eds) *Polar Bears: Proceedings of the 12th Working Meeting of the IUCN/SSC Polar Bear Specialist Group, 3–7 February, Oslo, Norway.* Occasional Paper of the IUCN Species Survival Commission, Gland, Switzerland, pp. 51–66.

Lunn, N.J., Atkinson, S., Branigan, M., Calvert, W., Clark, D., Doidge, B., Elliot, C., Nagy, J., Obbart, M., Otto, R., Stirling, I., Taylor, M., Vandal, D. and Wheatley, P. (2002) Polar bear management in Canada 1997–2000. In: Lunn, N.J., Schliebe, S. and Born, E.W. (eds) *Polar Bears: Proceedings of the 13th Working Meeting of the IUCN/SSC Polar Bear Specialist Group, 23–28 June 2001, Nuuk, Greenland.* Occasional Paper of the IUCN Species Survival Commission, Gland, Switzerland, pp. 41–52.

Lyster, S. (1993) *International Wildlife Law: An Analysis of International Treaties Concerned with the Conservation of Wildlife.* Cambridge University Press, New York.

Marsh, J. and Johnston, M.E. (1983) Conservation, tourism and development in the North: the case of northern Yukon. *Paper Presented at the 6th Annual Applied Geography Conference.* 12–15 October 1983, Toronto, Ontario.

Newsome, D., Dowling, R.K. and Moore, S.A. (eds) (2005) *Wildlife Tourism.* Multilingual Matters Ltd, Clevedon, UK.

Norton, D.W. and Gaylord, A.G. (2004) Drift velocities of ice floes in Alaska's Chukchi sea flaw zone: determinants of success by spring subsistence whalers in 2000 and 2001. *Arctic* 57, 347–362.

Notzke, C. (2006) *Stranger, the Native and the Land.* Captus University Publications, North York, Ontario, Canada.

Nuttall, M. (2005) Tourism. In: Nuttall, M. (ed.) *Encyclopedia of the Arctic.* Routledge, New York, pp. 2037–2040.

Obbart, M.E. and Walton, L.R. (2004) The importance of polar bear provincial park to the southern Hudson Bay polar bear population in the context of future climate change. In: Rehbein, C.K., Nelson, J.G., Beechey, T.J. and Payne, R.J. (eds) *Proceeding of the Parks Research Forum of Ontario, Annual General Meeting, May 4–6, 2004.* Lakehead University, Parks and Research Form of Ontario: Waterloo, Ontario, Canada, pp. 105–116.

Olsen, K. (2006) Making differences in a changing world: the Norwegian sámi in the tourist industry. *Scandinavian Journal of Hospitality and Tourism* 6(1), 37–53.

Ontario Ministry of Natural Resources (OMNR) (1977a) *Background Information: Polar Bear Provincial Park.* Ministry of Natural Resources, Toronto, Ontario, Canada.

Ontario Ministry of Natural Resources (OMNR) (1977b) *Planning Proposal: Polar Bear Provincial Park.* Ministry of Natural Resources, Toronto, Ontario, Canada.

Ontario Ministry of Natural Resources (OMNR) (1980) Polar bear provincial park: master plan. Ministry of Natural Resources, Toronto, Ontario, Canada.

Ontario Ministry of Natural Resources (OMNR) (1994) *Polar Bear Provincial Park Management Plan Review: Public Comments and Co-planning Team Responses Concerning the Summary of Issues and Options Discussion Paper (November 1991).* Toronto, Ontario, Canada.

Ontario Parks (2005) Polar Bear Provincial Park. Master Plan – Proposed Amendment – November 2005. Available at: http://www.ontarioparks.ca/english/planning_pdf/pola_plan_ammend.pdf

Ovsyanikov, N. (1998) *Polar Bears.* Raincoast Books, Vancouver, British Columbia.

Pagnan, J. (2003) Climate change impacts on Arctic tourism – a preliminary review. *Proceedings of the 1st International Conference on Climate Change and Tourism, Djerba, Tunisia, April 2003.* World Tourism Organization, Madrid, Spain.

Polar Bears International (PBI) (2006) Wrangel Island. Available at: http://www.polarbearsinternational.org/pbi-supported-research/wrangel-island/

Portman, T. (2004) Tundra struck: exploring Ontario's polar bear provincial park. Available at: http://www.cnf.ca/magazine/fall04/tundra.html

Prevett, J.P. (1989) Hudson Bay lowlands and Polar Bear Provincial Park. In: Theberge, J.B. (ed.) *Legacy: The Natural History of Ontario.* McLelland & Stewart Inc., Toronto, Canada, pp. 250–256.

Razzhivin, V. (2005) Wrangel Island. In: Nuttall, M (ed.) *Encyclopedia of the Arctic.* Routledge, New York, pp. 2196–2197.

Reid, D. (2003) *Tourism, Globalization and Development: Responsible Tourism Planning.* Pluto Press, London.

Schliebe, S., Wiig, O., Derocher, A. and Lunn, N. (2006) *Ursus maritimus.* 2006 IUCN Red List of Threatened Species. www.iucnredlist.org.

Scott, P.A. and Stirling, I. (2002) Chronology of terrestrial den use by polar bears in Western Hudson Bay as indicated by tree growth anomalies. *Arctic* 55(2), 151–166.

Slocum, R. (2004) Polar bears and energy-efficient lightbulbs: strategies to bring climate change home. *Environment and Planning* 22, 413–438.

Smith, L.V. (1997) The inuit as hosts: heritage and wilderness tourism in Nunavut. In: Price, M.E. and Smith, V.L. (eds) *People and Tourism in Fragile Environments*. Wiley, New York, pp. 33–50.

Stenhouse, G.B., Lee, L.J. and Poole, K.G. (1988) Some characteristics of polar bears killed during conflicts with humans in the Northwest Territories. *Arctic* 41, 275–278.

Stewart, E.J., Draper, D. and Johnston, M.E. (2005) A review of tourism research in the polar regions. *Arctic* 58(4), 383–394.

Struzik, E. (2004a) Search and rescue: paying for others' mistakes. *The Edmonton Journal* November 7, 2004.

Struzik, E. (2004b) Difficult births: creating new parks. *The Edmonton Journal* November 7, 2004.

Travel Alaska (2006) Bear viewing. Available at: http://travelalaska.com/Regions/TipsDetail. aspx?FAQID = 98&PageTitle = Bear%20Viewing

Travel Wild Expedition (2006) The polar bears and wildlife of Spitsbergen, July 5–17 and July 14–27, 2006. Available at: http://travelwild.com/Spitsbergen2006-TravelWild.pdf

United Nations Educational, Scientific and Cultural Organization (UNESCO) (1998) International Co-ordinating Council of the Programme on Man and the Biosphere (MAB). Fifteenth Session Periodic Review of the Status of Biosphere Reserves Designated More Than Ten Years Ago. Available at: http://unesdoc.unesco.org/images/0011/001143/114396Eo.pdf

United Nations Educational, Scientific and Cultural Organization (UNESCO) (2006) World HeritageSites; Natural System of Wrangel Island Reserve. Available at: http://whc.unesco. org/en/list/1023

Usher, A.J. (1993) Polar Bear Park and area: community-oriented tourism development in Ontario's Arctic. In: Johnston, M.E. and Haider, W. (eds) *Communities, Resources and Tourism in the North*, Volume 2. Lakehead University Centre for Northern Studies, Northern and Regional Studies Series, Thunder Bay, Ontario, Canada, pp. 17–34.

Viken, A. (2006) Tourism and Sámi Identity – An analysis of the tourism identity nexus in a Sámi community. *Scandinavian Journal of Hospitality and Tourism* 6(1), 7–24.

Weaver, D. (2002) The evolving concept of ecotourism and its potential impacts. *International Journal of Sustainable Development* 5(3), 251–264.

Wright, P. and McVetty, D. (2000) Tourism planning in the Arctic Banks island. *Tourism Recreation Research* 25(2), 15–26.

Ziaja, W. (2004) Spitsbergen landscape under 20th century climate change: Sorkapp Land. *Ambio* 33(6), 295–299.

21 Marine Wildlife and Tourism Management: Scientific Approaches to Sustainable Management

M. LÜCK AND J.E.S. HIGHAM

Providing experiences to tourists who engage with wild animals in the marine context has become a large industry, with the diversification from whale watching to the observation of many other species, such as dolphins and porpoises, seabirds, pinnipeds, sharks, penguins, polar bears, manatees and sea otters (e.g. Lück, 2003a; Hoctor, 2003; Sorice *et al.*, 2003; Dobson, 2006; Orsini *et al.*, 2006; Lemelin, 2008). In the introduction we stated that this book will probably pose more questions than answers. This seems to be the case at the end of this volume. For example, Higham and Hendry (Chapter 19) highlight the quite different impact perceptions as they relate to visitors on different viewing platforms. But in doing so they highlight the chronic lack of understanding of key impacts associated with different viewing platforms, not to mention the symmetric and asymmetric social impacts that exist at sites where multiple platforms exist in close proximity. In Chapter 13, Miller emphasizes that in order to manage marine protected areas (MPAs) appropriately, a number of challenges need to be addressed. He contends that the need for better integration across sectors and academic disciplines clearly exists. It is crucial for marine wildlife managers to take advantage of not only academic research, but also the practical experiences of the private sector and the anecdotes of tourists themselves.

The chapters that comprise this volume highlight the management challenges of marine wildlife tourism in a range of settings and viewing contexts. This includes not only the now well-established tourist experiences such as whale watching, but also more recent developments, such as the enormous interest in the observation of polar bears in Churchill, Manitoba (Canada), and growing interest in experiencing sharks in their natural habitat.

Sobel and Dahlgren (2004, p. 3) contend that '[p]eople who know the sea know something is wrong'. This statement is well supported by Ellis' (2003) volume entitled *The Empty Ocean,* which underlines the enormity of current environmental management challenges associated with marine environments. The authors of these two volumes make little direct reference to tourism, but to general threats to the oceans and all creatures living within them. Among the greatest threats are unsustainable fishing practices, the use of drift nets (also known as 'wall of death' nets, commonly upward of tens of kilometres in length) and gill nets, long line by-catches and the pollution of the seas, originating on land and on ships. Prior to the introduction of Exclusive Economic Zones (EEZs), which is the waters bordering a particular country, virtually anybody could fish anywhere. Even today, international waters are largely unregulated, and overfishing is a common practice (Ellis, 2003). Ellis (2003) provides innumerable examples of the decline of fish stocks due to these unsustainable practices, for example, cod, haddock and tuna. He contends that overfishing is the prime example of what Hardin (1968) refers to as the 'Tragedy of the Commons'. In fact, he argues that the marine environment underlines this concept at a more pressing level than terrestrial environments, despite illustrating his original discussion of the concept with land-based rather than marine examples (e.g. the overgrazing of common pool pastures). In order to avoid overfishing, countries have chosen to extend their EEZs; for example, Canada has recently extended its EEZ to 200 miles.

However, the results are not very satisfactory. First, fish and marine mammals do not know political boundaries, and often move between EEZs and international waters. The case of the migratory whale shark (*Rhincodon typus*) is a case in point. This animal, the world's largest fish species, is protected in Australian territorial waters where it is the subject of increasing interest on the part of conservationists and tourists alike. However, its migrations into Southeast Asian waters bring it into contact with fishing boats that value the whale shark for its fins rather than its conservation value. Second, the introduction or extension of an EEZ does not eliminate the problems of the commons, but merely turns international commons into national commons (Ellis, 2003). Slooten (in press) estimates that approximately 500,000–800,000 marine mammals are caught in gill nets as by-catch every year, which is another example of inappropriate fishing methods. Gill netting is by far the most common cause of death for not only marine mammals, but also for fish and seabirds (Slooten, in press).

Similar and equally pressing problems exist when one considers the pollution of marine environments. There are two main sources of oceanic pollution: first, commercial fishing, the cruise industry, recreational boating, commercial shipping and oil/gas offshore rigs. Second, land-based pollution, in particular, from human waste and refuse, industrial waste and agricultural run-off into rivers and streams and ultimately into the oceans. Berghan (1998) estimates that 70% of all rubbish found in the oceans is derived from land-based sources. Harriott (2004) presented the results of interviews with experts and professionals experienced in reef issues, and reported that they saw marine tourism as a much smaller threat to the Great Barrier Reef than other commercial activities, such as fishing and agriculture. Particularly, plastics and styrofoam are a major problem, because they float and can be windblown and carried on ocean currents over hundreds of miles. Berghan (1998) notes that many plastic bottles, plastic sheeting and styrofoam cups have even been found on remote Antarctic beaches. She compiled the 'Dirty Dozen', a ranking of the most common objects of marine refuse during a large Coastal Cleanup of the US beaches in 1996 (Table 21.1).

It is of course impossible to speculate on the specific origin of such volumes of rubbish recovered from coastal and marine environments. As such it is also impossible to attribute proportions of waste to tourism and non-tourism sources. No doubt much of this problem is attributable to inappropriate coastal urban and residential waste management. It is equally certain that a proportion of this waste originates from the irresponsible practices of recreationists and tourists, both on a personal level and from poor commercial practices. Irrespective of the apportioning of blame, the fact remains that the harm caused by both the deliberate and inadvertent disposal of human waste, industrial waste and non-recycled rubbish in the marine environment has, and will continue to contribute to the mortality and morbidity, and therefore the decline and disappearance of species of wild animals.

MPAs in various forms, including marine parks, marine reserves and sanctuaries, have been established in many countries over the past few decades (Hoyt, 2005). The extent to which animals are protected in these MPAs depends on the designation and national legislation. Sobel and Dahlgren (2004) contend that the establishment of MPAs has become increasingly widespread, but warn that the application of MPAs to marine conservation and fisheries is still in its infancy. Despite the overwhelming scientific evidence of various benefits of MPAs as a management tool, for example, the recovery of declining species, and educational opportunities to visitors (Lück, 2003b; Sobel and Dahlgren, 2004), they also note that the scientific study of marine reserves is complicated, and influenced by many factors, including:

Table 21.1. The 'Dirty Dozen' of the 1996 US Coastal Cleanup. (From Berghan, 1998.)

Rank	Debris item	Total number reported	% of total debris collected
1	Cigarette butts	608,759	16.20
2	Plastic pieces	240,820	6.41
3	Foamed plastic pieces	206,890	5.51
4	Plastic food bags/wrappers	205,762	5.48
5	Plastic caps/lids	179,103	4.34
6	Paper pieces	158,957	4.23
7	Glass pieces	140,667	3.74
8	Plastic straws	131,602	3.50
9	Metal beverage cans	130,134	3.46
10	Glass beverage bottles	127,633	3.40
11	Plastic beverage bottles	121,703	3.24
12	Foamed plastic cups	96,394	2.57
	Total 'Dirty Dozens'	2,348,424	62.08

- Control and replication issues (both spatial and temporal);
- Natural environmental and recruitment variability;
- The complicating effects of other management measures;
- Changes in fishing patterns or effort external to the reserves (Sobel and Dahlgren, 2004, p. 93).

This, however, should not detract from the establishment of MPAs, even if it is a temporary or precautionary measure until scientific evidence can be achieved to provide insights into likely consequences of tourism and other human activities. Hoyt (2005) argues that MPAs can be seen as such a precautionary approach. The precautionary principle was developed in the 1970s in Germany ('Vorsorgeprinzip'), and has since been adopted as part of environmental policies in over 40 countries (Fennell, 2006). It is based on the notion that in the absence of scientific proof, precautions need to be taken until scientific research can produce reliable data. In fact, the lack of scientific evidence is no reason to postpone action to avoid potential harm and damage, especially if irreversible, to the environment (Cicin-Sain and Knecht, 1998). According to VanderZwaag (1994, p. 7), the precautionary principle contains the following core elements:

- A willingness to take action (or no action) in advance of formal scientific proof;
- Cost effectiveness of action, that is, some consideration of proportionality of costs;
- Providing ecological margins of error;
- Intrinsic value of non-human entities;
- A shift in the onus of proof to those who propose change;
- Concern with future generations;
- Paying for ecological debts through strict/absolute liability regimes.

The bottom line for this approach is that, if in doubt due to the absence of scientific certainty, then actions should err on the side of caution, at least until rigorous insights are available (Hoyt, 2005). Bejder (2007, Perth, WA, personal communication) argues the case for shifting the burden of evidence from scientists proving that impact issues do exist, to tourism operators and management agencies demonstrating beyond reasonable doubt that effective mitigation measures are in place. In the case of marine wildlife tourism development, this would entail conservative guidelines and regulations regarding, for example, approach distances, times of interaction, duration of interaction, intensity of interaction (e.g. watching, feeding, swimming with, etc.), modification of the environment (e.g. erection of viewing platforms, hides, tracks, jetties, etc.) and many more (see, e.g. Lien, 2000). The burden of proving that wildlife interaction activities have minimal adverse effects on the species in question (or, more realistically, 'acceptable' or sustainable impacts) should fall, at least to some degree, upon the providers of commercial activities (Cicin-Sain and Knecht, 1998).

At a much larger scale, and not encompassing marine wildlife tourism alone, has been the introduction of Integrated Coastal Management (ICM) policies in the late 20th century. ICM is closely linked to the United Nations marine regulatory regime (the Law of the Sea) and the global call for sustainable development, following the 1992 Earth Summit in Rio de Janeiro (Nichols, 1999). Cicin-Sain and Knecht (1998, p. 39) define ICM as 'a continuous and dynamic process by which decisions are made for the sustainable use, development, and protection of coastal and marine areas and resources.' ICM is a six-stage process (Fig. 21.1), and recognizes the unique interface of land-based and water-based resources, and aims for a sustainable management regime of these.

In particular, it aims to overcome the fragmented regulations and policies governing the management of both land and sea. It is multipurpose oriented, that is, it analyses the potential impacts of development, conflicts and relationships between user groups, physical and biological processes and implications, and promotes the streamlining of sectoral, coastal and ocean activities (Cicin-Sain and Knecht, 1998). The ideals of ICM are laudable, and intend to overcome significant problems caused by a plethora of often conflicting multinational and multiple-agency regulations. However, there has also been some criticism of ICM. Nichols (1999, p. 388) warns that 'by promoting the overhaul of existing social and spatial organization in coastal zones, and by asserting the primacy of resource access for modern economic interests, ICM may introduce more rather than less social conflict and ecological degradation'. Particularly in developing countries, ICM is often growth-centred rather than focused on environmental conservation. In the case of marine wildlife tourism, all too often the economic interests of the tour operators prevail over conservation concerns associated with focal animals and species. Rigorous scientific research, as illustrated in many

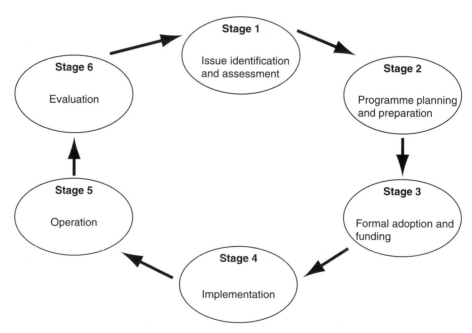

Fig. 21.1. The six stages of an ICM process. (From Cicin-Sain and Knecht, 1998, p. 58.)

chapters of this volume, needs to be taken into consideration when planning and regulating interactions with and actions around marine wildlife.

The challenges of implementing and managing marine wildlife activities often appear to be unsurmountable due to the many stakeholders with different (and often opposing) interests involved in the management of marine wildlife. Forestell and Kaufman (1993) argue that it is a misnomer to manage wildlife, as it is people interacting with wildlife who should be the focus of visitor management. Thus, an environment-focused approach is to be favoured over an anthropocentric, economic-based approach. Such an approach is not necessarily to the disadvantage of the providers of wildlife interaction experiences. Many of the preconceptions of various stakeholders are based on misinformation, and on the fear of losing the viability of their operation. Taking these fears into account, it could be beneficial for resource managers to appropriately 'sell' conservation to tour operators and tourists/recreationists. When visitors and operators understand the widespread benefits of conservation strategies, they will be much more receptive to such guidelines and, as a consequence, much more likely to support them actively.

This book seeks to raise and critically address issues that relate to marine wildlife and tourism management. As such the chapters in this volume collectively address a wide range of legislative, conservation, impact and management issues. First, it is important to understand the changing patterns of tourist demand in respect to accessing and experiencing wild animal populations in the marine environment (Part I). With significant developments in technology, the marine environment generally, and marine species specifically, have become much more accessible in a short span of time (Orams, 1999). For example, the development of the scuba apparatus by Jacques Cousteau and Emile Gagnon in the 1940s opened the underwater world to a large audience (Martinez, in press). Similarly, while vast areas of the Great Barrier Reef could previously be safely subject to de facto management because they were beyond the range of high volume day trippers, this is no longer the case. New marine technologies have provided high volume day visitors with safe, comfortable and fast access to vast expanses of the Great Barrier Reef Marine Park.

It has been argued that a new type of tourist emerged in the 1980s and 1990s, being much more engaged with the host countries, cultures and natural environments (Krippendorf, 1986; Poon, 1994). Increasing demand for marine wildlife experiences is testament to this development, as illustrated by Dobson (Chapter 3), Dearden et al. (Chapter 4) and Lemelin (Chapter 5). The search for the benefits of participation in marine wildlife-viewing activities has been demonstrated in Chapter 2 by Zeppel and Muloin. In the case of whale watching, Malcolm and Duffus found that there are various degrees of specialization among tourists engaging in these activities, from the generalist tourist who books a whale-watching tour as just one part of a wider holiday experience, to the very knowledgeable whale-watching enthusiast who travels to certain places particularly for the whale-watching opportunities (Chapter 6). Both technological advances, especially in media and information technology, and increased interest in the marine environment have also led to an increasing demand to see marine wildlife at aquaria and marine parks, especially among certain market groups. The fascination with fish and marine invertebrates has even triggered an ever-growing number of aquaria in private homes (Chapter 7).

Secondly, the potential positive and negative impacts of marine wildlife viewing activities are discussed in Part II. There is no doubt that tourism oper-ations can provide economic benefits to commercial operators and their employees, as well as employment and economic development opportunities to the communities within which they are situated (Cater and Cater, Chapter 8). However, there is growing concern that rapidly increasing demand for wildlife-viewing experiences also places increasing pressure on the natural environment, including focal animal populations. Some of these impacts, and the call for more rigorous management of respective sites and activities, are discussed in Part II, with specific attention paid to the impacts of tourism (and other human activities) on penguins (Seddon and Ellenberg, Chapter 9; Shelton and McKinley, Chapter 12), pinnipeds (Newsome and Rodger, Chapter 10) and marine mammals (Lusseau, Chapter 11). While there is con-sensus among these authors that humans may seriously impact populations of wild animals, it is also apparent that the impacts of human activities can vary dramatically between different sites, different times of the day/month/year, different species and even different individuals within a population. Futhermore, we have much to learn about the impacts of tourism as they relate to (and in some cases pale into insignificance alongside) other causes of anthropogenic impact. Only with a clear understanding of these impact issues as they exist in similar but significantly different contexts, will managers be sufficiently well placed to actually manage marine wildlife-tourism phenomena.

Thirdly, increasing demand for marine wildlife interactions brings with it growing concerns for the sustainability of the marine environment in general and focal species in particular. In order to mitigate the potential negative impacts of tourist activities, a range of policy, legislative and management tools has been developed. An understanding of marine wildlife and tourism manage-ment is incomplete without careful consideration of ethical, legislative and management arrangements as they exist in various contexts. Miller (Chapter 13) investigates the challenges that accompany the implementation of MPAs, while Garrod discusses ethical issues associated with marine wildlife tourism (Chapter 14). Forestell and Maher (Chapters 15 and 16) pay attention to the specific examples of whale-watching regulations and resource management in Antarctica, respectively.

Lastly, Part IV addresses tourism management as it relates to marine wild-life. This section highlights the importance of being able to give priority to tourism management challenges, and this involves differentiating between tour-ism and other anthropogenic impacts on marine animals. Constantine and Bejder (in Chapter 17) call for a paradigm shift in the management of the whale-watching industry, away from an anthropocentric focus to an ecocentric management regime. This also requires managers to distinguish between the management of direct tourism impacts and incidental impacts, such as the development of infrastructure, regardless of whether new infrastructure devel-opments are intended to serve tourism or not. Lück (Chapter 18) argues that structured interpretation on marine wildlife tours can be a valuable management tool, helping the conservation efforts of wildlife managers. The management of wildlife viewing experiences is also dependent on the respective viewing platform. Higham and Hendry (Chapter 19) illustrate this using the example of boat-based, land-based and kayak-based orca watching in the San Juan Islands, USA. Lemelin and Dyck, in Chapter 20, review management strategies of polar bear tourism in various parts around the world noting the full range of approaches from the complete absence of management intervention to strict regulation.

The underlying purpose of this book, as established in Chapter 1, is to provide rigorous scholarly insights drawn from the natural and social sciences into the management of tourist interactions with wild animal populations in the marine environment. It has been argued from the outset that the sustainable management of marine wildlife tourism phenomena is dependent on rigorous scientific research, and management responsiveness to good science. The chapters herein provide a range of insights into marine wildlife and tourism management drawn from various scientific disciplines. They contribute to answering important questions and, in doing so, raise additional questions that relate to this subject.

It is hoped, therefore, that most immediately this book will contribute to two important ends. One is the development and implementation of effective and responsive management regimes to oversee sustainable tourism development in marine contexts, and specifically in relation to tourist interactions with wild animals in coastal and marine environments. Secondly, it is hoped that this book both highlights and adds momentum to the critical role played by natural and social scientists in offering guidance to resource managers, policy makers, government agencies, non-governmental organizations, tourism operators and other stakeholders, including tourists themselves, who contribute to the all-too-challenging matter of sustainable tourism in the marine environment. The ultimate aim is comprehensive and rigorous scientific approaches to the sustainable management of wildlife-based tourism in coastal and marine environments.

References

Berghan, J. (1998) *Marine Mammals of Northland*. Department of Conservation, Wellington, New Zealand.

Cicin-Sain, B. and Knecht, R.W. (1998) *Integrated Coastal and Ocean Management: Concepts and Practices*. Island Press,Washington, DC.

Dobson, J. (2006) Sharks, wildlife tourism, and state regulation. *Tourism in Marine Environments* 3(1), 15–24.

Ellis, R. (2003) *The Empty Ocean*. Island Press, Washington, DC.

Fennell, D.A. (2006) *Tourism Ethics*. Channel View Publications, Clevedon, UK.

Forestell, P.H. and Kaufman, G.D. (1993) Resource managers and field researchers: allies or adversaries? In: Postle, D. and Simmons, M. (eds) *Encounters with Whales'93*. Great Barrier Reef Marine Park Authority, Townsville, Queensland, Australia, pp. 17–26.

Hardin, G. (1968) The tragedy of the commons: the population problem has no technical solution; it requires a fundamental extension in morality. *Science* 162 (December), 1243–1248.

Harriott, V.J. (2004) Marine tourism impacts on the great barrier reef. *Tourism in Marine Environments* 1(1), 29–40.

Hoctor, Z. (2003) Community participation in marine ecotourism development in West Clare, Ireland. In: Garrod, B. and Wilson, J.C. (eds) *Marine Ecotourism: Issues and Experiences.* Channel View Publications, Clevedon, Queensland, Australia, pp. 171–176.

Hoyt, E. (2005) *Marine Protected Areas for Whales, Dolphins and Porpoises: A World Handbook for Cetacean Habitat Conservation.* Earthscan, London.

Krippendorf, J. (1986) The new tourist – turning point for leisure and travel. *Tourism Management* 7(3), 131–135.

Lemelin, H.R. (in press) Polar bears. In: Lück, M. (ed.) *Encyclopedia of Tourism and Recreation in Marine Environments.* CAB International, Wallingford, UK.

Lien, J. (2000) *The Conservation Basis for the Regulation of Whale Watching in Canada by the Department of Fisheries and Oceans: A Precautionary Approach.* Department of Fisheries and Oceans, St John's, Newfoundland, Canada.

Lück, M. (2003a) *Environmentalism And On-Tour Experiences Of Tourists On Wildlife Watch Tours In New Zealand: A Study Of Visitors Watching And/Or Swimming With Wild Dolphins.* PhD thesis, University of Otago, Dunedin, New Zealand.

Lück, M. (2003b) Education on marine mammal tours as agent for conservation – but do tourists want to be educated? *Ocean and Coastal Management* 46(9/10), 943–956.

Martinez, E. (in press) Cousteau, Jacques-Yves (1910–1997). In: Lück, M. (ed.) *Encyclopedia of Tourism and Recreation in Marine Environments.* CAB International, Wallingford, UK.

Nichols, K. (1999) Coming to terms with integrated coastal management: problems of meaning and method in a new arena of resource regulation. *The Professional Geographer* 51(3), 388–399.

Orams, M.B. (1999) *Marine Tourism: Development, Impacts and Management.* Routledge, London.

Orsini, J.-P., Shaughnessy, P.D. and Newsome, D. (2006) Impacts of human visitors on Australian sea lions (*Neophoca cinerea*) at Carnac Island, Western Australia: implications for tourism management. *Tourism in Marine Environments* 3(2), 101–116.

Poon, A. (1994) The new tourism revolution. *Tourism Management* 15(2), 91–92.

Slooten, E. (in press) Gillnets. In: Lück, M. (ed.) *Encyclopedia of Tourism and Recreation in Marine Environments.* CAB International, Wallingford, UK.

Sobel, J. and Dahlgren, C. (2004) *Marine Reserves: A Guide to Science, Design, and Use.* Island Press,Washingon, DC.

Sorice, M.G., Shafer, S.C. and Scott, D. (2003) Managing endangered species within the use/preservation paradox: understanding and defining harassment of the West Indian Manatee. *Coastal Management* 31, 319–338.

VanderZwaag, D. (1994) *CEPA and the Precautionary Principle/Approach.* Minister of Supply and Services, Hull, Quebec, Canada.

Index